Distribution Power Systems and Power Quality

Distribution Power Systems and Power Quality

Special Issue Editor

Birgitte Bak-Jensen

MDPI • Basel • Beijing • Wuhan • Barcelona • Belgrade • Manchester • Tokyo • Cluj • Tianjin

Special Issue Editor
Birgitte Bak-Jensen
Aalborg University
Denmark

Editorial Office
MDPI
St. Alban-Anlage 66
4052 Basel, Switzerland

This is a reprint of articles from the Special Issue published online in the open access journal *Energies* (ISSN 1996-1073) (available at: https://www.mdpi.com/journal/energies/special_issues/distribution_power_system).

For citation purposes, cite each article independently as indicated on the article page online and as indicated below:

LastName, A.A.; LastName, B.B.; LastName, C.C. Article Title. *Journal Name* **Year**, *Article Number*, Page Range.

ISBN 978-3-03936-250-9 (Pbk)
ISBN 978-3-03936-251-6 (PDF)

Contents

About the Special Issue Editor

Birgitte Bak-Jensen (senior IEEE member 2012) received her MSc degree in Electrical Engineering in 1986 and a PhD in Modelling of High Voltage Components in 1992; both from the Institute of Energy Technology, Aalborg University, Denmark. From 1986–1988, she worked at Electrolux Elmotor A/S, Aalborg, Denmark, as an Electrical Design Engineer. She is now Professor MSO in the Intelligent Control of the Power Distribution System at the Institute of Energy Technology, Aalborg University, where she has worked since August 1988. Her fields of interest are mainly related to the operation and control of the distribution network grid, including power quality and stability in power systems, and taking the integration of dispersed generation and smart grid issues like demand response into account. Furthermore, multi-energy systems, taking the interaction between the electrical grid and i.e. the heating and transport sector into account, is a key area of her interest. She has participated in many projects concerning the control and operation of small dispersed generation units in the distribution network operated in connected and islanded mode, and the utilization of demand side management, for instance, using electrical vehicles, electrical boilers or heat pumps as energy storages, used for levelling out fluctuations from renewable power units. She is now convenor of Cigre WG C6/1.33 on multi-energy systems.

Preface to "Distribution Power Systems and Power Quality"

Today, a lot of renewable power generation units, such as wind power systems, photo-voltaic and small biomass fired combined heat and power plants, are integrated into the power system. The first two generate power depending on weather conditions, and therefore have fluctuating power production. Often, the power plants produce power in an on–off controlled way, dependent on heat demand, which also leads to power fluctuation. Furthermore, a lot of the new power generation units are equipped with electronic power converters, which may inject harmonics into the power system. This can also affect power quality. Moreover, at distribution level, the hosting capacity of the lines is not only affected by new, small power generation units, which may lead to the voltage rising above the limit, but also new, large loads which are seen in the grid, such as electrical vehicles and heat pumps, which might lead to voltages below the lower limit. These load units might also cause harmonic injections, together with other converter- and rectifier-based loads in the grid. Another concern is the reliability of such systems; some claim that, in the future, there will be less interruptions, due to higher possibilities for ancillary services from all small units, but others claim that the integration of new units will lead to more interruptions, since they will replace some of the central power plants. Furthermore, the protection system might be affected by reverse power flow and shifting short circuit level.

Therefore, this Special Issue focuses on the hosting capacity of distribution grids, how to counteract voltage fluctuations and harmonics, and how to ensure the reliability and stability of the future power system, with a special focus on distribution systems with high dispersed power generation. In the papers, you will find different applied methods to analyze the grid behavior using, for instance, Monte Carlo simulations, piecewise bound constrained optimization, S-transform and probabilitic neural networks and wavelet methods.

Birgitte Bak-Jensen
Special Issue Editor

Article

Reliability Assessment of Power Systems with Photovoltaic Power Stations Based on Intelligent State Space Reduction and Pseudo-Sequential Monte Carlo Simulation

Wenxia Liu, Dapeng Guo *, Yahui Xu, Rui Cheng, Zhiqiang Wang and Yueqiao Li

School of Electrical and Electronic Engineering, North China Electric Power University, Beijing 102206, China; liuwenxia001@163.com (W.L.); 18811384432@163.com (Y.X.); ncepu_chengrui@126.com (R.C.); wwwgode@163.com (Z.W.); lyqiao@ncepu.edu.cn (Y.L.)
* Correspondence: 18739942918@163.com; Tel.: +86-187-3994-2918

Received: 8 April 2018; Accepted: 22 May 2018; Published: 3 June 2018

Abstract: As the number and capacity of photovoltaic (PV) power stations increase, it is of great significance to evaluate the PV-connected power systems in an effective, reasonable, and quick way. In order to overcome the challenge of PV's time-sequential characteristic and improve upon the computational efficiency, this paper presents a new methodology to evaluate the reliability of the power system with photovoltaic power stations, which combines intelligent state space reduction and a pseudo-sequential Monte Carlo simulation (PMCS). First, a non-aggregate Markov model of photovoltaic output is established, which effectively retains some time-sequential representation of the PV output. Then, the differential evolution algorithm (DE) is introduced into the sampling stage of PMCS to carry out an intelligent state space reduction (ISSR). By using the DE algorithm, success states are searched out and removed, thus the state space is reduced and formed with a high density of loss-of-load. Hence, unnecessary samplings are avoided, which optimizes the PMCS sampling mechanism and improves the computational efficiency. Finally, the proposed method is tested in the modified IEEE RTS-79 system. The results indicate that this new method has a better computational efficiency than the time-sequential Monte Carlo simulation method (TMCS) and pure PMCS. In addition, the effectiveness and feasibility of this method are also verified.

Keywords: photovoltaic power stations; power systems reliability; non-aggregate Markov model; pseudo-sequential Monte Carlo simulation; intelligent state space reduction

---

1. Introduction

As photovoltaic (PV) power generation is one of the most important renewable energies, grid-connected photovoltaic power stations have aroused attention around the world and have been developed and utilized rapidly. With the increasing penetration of PV in power systems, the power system faces the impact of random fluctuations of PV output. Therefore, it is necessary to make accurate assessments of the reliability of PV-connected power systems. The reliability indices are of great significance for the power system to plan its expansion, arrange power generation, and energy trading [1–3]. However, the time-sequential characteristics and fluctuations of PV output will increase the computational amount in any reliability assessment. Therefore, it is particularly important to evaluate the reliability of PV-connected power systems in an effective, reasonable, and quick way.

At present, the reliability assessment methods for power systems are generally divided into the analytical method and the Monte Carlo simulation (MCS) method. MCS can get rid of the constraints of the system scale and is particularly suitable for large-scale composite power systems [4,5]. In addition, MCS includes two types: time-sequential MCS and non-sequential MCS. However, the characteristics of PV makes the reliability assessment work face greater challenges. On one hand, considering the model of a PV power plant in a simulation method, the computational amount will become extremely complicated when a PV has a large number of states. On the other hand, due to the continuous development of modern power systems and the increasing improved reliability level of the system as well as its components, the computational efficiency of MCS is gradually reduced.

In order to obtain reliability indices with high efficiency, a lot of research has been conducted at home and abroad. Common methods include the stratified uniform sampling method, the importance sampling method, and state space reduction. In the stratified sampling method [6,7], the sampled space is divided into several sub-spaces, sampled separately, and then the indices of these sub-spaces are integrated. Thus, the variance is reduced by reasonably allocating the proportion of the sampling results in each sub-space. In [8], an important sampling method based on the optimal multiplier is proposed, and the optimal multiplier is continuously reconstructed by the component state each time the system failure occurs. At last, the construction of the importance distribution function is completed. In [9], the cross entropy algorithm is used in importance sampling, and the optimal probability model of system components is established, based on the cross entropy algorithm, to reduce the variance coefficient of the sample space. It is worth mentioning that state space reduction is an effective methodology by which most of the success states can be reduced out from the original state space by a certain sampling mechanism. And the remaining state space with a higher density of loss-of-load states allows the MCS to obtain more loss-of-load states in sampling, which in turn, speed up the convergence of the variance coefficient. Based on state decoupling, Mitra and Singh et al. proposed a state space reduction method in 1996, which achieved the reduction of the state space [10–12]. In recent years, on this basis, scholars at home and abroad have fully exploited the fast and random search ability of this intelligent algorithm. Therefore, the method of the Intelligent State Space Reduction (ISSR) is formed systematically. In [13–15], the genetic algorithm and the binary particle swarm both are used in state space reduction to speed up the convergence of MCS. In [15], the performance comparison of the intelligent state space reduction is carried out under different heuristic algorithms. In the comparison, not only are the proposed genetic algorithm and binary particle swarm algorithm considered, but the mutex binary particle swarm and binary particle swarm optimization are also involved.

However, the methods mentioned above are all applied to the framework of the non-sequential Monte Carlo method. Taking into account that a large scale of renewable energies and other components connect to the grid, the non-sequential Monte Carlo method is no longer applicable due to the time sequential properties of the components and correlation with the adjacent system states cannot be depicted. Therefore, the improved sequential Monte Carlo method will have a wider application prospect. As has been confirmed, the parallel computation technique and pseudo-sequential Monte Carlo simulation can effectively improve the computational efficiency of TMCS [16]. The parallel computation technique is the parallel computation and information interaction between multiple computers, analyzing and calculating the power flow for the system states at each time section. By using this technique, the computational time is reduced and in the meantime it depends on the computer hardware equipment. The pseudo-sequential Monte Carlo simulation (PMCS) [17] is the combination of the sequential and non-sequential Monte Carlo method. To be specific, the loss-of-load states are sampled randomly by the non-sequential Monte Carlo method, followed by constructing the sub-sequences of the loss-of-load states via the time-sequential Monte Carlo method. Only partial states need to be analyzed for the time-sequential information, which improves the computational efficiency and is thus called a "pseudo-sequence". Although the pseudo-sequential MCS can improve computational efficiency, it is still at a distinct disadvantage when compared with non-sequential MCS. In [18], the state transition technique is applied to the pseudo-sequential simulation and this technique is used to speed up the formation of the sub-sequence of the loss-of-load states. However, it is shown that the time-consumption of the pseudo-sequential simulations is mainly due to a large number of ineffective states (success state) that are sampled and evaluated in the non-sequential process [19].

The adoption of a more advanced state sampling mechanism will further improve the computational efficiency of PMCS.

In view of all the above considerations, this paper proposes a kind of PMCS method based on an ISSR. First, a non-aggregated Markov model of photovoltaic power generation is built to make it appropriate to the process of a pseudo-sequential simulation. Secondly, the differential evolution algorithm is introduced in the process of intelligent state space reduction, so the success states can be quickly sought and the set of success states can be established. By this way, the sampling mechanism is optimized by ISSR, which greatly increases the probability of sampling loss-of-load states and reduces the amount of work in the states' evaluation. Therefore, the improvement of the existing PMCS is realized.

2. Non-Aggregate Markov Model of Photovoltaic Output

As a result of the photovoltaic power out being time varying, the best option to assess the reliability of a PV-connected power system is by using the time-sequential Monte Carlo simulation. However, a huge amount of CPU time is needed for such a detailed simulation, which can make the evaluation unfeasible for large and complex systems. Considering this difficulty, in order to retain the time-sequential characteristics of PV output as much as possible, a non-aggregate Markov model of the photovoltaic power generation is proposed here, which can make it better applied to the evaluation method that is going to be proposed in the following sections.

As is shown in Figure 1, in the photovoltaic output model, the total hours of a year T is divided into Q intervals with the same length ΔT. For the interval i, the photovoltaic output P_i takes the mean value of statistical data during the interval i. Then, according to the time sequence of the PV output curve, all the PV output states are linked in chronological order. In this model, a constant transition rate of $\lambda = \frac{1}{\Delta T}$ between two connected states is adopted. Thus, a non-aggregate Markov model of photovoltaic power generation is formed.

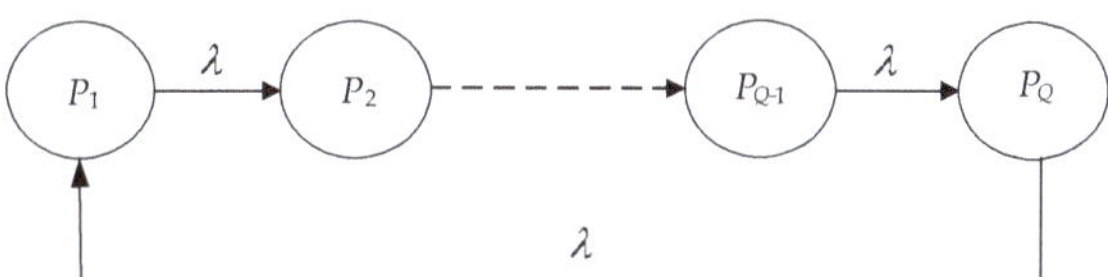

Figure 1. Non-aggregate Markov model of photovoltaic output.

3. Power System Reliability Evaluation Based on Pseudo-Sequential Monte Carlo Simulation (PMCS)

3.1. Basic Theory of PMCS

PMCS is a combination of the sequential Monte Carlo simulation (TMCS) and the non-sequential Monte Carlo simulation, which also maintains the flexibility and accuracy of TMCS while speeding up the system reliability evaluation. Compared with the TMCS, PMCS only takes into account the sequential information of the sub-sequences of the loss-of-load states, which contributes to the reliability indices in the simulation process. In PMCS, the system states are randomly sampled based on non-sequential MCS. If the sampled state is in the loss-of-load state section, then mark this section as the starting point. Based on the state transition equation of the loss-of-load states set, the time duration is respectively extended backward and forward from the starting point until to a certain success system state is achieved, thus forming the subsequence of the loss-of-load state.

The subsequences of the loss-of-load states are formed via forward and backward simulation, which is shown in Figure 2, and the procedures are as follows:

(1) Forward time-sequential simulation: starting from the selected loss-of-load state X_s, the state transition process continuously goes on until it reaches a success state. The probability for the state transition from X_s to X_t is expressed as:

$$P_{st} = \frac{f_{st}}{f_s^{out}} = [P(X_s)\lambda_{st}] / [P(X_s)\sum_{i=1}^{Ms}\lambda_{si}] \tag{1}$$

where f_{st} is the frequency of system state X_s transferring to X_t; f_s^{out} is the frequency of departure from state X_s; $P(X_s)$ is the occurrence probability of the state X_s, λ_{st} is the transition rate of the component whose state changes during the transferring process from X_s to X_t; M_s is the number of states which the system can turn into after leaving the state X_s.

(2) The time-sequential backward simulation: starting from the selected loss-of-load state X_s, continue the state transition process of backwards until success state is found. The probability of the state transition from X_t to X_s is:

$$P_{rs} = \frac{f_{rs}}{f_s^{in}} = [P(X_r)\lambda_{rs}] / [\sum_{i=1}^{Mr} P(X_i)\lambda_{is}] \tag{2}$$

where f_{rs} is the frequency that the system state X_r transferring to X_s; f_s^{in} is the frequency of arriving at state X_s; $P(X_i)$is the occurrence probability of the state X_r; λ_{is} is the transition rate of the state changing component whose state changes during the transferring process from X_i to X_s; M_r is the number of states that the system can arrive at the state X_s.

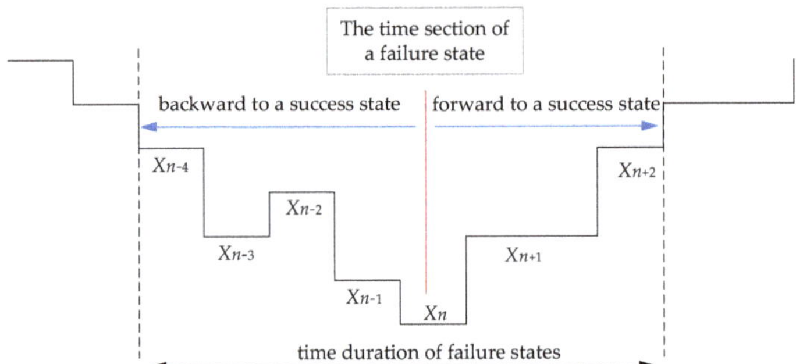

Figure 2. The schematic diagram of forward and backward method used in pseudo-sequential Monte Carlo simulation (PMCS).

For the failure subsequence formed by the forward/backward simulation, the total time expectation of the failure duration can be expressed as:

$$E[D_s] = \sum_{i \in s} E[D_i], \tag{3}$$

where

$$E[D_i] = 8760 / \left[\sum_j \lambda_j\right] \tag{4}$$

where $E[D_i]$ is the time expectation of failure duration in the *i*th system state within the failure subsequence, and λ_j is the transition rate.

3.2. Computation of PMCS Reliability Indices

During the simulation process of PMCS, only the failure sequences are taken into account. Therefore, in order to decrease the error deviation, it is necessary to force the reliability indices to map back to the original state space. The basic principle of computing the PMCS reliability indices is to convert the reliability indices based on the failure subsequence into those based on the common sampled states. In the PMCS, the expected values of LOLP (Loss of Load Probability) and EENS (Expected energy not supplied) can be expressed as [19]:

$$E[H_{LOLP}] = \frac{1}{N} \sum_{i=1}^{N} H_{LOLP}(X_i),$$
(5)

$$E[H_{EENS}] = \frac{1}{N} \sum_{i=1}^{N} H_{EENS}(X_i)$$
(6)

where N is the overall times of non-sequential sampling. $H_{LOLP}(X_i)$ and $H_{EENS}(X_i)$ are the test results of sampled state X_i corresponding to the reliability indices, which are given as follows:

$$H_{LOLP}(X_i) = \begin{cases} 1 & X_i \in X_f \\ 0 & X_i \notin X_f \end{cases},$$
(7)

$$H_{EENS}(X_i) = \begin{cases} \dfrac{\sum\limits_{S_j \in M_i} P_S(S_j) D(S_j)}{\sum\limits_{S_j \in M_i} D(S_j)} & X_i \in X_f \\ 0 & X_i \notin X_f \end{cases},$$
(8)

where M_i is the sub-sequence generated from loss-of-load states; $P_S(\cdot)$ is the load curtailment of a certain state; $D(\cdot)$ is the duration of a certain state S_j; X_f is the set of loss-of-load states.

4. Pseudo-Sequential Monte Carlo Simulation Based on Intelligent State Space Reduction

4.1. The Concept of Intelligent State Space Reduction

For the power system, the vast majority of system states are success states, while the loss-of-load states just account for a small proportion (The distribution of the power state space is shown in Figure 3). However, the success states contribute less to the reliability indices calculation, resulting in a large number of invalid samples during the sampling process.

Figure 3. The constituents of the system state space.

The ISSR is an effective method to facilitate the sampling of loss-of-load states. The first step is to guide the generation evolution via the intelligent algorithm, and in the process of population generation, the success states are quickly searched and stored in the set of success states. Then the set of success states is moved out of the original overall state space, and as a result of which, due to the remaining state space having a higher density of loss-of-load states, the probability of loss-of-load states to be sampled is greatly increased. With the same convergence accuracy, compared with traditional Monte Carlo sampling, this approach features fewer samples needed and less time-consumption. The sketch of the ISSR is shown in Figure 4.

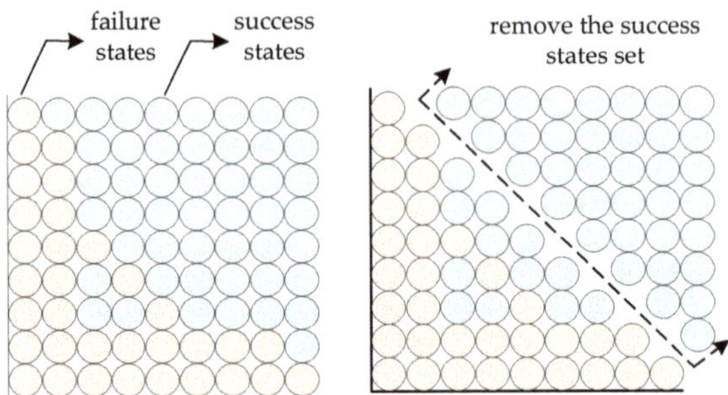

Figure 4. The schematic diagram of Intelligent State Space Reduction (ISSR) algorithm.

4.2. The Intelligent State Space Reduction Based on Differential Evolution Algorithm

The differential evolution algorithm is a heuristic random search algorithm based on population differences, attracting much more attention because of its simple principle, less control parameters, and strong robustness. The operation flow of DE is similar to that of other evolution algorithms, including mutation, crossover, and selection. A differential strategy is used for DE's mutation operation, that is, by using the differential vectors between individuals within a generation to interrupt the individuals, the mutation of individuals can be achieved. DE's mutation operation effectively utilizes the population distribution to improve the search ability, and in this way, the deficiency of mutation in the Genetic Algorithm is overcome. Therefore, this paper adopts DE to guide the generation evolution, thus completing the rapid search for success states. Let $X_{i,t}$ denote the individual i (i.e., the system state) in generation t, which is expressed as follows:

$$X_{i,t} = (x_{i,t}^1, x_{i,t}^2, \cdots, x_{i,t}^n),\ i = 1, 2, \cdots, M, \tag{9}$$

$$x_{i,t}^j = \left\{ \begin{array}{ll} 0, & success\ state \\ 1, & failure\ state \end{array} \right. \quad j = 1, 2, \ldots, n, \tag{10}$$

where $x_{i,t}^j$ indicates the state of component j in the ith individual, tth generation. 0 represents success state, 1 represents failure state, n is the number of system components, and M indicates the size of the generation population.

For all individuals in the generation population, it is important to set the appropriate fitness function. In this paper, referring to previous work, the fitness function $Fit_{(k)}$ is defined as follows [13]:

$$Fit_{(k)} = Copy_k \times P_k \times E_k, \tag{11}$$

where $Copy_k$ represents the number of all possible permutations of the system state k, the generator set can be divided into m groups according to its rated capacity, G_j represents the total number of generators in the jth group, and O_j is the total number of normal working generators in the jth group, which can be represented by:

$$Copy_k = \left[\begin{array}{c} G_1 \\ O_1 \end{array} \right] \cdots \left[\begin{array}{c} G_j \\ O_j \end{array} \right] \cdots \left[\begin{array}{c} G_m \\ O_m \end{array} \right], \tag{12}$$

where P_k represents the probability that the system state k occurs, which can be represented by:

$$P_k = \prod_{i=1}^{n} p_i, \tag{13}$$

$$p_i = \begin{cases} 1 - FOR_i, & Normal\ working\ component\ i \\ FOR_i, & Failed\ component\ i \end{cases}, \tag{14}$$

where: FOR_i represents the unavailability of component i; C_k represents the total power generation capacity in state k generators set, and U_k represents the actual power generation capacity in state k generators set after the optimal power flow (OPF); E_k is the surplus power supply in state k, which is expressed as:

$$E_k = \begin{cases} C_k - U_k, & success\ states \\ U_k - C_k, & failure\ states \end{cases}, \tag{15}$$

The fitness function will guide the system to increase the total power generation capacity and circuit capacity, which can further facilitate the intelligent search for the success states. The intelligent algorithm aims to search out more success states in a short period of time rather than to solve an optimization problem. Therefore, the stopping criterion of ISSR is supposed to be the number of generations.

The steps for the state space reduction based on DE are as follows:

Step 1: Generate the first generation of population according to the unavailability of individual components, and the fixed size of population is M.

Step 2: Identify each individual in the population and judge whether it is a success state; if so, store the individual in the set of success states.

Step 3: Individual evaluation. The fitness function values for each individual $X_{i,t}$ are calculated by Equation (11).

Step 4: Mutation operation. For each individual $X_{i,t}$ in the population, the three mutually different integers $r_1, r_2, r_3 \in \{1, 2, \ldots, M\}$ are randomly generated, and the four numbers r_1, r_2, r_3, and i are required to be different from each other. Since each individual is represented by a binary bit string, the logical operation is adopted instead of the arithmetic operation to ensure that each individual bit string in the evolution generations can only be 0 or 1. As "$\oplus$" is used to indicate "exclusive OR" operation, "$\otimes$" indicates "and", and "$+$" indicates "or", finally the mutation individual $V_{i,t}$ is produced according to the Equation (16):

$$V_{i,t} = X_{r1,t} + F \otimes (X_{r2,t} \oplus X_{r3,t}), \tag{16}$$

where the mutation factor F is a randomly-generated binary bit string.

Step 5: Cross operation. For the mutation individual $V_{i,t}$ and the target $X_{i,t}$ in the population, based on Equation (17), the test individual is $U_{i,t} = (u^1_{i,t}, u^2_{i,t}, \ldots, u^n_{i,t})$. In order to ensure the evolution of the individuals, first of all, make sure that at least one in the $U_{i,t}$ is attributed by $V_{i,t}$ and the others are attributed either by the $V_{i,t}$ or by the $X_{i,t}$, which is determined by the crossover probability CR.

$$u^j_{i,t} = \begin{cases} v^j_{i,t}, & if\ rand_j \leq CR\ or\ j = j_{rand} \\ x^j_{i,t}, & otherwise \end{cases}, \tag{17}$$

where $rand_j$ is an evenly distributed real number randomly chosen between [0,1], and j_{rand} is a random integer of [1, 2, ..., n].

Step 6: Selection operation. The "greedy selection" strategy is adopted in this operation. The test individual $U_{i,t}$ and the target individual $X_{i,t}$ are made to compete with each other, and the one with better fitness value is selected as the individual of the generation $t + 1$.

$$X_{i,t+1} = \begin{cases} U_{i,t} & Fit(U_{i,t}) < Fit(X_{i,t}) \\ X_{i,t} & Fit(U_{i,t}) \geq Fit(X_{i,t}) \end{cases} \quad i = 1, 2, \cdots, M, \tag{18}$$

Step 7: Determine if the convergence criteria are met, that is, whether the fixed generations have been reached or not. If not, return to step 2; otherwise, stop the process of intelligent state space reduction.

4.3. The Evaluation Process of the PMCS Based on the Intelligent State Space Reduction

In view of the rare occurrence of loss-of-load events in the power system, it always takes much time to sample and evaluate the loss-of-load states in the non-sequential process of the PMCS. Therefore, in this part, a PMCS method based on ISSR is introduced to improve the probability of sampling the loss-of-load states and accelerate the computational speed. The simulation process can be divided into two parts. The first part is the process of intelligent state space reduction, which establishes the set of success states via ISSR. And the second part comes to the computational process of reliability indices via PMCS. In the computational process, firstly, the system states are randomly sampled by the non-sequential Monte Carlo simulation to search for the loss-of-load states in the reduced state space. For a sampled loss-of-load state, on the one hand, the loss-of-load state subsequence is determined by the forward/backward simulation until arriving at a success state. On the other hand, the point-in-time needs to be randomly sampled, at which point the loss-of-load event occurs. Once it is determined, in the duration of the loss-of-load state subsequence, the power generation of renewable energy can be obtained according to its time-sequential power curve. The flow chart of this algorithm is shown in Figure 5.

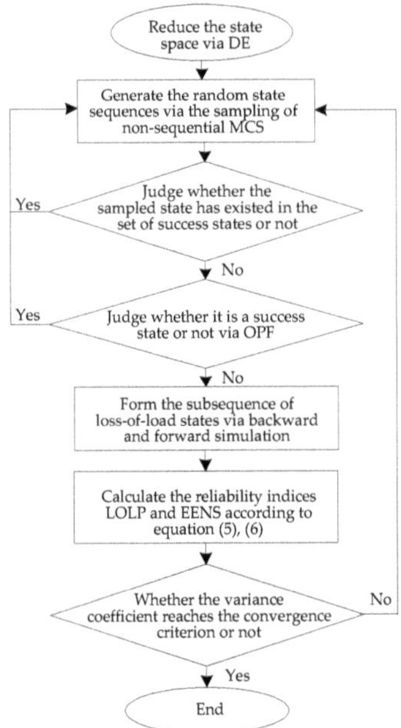

Figure 5. The reliability evaluation process by using PMCS based on the ISSR.

5. Case Study

In this paper, the methodology was implemented in a MATLAB platform and all computations were performed on a 64-bit Windows 7 system with an Intel i7-2600 CPU (4 cores at 3.4 GHz), 4 GB RAM.

A modified IEEE RTS-79 is taken as the test system, which has a total installed capacity of 3405 MW and a total peak load value of 2850 MW, consisting of 24 nodes, 38 lines, 32 generators, and a compensator. It is assumed that the size M of each generation population is 300, the crossover probability CR is 0.5, and the variation factor F is a randomly generated binary string (the probability for each bit to generate 0 or 1 is equal). At node 16, a PV power station is added, which has a total installed capacity of 150 MW, including 500 PV units with a capacity of 300 kW each. Figure 6 shows the real-time power curve and a non-aggregated Markov model for an individual PV unit in a typical day as well as in a certain region of northwest China. The real-time power curve is obtained from an individual photovoltaic unit in a PV power station, which is located at 34°16′ N, 108°54′ E. The angle of inclination of the photovoltaic modules is 19° southeast and the angle of orientation is 26°. The PV output value used in this paper is taken from the actual output data of the PV rooftop power station on 1 February 2018, and the acquisition step length is 5 min. For the non-aggregated Markov model of PV output, T is 24 h, ΔT is 1h. It can be seen from Figure 6, the non-aggregated Markov model simplifies the live power curve, and meanwhile, it retains some time-sequential characteristic of PV output.

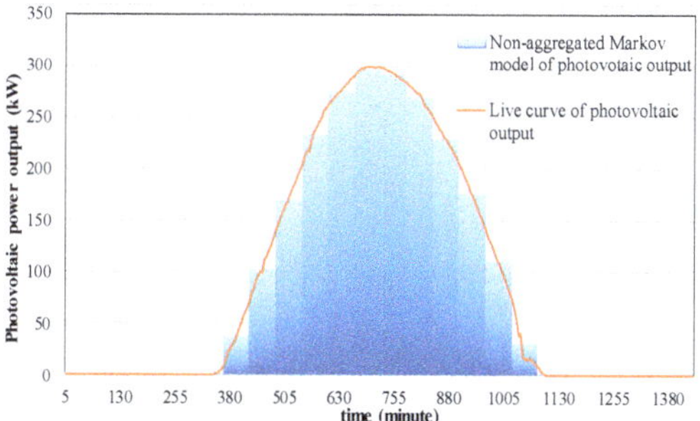

Figure 6. The actual power curve and non-aggregate Markov model of photovoltaic power output.

5.1. The Effects of DE on Generation Superiority

The purpose of the ISSR is to search for more success system states, so this paper regards the number of the success states in each generation as the reference criterion to measure the excellence of the population. In the process of reduction, the number of the success states in each generation changes along with the generation, which is as the curve shows below.

Figure 7 indicates that the number of success system states presents an upward trend along with the generations, which proves that the DE has successfully optimized the generations and achieved the generation evolution.

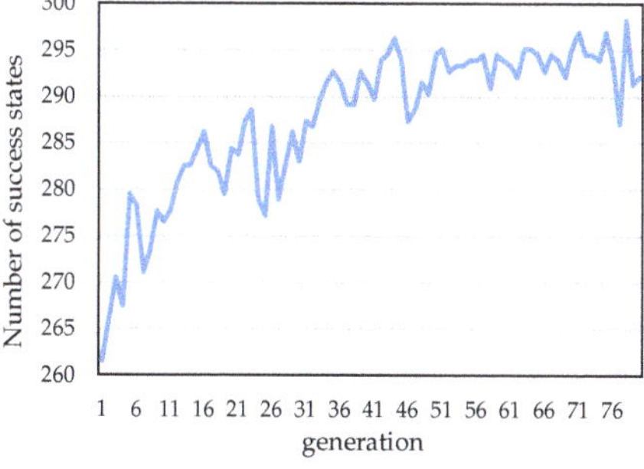

Figure 7. The trend graph that the number of success states change along with the generation.

5.2. The Effects of Generations on Computational Efficiency

The pseudo-sequential Monte Carlo simulation based on the ISSR can be generally divided into the process of ISSR and the reliability indices computation. The number of the generations, on the one hand, influences the scale of the individual states to be assessed in the process of ISSR. On the other hand, the density of the loss-of-load states in the reduced state space is also changed, which influences the computational efficiency of the reliability indices. The EENS variance coefficient of 5% is taken as the convergence index, and the computational time is compared and analyzed when the set number of generations is 50, 60, 70, and 80. The computational time change alongside the generations is shown in Table 1.

Table 1. The computational time under different generation numbers.

Generation Number	50	60	70	80
ISSR time/s	795.22	952.53	1104.4	1259.2
computation time/s	3760.3	3179.7	2038.2	1968.4
Total time/s	4555.6	4132.2	3142.5	3227.6

As it is shown in Table 1, the ISSR time increases linearly along with the increase of generations, while the computational time for the reliability indices decreases. This is because in the process of ISSR it is necessary to evaluate all individual states in the generations by calling OPF. Because the evaluation time spent on each individual is of little difference, as a whole, the spending time shows a linear increase. The more the generations, the larger the scale of the success states set will be, and the higher the density of the loss-of-load states in the reduced state space will be. Therefore, the variance convergence becomes faster with more generations, thus shortening the computational time of the reliability indices. When a generation increases from 60 to 70, the computational time of the reliability indices decreases the most. However, from 70 to 80, the decrease becomes the smallest. This is mainly because during the multiplying process from 60 to 70 generations, some success states newly emerge with larger probability. In this case, the set of success states includes the vast majority of the success states. Thus, the density of the loss-of-load states in the reduced state space increases greatly and the computational time decreases dramatically. However, during the multiplying process from 70 to 80, the generations have almost completely evolved. The diversity of generations and the scale of the success states have not improved so greatly. Therefore, compared with 70 generations, the decrease of the computational time is not obvious. When it is 70 generations, the total time is 3142.54 s, which is the optimum process of state space reduction and computation of the reliability indices.

5.3. Algorithm Comparison

In this section, the EENS variance coefficient of 5% is taken as the convergence index, and the proposed method is compared with traditional PMCS and TMCS. In the proposed method, with 70 chosen as the number of generations, the process of state space reduction as well as the computation of reliability indices is in the optimum. The results of the computation are shown in Table 2. According to (5) and (6), for PMCS and the new algorithm, reliability indices should be updated once after each time of the non-sequential sampling process. With LOLP and EENS obtained via TMCS made as the benchmark, the convergence processes of LOLP and EENS, as well as their variance change curves, are shown in Figures 8 and 9.

Table 2. Comparison of indices with different algorithms.

Algorithm	LOLP	EENS/MWh	Computation Time/s
TMCS	0.0400	5.5501	41,107
PMCS	0.0405	5.5207	5481.0
New algorithm	0.0395	5.5967	2038.2

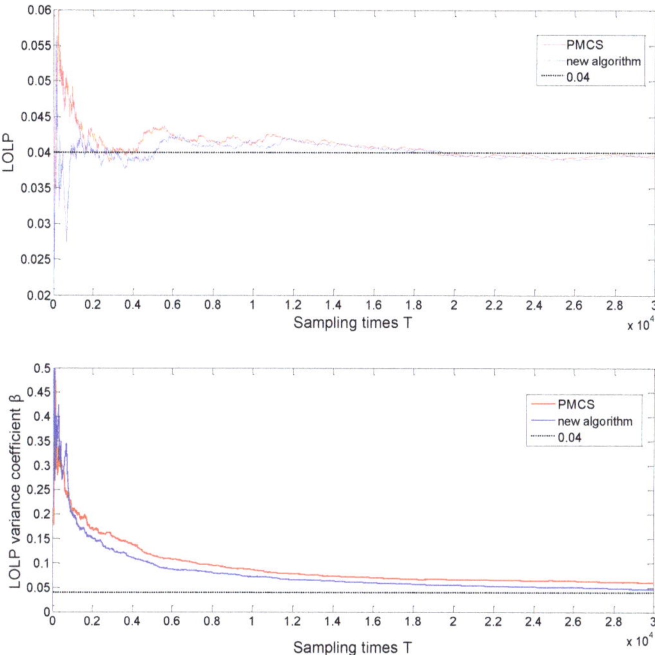

Figure 8. Convergence process of the Loss of Load Probability (LOLP) index in different algorithms.

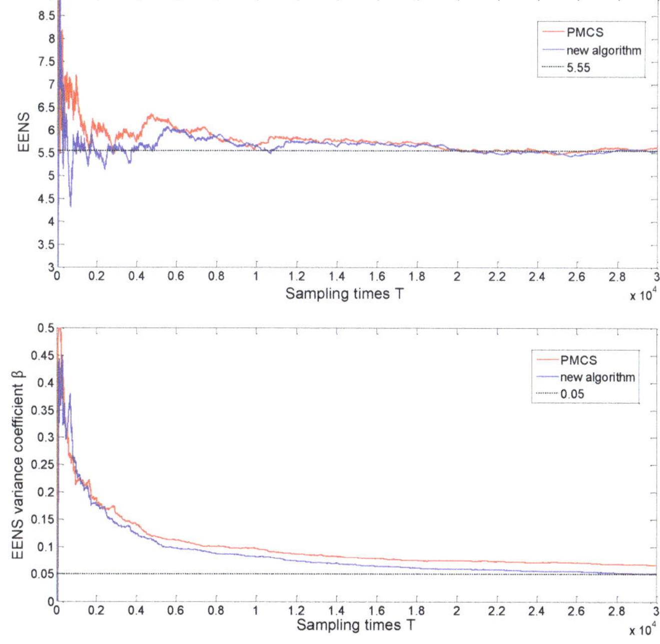

Figure 9. Convergence process of the expected energy not supplied (EENS) index in different algorithms.

Figures 8 and 9 indicate that the LOLP and EENS indices calculated by the PMCS and the new algorithm are almost equal to that calculated by the TMCS, which verifies the effective feasibility of PMCS and the new algorithm proposed in this paper. Table 2 shows that, the total computational time of the new algorithm is 3142.54 s, including the time of 2038.17 s spent for the computation of reliability indices and the time of 1104.37 s spent for the ISSR (see Table 1). When compared with the TMCS and PMCS, the computational efficiency of the new algorithm increases by 13.08 times and 1.74 times respectively. The PMCS only considers the time-sequential information of the loss-of-load subsequence, so its computational efficiency is higher than that of TMCS. As for the new algorithm, the state space with a high density of loss-of-load is established by ISSR, which greatly reduces the computational time of reliability indices and the total computation efficiency has been improved.

The subsequent work further verifies the improvement of ISSR to PMCS in the new algorithm. In the following analysis, T indicates the total number of non-sequential random samplings in PMCS, and D indicates the number of samplings in which the sampled states belong to the set of success states. For the remaining T-D samplings, the density of loss-of-load states is much higher via ISSR, that is, the majority of the states are loss-of-load states. By contrast, the traditional PMCS tends to a much lower density of loss-of-load states, that is, most of them are the success states. Here the number of non-sequential samplings of loss-of-load states is represented by S, and so the density of the loss-of-load states is represented as S/(T-D). The comparing results are shown in Table 3.

Table 3. Comparison of sampling amounts with traditional PMCS and the ISSR-based PMCS.

Content	T	S	D	T-D	S/(T-D)
PMCS	66,913	2707	0	66,913	4.04%
New algorithm	30,964	1223	28,134	2830	43.43%

As can be seen from Table 3, in the random sampling process via traditional PMCS, D equals 0. This is so because the set of success states is not established, and thus it is necessary to evaluate all the sampled states. In this case, its density of the loss-of-load states is 4.04%. But in the new algorithm, there is no need to evaluate the states in the set of success states by calling OPF. The results of LOLP and EENS both are 0, the total number of times for the non-sequential sampling is 30,964, and 28,134 of them are for the success states. Therefore in the new algorithm, only 2830 samplings are used for the state evaluation, hence the computational time for evaluating success states is greatly shortened. In the new algorithm, the density of loss-of-load events in the reduced state space is up to 43.22%, which is more than ten times that of using a traditional PMCS. On the basis of PMCS, the introduction of ISSR effectively optimizes the sampling mechanism of non-sequential simulation, further improving the efficiency of sampling the loss-of-load states and accelerating the computation speed.

In the above study, the modified IEEE RTS-79 system is a calculation example mainly used to verify the computational efficiency of this algorithm. In practice, this study has been supported by the practical project and is successfully applied to the electric power system of Zhejiang province in China. According to the original data provided by Zhejiang Electric Power Company, Zhejiang's 500 kV and main 220 kV grids have a total installed capacity of 190,684 MW and a total peak load value of 24,732 MW, consisting of 126 nodes, 85 load points, 197 lines, and 84 generators. And the system spare capacity is 1852 MW. This paper uses the algorithm to evaluate and analyze the reliability of the actual system. Table 4 shows the reliability indices obtained as well as the computation time via the different algorithms.

Table 4. The reliability indices of practical system obtained via different algorithms.

Algorithm	LOLP	EENS(MWh/Year)	Computation Time/s
TMCS	0.0213	20,992	7 h 43 min
PMCS	0.0218	21,140	4 h 18 min
New algorithm	0.0216	21,385	2 h 24 min

From Table 4, it can be seen that the new proposed algorithm has significant advantages in the reliability evaluation of practical complex systems and its computational efficiency is much higher than the TMCS and PMCS. Therefore, the proposed method has achieved important application value.

6. Conclusions

This paper proposes a fast reliability evaluation method based on the pseudo-sequential Monte Carlo simulation and intelligent state space reduction, and this proposal is tested in the modified IEEE RTS-79 system. The following conclusions are obtained: (1) Compared with the TMCS and traditional PMCS in reliability indices calculation, the proposed algorithm is performed with higher precision and computational efficiency; (2) The ISSR technique optimizes the traditional PMCS sampling mechanism to a large extent, reducing the times of invalid states sampling. But in the state evaluation process, OPF is still adopted for computation. If the state space can be directly divided by artificial intelligence technology, such as in support vector machines or neural networks, the speed of the reliability evaluation will be further accelerated; (3) The computational time decreases with the increase of the number of normal states reduced, while the computational time for state reduction increases. Thus, as for the use of ISSR, future research will focus on how to reasonably choose the optimal level of state space reduction and harmonize the contradiction between the pruning time and simulation time.

Author Contributions: Conceptualization, D.G. and Y.X.; Methodology, R.C.; Software, Z.W.; Validation, D.G., Y.X. and R.C.; Formal Analysis, D.G.; Investigation, Y.X.; Resources, Y.L.; Data Curation, Y.X.; Writing-Review & Editing, Y.X.; Funding Acquisition, W.L.

Acknowledgments: This work has received funding from the National Key R&D Program of China (2017YFB0903100). The National Key R&D Program of China covered the cost to publish in open access.

Conflicts of Interest: The authors declare no conflict of interest.

References

1. Mosadeghy, M.; Yan, R.; Saha, T.K. Impact of PV penetration level on the capacity value of South Australian wind farms. *Renew. Energy* **2016**, *85*, 1135–1142. [CrossRef]
2. Zhang, P.; Wang, Y.; Xiao, W.; Li, W. Reliability Evaluation of Grid-Connected Photovoltaic Power Systems. *IEEE Trans. Sustain. Energy* **2012**, *3*, 379–389. [CrossRef]
3. Zhou, P.; Jin, R.Y.; Fan, L.W. Reliability and economic evaluation of power system with renewables: A review. *Renew. Sustain. Energy Rev.* **2016**, *58*, 537–547. [CrossRef]
4. Da Silva, A.M.L.; de Resende, L.C.; da Fonseca Manso, L.A.; Miranda, V. Composite Reliability Assessment Based on Monte Carlo Simulation and Artificial Neural Networks. *IEEE Trans. Power Syst.* **2007**, *22*, 1202–1209. [CrossRef]
5. Bakkiyaraj, R.A.; Kumarappan, N. Optimal reliability planning for a composite electric power system based on Monte Carlo simulation using particle swarm optimization. *Int. J. Electr. Power Energy Syst.* **2013**, *47*, 109–116. [CrossRef]
6. Alban, A.; Darji, H.A.; Imamura, A.; Nakayama, M.K. Efficient Monte Carlo Methods for Estimating Failure Probabilities. *Reliab. Eng. Syst. Saf.* **2017**, *165*, 376–394. [CrossRef]
7. Nanou, S.I.; Tzortzopoulos, O.D.; Papathanassiou, S.A. Evaluation of an enhanced power dispatch control scheme for multi-terminal HVDC grids using Monte-Carlo simulation. *Electr. Power Syst. Res.* **2016**, *140*, 925–932. [CrossRef]

8. Gubbala, N.; Singh, C. Models and considerations for parallel implementation of Monte Carlo simulation methods for power system reliability evaluation. *IEEE Trans. Power Syst.* **1995**, *10*, 779–787. [CrossRef]

9. Sadeghi, M.; Kalantar, M. Multi types DG expansion dynamic planning in distribution system under stochastic conditions using Covariance Matrix Adaptation Evolutionary Strategy and Monte-Carlo simulation. *Energy Convers. Manag.* **2014**, *87*, 455–471. [CrossRef]

10. Mitra, J.; Singh, C. Incorporating the DC load flow model in the decomposition-simulation method of multi-area reliability evaluation. *IEEE Trans. Power Syst.* **1996**, *11*, 1245–1254. [CrossRef]

11. Singh, C.; Mitra, J. Composite system reliability evaluation using state space pruning. *IEEE Trans. Power Syst.* **2002**, *12*, 471–479. [CrossRef]

12. Mitra, J.; Singh, C. Pruning and simulation for determination of frequency and duration indices of composite power systems. *IEEE Trans. Power Syst.* **1999**, *14*, 899–905. [CrossRef]

13. Green, R.C.; Wang, L.; Singh, C. State space pruning for power system reliability evaluation using genetic algorithms. In Proceedings of the 2010 IEEE Power and Energy Society General Meeting, Providence, RI, USA, 25–29 July 2010; pp. 1–6.

14. Green, R.C.; Wang, L.; Alam, M.; Singh, C. State space pruning for reliability evaluation using binary particle swarm optimization. In Proceedings of the 2011 IEEE Power Systems Conference and Exposition, Phoenix, AZ, USA, 20–23 March 2011; pp. 1–7.

15. Green, R.C.; Wang, L.; Wang, Z.; Alam, M.; Singh, C. Power system reliability assessment using intelligent state space pruning techniques: A comparative study. In Proceedings of the 2010 IEEE International Conference on Power System Technology, Hangzhou, China, 24–28 October 2010; pp. 1–8.

16. Kadhem, A.A.; Wahab, N.I.A.; Aris, I.; Jasni, J.; Abdalla, A.N. Computational techniques for assessing the reliability and sustainability of electrical power systems: A review. *Renew. Sustain. Energy Rev.* **2017**, *80*, 1175–1186. [CrossRef]

17. Jco, M.; Pereira, M.V.F.; Da Silva, A.L. Evaluation of reliability worth in composite systems based on pseudo-sequential Monte Carlo simulation. *IEEE Trans. Power Syst.* **1994**, *9*, 1318–1326.

18. Mello, J.C.O.; Da Silva, A.L.; Pereira, M.V.F. Efficient loss-of-load cost evaluation by combined pseudo-sequential and state transition simulation. *IET Proc. Gen. Trans. Distrib.* **1997**, *8*, 147–154. [CrossRef]

19. Zhang, P.; Li, W.; Li, S.; Wang, Y.; Xiao, W. Reliability assessment of photovoltaic power systems: Review of current status and future perspectives. *Appl. Energy* **2013**, *104*, 822–833. [CrossRef]

Article

Proposals for Enhancing Frequency Control in Weak and Isolated Power Systems: Application to the Wind-Diesel Power System of San Cristobal Island-Ecuador

Danny Ochoa * and Sergio Martinez

Department of Electrical Engineering, E.T.S.I. Industriales, Universidad Politecnica de Madrid, 28006 Madrid, Spain; sergio.martinez@upm.es
* Correspondence: danny.ochoac@alumnos.upm.es; Tel.: +34-608-892-552

Received: 19 February 2018; Accepted: 10 April 2018; Published: 12 April 2018

Abstract: Wind-diesel hybridization has been emerging as common practice for electricity generation in many isolated power systems due to its reliability and its contribution in mitigating environmental issues. However, the weakness of these kind of power systems (due to their small inertia) makes the frequency regulation difficult, particularly under high wind conditions, since part of the synchronous generation has to be set offline for ensuring a suitable tracking of the power demand. This reduces the power system's ability to absorb wind power variations, leading to pronounced grid frequency fluctuations under normal operating conditions. This paper proposes some corrective actions aimed at enhancing the frequency control capability in weak and isolated power systems: a procedure for evaluating the system stability margin intended for readjusting the diesel-generator control gains, a new wind power curtailment strategy, and an inertial control algorithm implemented in the wind turbines. These proposals are tested in the San Cristobal (Galapagos Islands-Ecuador) hybrid wind-diesel power system, in which many power outages caused by frequency relays tripping were reported during the windiest season. The proposals benefits have been tested in a simulation environment by considering actual operating conditions based on measurement data recorded at the island.

Keywords: wind-diesel power systems; power system modeling; variable-speed wind turbine; frequency control; stability analysis; wind power curtailment; inertial control

1. Introduction

Historically, stand-alone diesel power generation has shown to be the most suitable solution for electrification in remote populated areas and islands, due to its reliability and its quick installation [1]. However, burning diesel fuel for that purpose might be expensive and contribute to the emission of greenhouse gasses and to local pollution. In order to tackle these drawbacks, many countries have raised policies and incentives aimed at promoting the integration of alternative power generation technologies, mainly based on the use of renewable energy resources. In this sense, wind-diesel hybridization has been emerging as common practice in many weak and isolated power systems in last years [2]. In such kind of systems, diesel generators are able to correct the power imbalances caused by the load variation as well as the fluctuating nature of wind power. Nevertheless, during high wind condition periods, the grid frequency control could be a challenging task, which deserves special treatment compared to that given to large power systems. In many cases, the system operator could force to limit part of the available wind power in order to preserve an adequate frequency regulation margin to face any frequency event, for example, caused by a sudden loss of power generation [3–5]. This procedure, also called wind power curtailment, is necessary since modern wind turbines incorporate a grid connection interface based on power electronics, and do not provide inertia to the system as the conventional synchronous generators actually do in a natural manner. When the penetration of wind energy increases, the inertial response of the system tends to degrade progressively, which represents a threat for the power supply continuity [6]. Real situations that confirm the aforementioned technical difficulties took place in the San Cristobal Island hybrid wind-diesel power system (SCHPS), located at the Galapagos Islands-Ecuador, where the high participation of wind energy in the load sharing caused partial and total power outages during the windiest season of 2015. These events were caused by the tripping of certain frequency relays installed to protect the generation units. Given the seriousness of that situation, it has been found interesting to consider such power system as a case of study in this work.

Nowadays, it is possible to find in the literature many scientific contributions aimed to face the technical issues associated with the significant integration of wind turbines into small power systems. For instance, works presented in [6–11] propose the use of additional energy storage systems such as batteries, ultra-capacitors, and flywheels, in order to absorb the fluctuating nature of the power injection from wind and, therefore, to reduce the induced frequency deviations in the grid. Although these solutions have proven to be the most effective, from a technical perspective, the costs associated with their installation and maintenance could constitute a drawback to consider. On the other hand, there are also control strategies that focused on taking advantage of the available kinetic resources existing in the wind turbines in order to enable the frequency support provision while avoiding additional hardware installation. These contributions, as in [12–16], are conceived to provide a variable speed wind turbine with a suitable inertial response to support the grid frequency in normal operating conditions and in case of contingency. Since these strategies force the wind turbine to operate under non-optimal conditions, the cost of the wind energy waste could represent a clear disadvantage. However, many grid codes consider the possibility of sacrificing part of the generated energy in order to safeguard the quality, stability, reliability and continuity of the power supply [17,18]. This paper presents some proposals aimed at enhancing the frequency response of a weak and isolated power system with a significant share of wind energy by using the approach offered by the second group of solutions, that is, by leveraging the existing equipment installed in the system.

The first contribution of this paper is to present a procedure for evaluating the stability margin of the power system based on the classical root-locus analysis in order to foresee the consequences of manipulating certain parameters of the diesel-engine speed governors in the grid frequency response. From the previous analysis, an adjustment of the speed governor gains towards improving the frequency support capability from conventional generators is suggested. The proposed procedure is based on the conventional generation only, since variable-speed wind turbines do not provide, in principle, inertial response to the grid. Next, to go beyond this approach, the solutions proposed below are intended to make it possible for the wind turbine to mitigate its negative effects to the grid, and are designed to be implemented in its power controllers.

The second improvement proposal, which constitutes the main contribution of this paper, consists in a new wind power curtailment strategy with better benefits compared to those strategies based on the pitch angle regulation. This approach proposes the combined use of electromagnetic and mechanical variables to perform an effective wind power limitation. This results in an almost constant power injection to the grid when the available wind power exceeds a maximum limit established by the system operator, leading to a significant reduction of the grid frequency fluctuations under wind curtailment conditions.

Complementarily, in order to get a milder evolution of the grid frequency in those intervals where the wind energy is delivered under the criterion of optimal conversion (non-curtailment condition), the implementation of an inertial control based strategy in the wind turbine controllers is presented as a third corrective action. Such strategy enables the participation of the variable-speed wind turbines in the grid frequency support in cooperation with the conventional synchronous generation under normal operating conditions and under contingency scenarios.

This paper is organized as follows: Section 2 introduces the description and modeling of the SCHPS along with its implementation in MATLAB/Simulink®. All the proposals devised to enhance the frequency control tasks by the test power system when it is subjected to normal operating conditions are presented in Section 3. In order to assess the effectiveness of the improvement proposals under more critical operating conditions, Section 4 presents a detailed study of the test system when it experiences a frequency event caused by a sudden loss of wind power. Finally, the most relevant conclusions are established in Section 5.

2. Power System under Study

2.1. Description and Modeling

Figure 1a shows the configuration of the hybrid wind-diesel power system of San Cristobal Island. The conventional power plant consists of three 813 kVA-synchronous generation units (DG_1, DG_2, and DG_3), each coupled to a diesel engine. These units inject the produced energy to a three-phase 13.2 kV bus of the distribution substation (DSS) located in the courtyard of the diesel power plant. From the DSS, three primary feeders distribute the distribution of the generated energy at different consumption points of the island. The San Cristobal wind farm, located 12 km from the DSS, is composed of three 800 kW-full converter wind turbines (WT_1, WT_2, and WT_3) whose delivered energy is dispatched to the DSS through a 13.2 kV transmission line. Table 1 provides some relevant parameters from the nameplate of individual generation units installed at SCHPS.

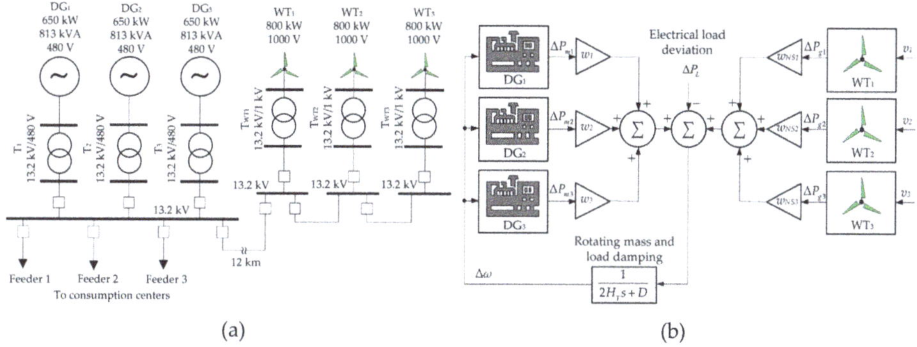

(a)   (b)

Figure 1. Wind-diesel hybrid power system of San Cristobal Island: (**a**) Layout diagram; (**b**) Load frequency control scheme.

Table 1. Dataset of generation units.

Wind Turbine		Diesel-GENSET	
Model	MADE A-59	Model	CAT-3512 DITA
Type	IV-Full converter	Type	-
Rated power	800 kW	Rated power	650 kW
Generator speed range	750–1650 RPM	Capacity	813 kVA
Gearbox ratio	1:66.185	Output voltage	480 V $\pm$ 5%
Synchronous speed	1500 RPM	Frequency	60 Hz
Converter output voltage	1000 V $\pm$ 10%	Synchronous speed	1200 RPM
Converter output frequency	60 Hz	Equivalent constant of inertia (base 3 × 813 kVA)	0.4208 s

The load frequency control (LFC) scheme that represents the short-term behavior of frequency and power of the SCHPS is depicted in Figure 1b. That scheme is derived from the swing equation of a synchronous generator (1) [19]. In this equation, that considers small deviations of power and rotor speed, the inertial characteristics of the synchronous generators are described by the equivalent constant of inertia, H_T. The electric load is characterized by its deviation ΔP_L from its pre-disturbance value and its damping effect, D. $\Delta\omega$ is the equivalent synchronous rotor speed deviation, which is equivalent to the grid frequency deviation, Δf, if both are expressed in per unit. ΔP_m is the mechanical power deviation introduced by the speed governor in response to a grid frequency change Δf, and ΔP_g denotes the deviation of the active power injected by the wind turbine. v is the horizontal component of wind speed incoming to the wind rotor.

$$\Delta P_{m[pu]} - \Delta P_{L[pu]} = 2H_{T[s]}\frac{d\Delta\omega_{[pu]}}{dt} + D\Delta\omega_{[pu]} \tag{1}$$

The factors w_i are the energy share weights and represent the ratio between the i-th synchronous generation unit capacity and a specified base power. Complementarily, w_{NSi} are the energy share weights from the non-synchronous generation units (variable-speed wind turbines, in this case) whose definition is similar to w_i. The use of these factors allows the expression of all the power deviation signals from the generation units in the same per unit base chosen for the whole power system. However, as it is expressed in (2), only the factor w_i must be considered for the evaluation of H_T. In this definition, H_i is the constant of inertia of each synchronous generation unit computed in its own base power, and m is the total number of online synchronous generators.

$$H_T = \sum_{i=1}^{m} w_i H_i \tag{2}$$

The model that represents the time-domain behavior of the diesel generation unit has been taken from [20] and schematized in Figure 2, where ω_r is the rotor shaft speed, T_m denotes the mechanical torque, and P_m is the mechanical power. T_{max} and T_{min} are the mechanical torque limits, P_{m0} is the pre-disturbance value of P_m, and ΔP_m is the mechanical power deviation.

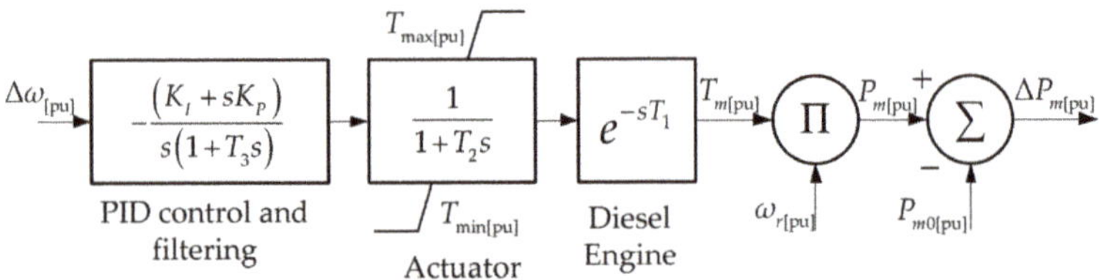

Figure 2. Simplified model of the diesel generation unit.

In addition, the general scheme of the wind turbine model installed at the San Cristobal wind farm is shown in Figure 3a. A three-blade wind rotor is coupled to a two pole-pairs synchronous generator throughout a gearbox. The stator windings of the electrical generator are connected to the grid via a back-to-back power electronics converter (PEC). The decoupling provided by the DC-link allows the wind turbine to operate with variable speed regardless the grid frequency value. The PEC consists of two converters: Machine Side Converter (MSC) and Grid Side Converter (GSC) [21]. The MSC controls the active power delivered by the synchronous generator following a maximum power point tracking (MPPT) characteristic, which is devised to maximize the mechanical power extraction by the turbine, while the GSC regulates both the DC-link voltage and the output reactive power at the grid connection point [22]. Finally, the rotor of the synchronous generator is excited by an external DC power supply.

In order to represent the short-time dynamics of the variable speed wind turbine, the simplified electromechanical model proposed in [23] and depicted in Figure 3b has been used. This model, originally designed for a doubly-fed induction generator based wind turbine, can be applied to represent a wind turbine with a synchronous generator via full converter due to the similarities between their mechanical topologies and because, within the time frame considered in the load frequency control studies, the electromagnetic time constants are negligible compared to the mechanical ones. This fact allows to represent the dynamics of the power electronic converter and the electrical generator by a first order transfer function with time constant τ_C. In Figure 3b, T_t and T_{em} are the mechanical and the electromagnetic torque, respectively. P_g is the total output active power, ω_t and ω_g are the angular speed of the turbine and the electric generator, and β denotes blade pitch angle. P_{g0} and ΔP_g are the pre-disturbance value and the deviation of the output active power, respectively. The variables with the superscript * represent their set-point values. The rest of parameters of both models are defined in the Appendix A.

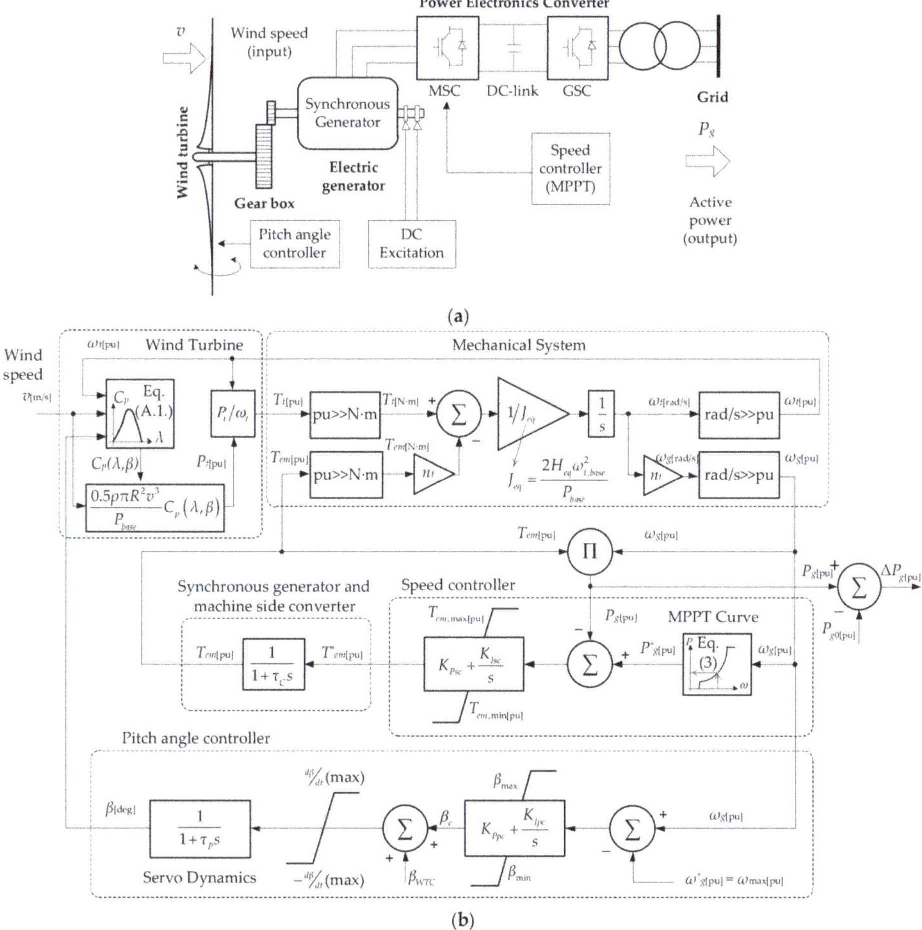

Figure 3. Wind turbine installed at the San Cristobal wind farm: (**a**) General layout; (**b**) Simplified electro-mechanical model.

Based on commercial datasheets and real data recorded in field tests performed at the SCHPS, all the parameters of those models have been assigned and tuned in order to represent, in a reliable way, the actual power system performance. Figure 4 shows the power curves particularized for the commercial wind turbine model. In this illustration, the MPPT curve is also plotted, whose analytical representation is expressed in (3). In this equation, K_{opt} is the optimization constant which depends on constructive characteristics of the wind rotor, and P_{max} is the maximum active power to be delivered by the wind turbine. The different operating zones of the wind turbine along with their associated parameters are graphically defined in Figure 4a. Appendix A contains all the numerical data necessary for the evaluation of (3).

$$P_t = \begin{cases} \dfrac{K_{opt}\omega_0^3}{(\omega_0-\omega_{min})}\left(\omega_g - \omega_{min}\right), & \omega_{min} \leq \omega_g < \omega_0 \\[2mm] K_{opt}\omega_g^3, & \omega_0 \leq \omega_g \leq \omega_1 \\[2mm] \dfrac{P_{max}-\overbrace{K_{opt}\omega_1^3}^{P_{v(base)}}}{(\omega_{max}-\omega_1)}\left(\omega_g - \omega_{max}\right) + P_{max}, & \omega_1 < \omega_g \leq \omega_{max} \end{cases} \tag{3}$$

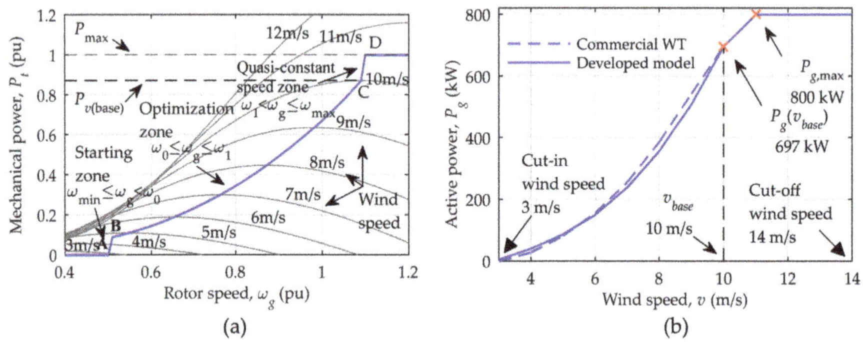

Figure 4. Wind turbine power curves: (**a**) ω-P chart; (**b**) Power-wind speed curve.

The models depicted in Figures 2 and 3 were validated by considering some real contingency situations in the studied power system. For instance, Figure 5 shows the time domain behavior of some variables recorded in the SCHPS as a consequence of a sudden loss of 700 kW of wind energy caused by the intentional disconnection of wind turbines WT_1 and WT_2. Such disturbance was introduced elapsed 18 s from the beginning of data recording. Prior to the contingency, the electric load was supplied only by the machines: DG_1, DG_2, WT_1, and WT_2, whose active power values are provided in Table 2. During the time frame covered by the data acquisition, the actual demanded power hardly experiences any variation, hence, this variable will be kept unchanged along the time domain simulation ($\Delta P_L = 0$).

Table 2. SCHPS real contingency scenario.

Inertial Characteristics of Power System	
H_T	D
2×0.4208 s	0.5 pu
Pre-Disturbance Values	
Wind	Diesel
$P_{g1(0)} = 370$ kW ($v_1 = 8.1$ m/s)	$P_{m1(0)} = 240$ kW
$P_{g2(0)} = 330$ kW ($v_2 = 7.8$ m/s)	$P_{m2(0)} = 260$ kW

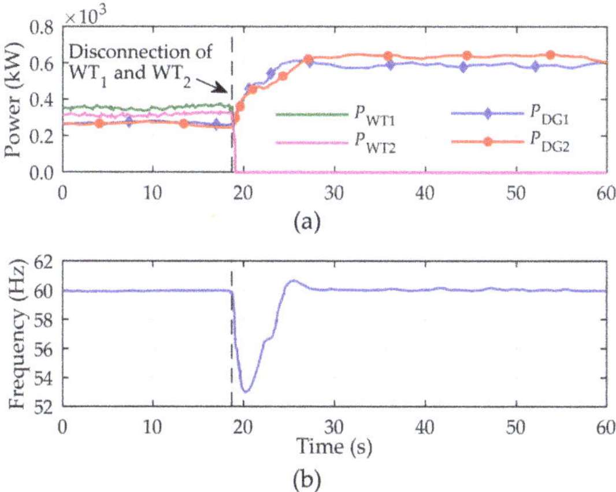

Figure 5. Real contingency scenario at the SCHPS: (**a**) Power generation; (**b**) Grid frequency.

For the implementation of the LFC-scheme in MATLAB/Simulink®, the set of values assigned to the parameters of the speed governor of diesel-generation units has been replicated on the three thermal machines, for simplicity. Therefore, an identical response of the power and rotor speed is expected for DG_1–DG_3 under both normal operation and contingency. A similar reasoning has been applied to the model parameters of the wind turbines WT_1–WT_3. The base power for per unit calculation is chosen as the total installed capacity of the diesel power plant (S_{base} = 3 × 813 kVA), so, the energy share factors will be $w_1 = w_2 = w_3 = 1/3$ and $w_{NS1} = w_{NS2} = w_{NS3} = 0.328$. Table 2 presents relevant information for the representation of this specific contingency scenario, where the generators DG_3 and WT_3 remain offline. The time domain behavior of the variables of interest recorded in the real power system and that obtained in simulation are compared in Figure 6. Note that, it exists a high correlation, in amplitude and time, between the response of grid-frequency and total active power of the wind turbine. The similarities between the curves obtained by simulation and from real measurements allow to validate the SCHPS model presented in this section.

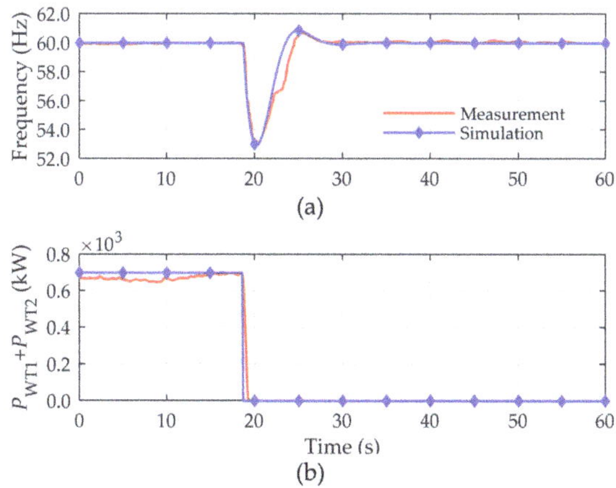

Figure 6. SCHPS model validation: (**a**) Grid frequency; (**b**) Wind farm generated power.

2.2. Operation and Control

According to the proceeding manual of the SCHPS, the total active power produced by the wind farm, P_{WF}, is regulated by a wind farm controller (WFC) structure as it is shown in Figure 7a. Two criteria are used for dispatching the energy production of the wind farm: ecological and technical. For addressing the first, the active power reference signal, P^*_{WF}, is set so that the power demand can be supplied by both the wind farm and the diesel power plant by burning the least possible fossil-fuel. For preserving the efficiency and lifespan of the diesel-engines, it is desirable that they operate above their minimum mechanical power, $P_{m,min}$, defined about 25% from its rated value. The total number of online diesel generators is denoted by n_{DG}.

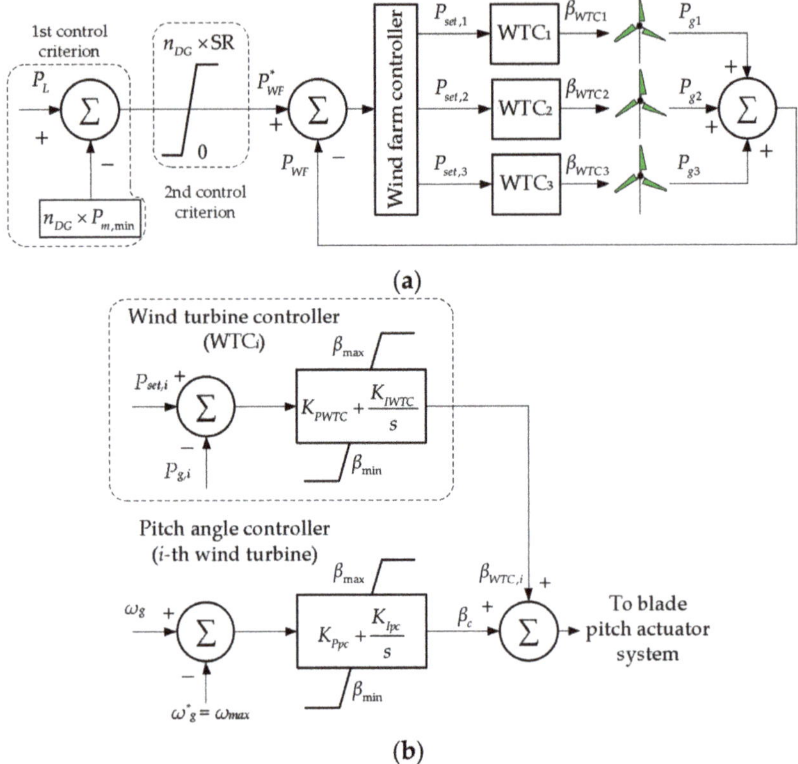

Figure 7. Wind farm controller structure: (**a**) General scheme; (**b**) Implementation of the WTC.

For meeting the second criterion, it is expected that each online diesel generator must be able to withstand a sudden loss of wind power by using its spinning reserve (SR) to preserve the power system stability. In the SCHPS, the power reserve is calculated as: $SR = (P_{nom} - P_{m,min})$. For implementing this criterion, a saturation block is added to the control scheme and its output is applied to the WFC. The WFC sends a maximum active power reference, P_{set}, to the i-th wind turbine controller (WTC). The WTC uses the pitch angle, β_{WTC}, as a control variable in order to limit the wind power captured by the i-th turbine. This signal is applied to the blade pitch angle controller of each wind turbine as Figure 7b illustrates. If the actual total wind power production is below its maximum specified value, each wind turbine is required to operate according to a MPPT-characteristic to extract the optimal power from wind.

2.3. Simulation of the SCHPS under Normal Conditions

The historical dispatching records of the SHSC reveal that between June and September there is a surplus of wind energy production every year [24]. In a particular case, high favorable wind conditions were registered in the San Cristobal wind farm on August 2015. This condition led wind turbines to cover the energy needs of the island up to 66% under the maximum demand scenario ($P_{L\max}$ = 2424 kW). During the same period, several unplanned power outages were reported. These were mainly caused by the trip of certain frequency protection relays both in the wind farm and in the diesel power plant, by causes still unknown by the operators. Given the technical seriousness of these events, such scenario will be taken as the main case of study in this work. Tables 3 and 4 list relevant information concerning to the real status of the SHSC for the scenario described above, that approximately corresponds to the highlighted point in the wind speed profile shown in Figure 8. This illustration was generated by using real data measured at the hub height of the wind turbines (51 m) under gusty conditions.

Table 3. SCHPS power production *.

Magnitude	Wind			Diesel		
	WT$_1$	WT$_2$	WT$_3$	DG$_1$	DG$_2$	DG$_3$
Wind speed (m/s)	10.0	10.9	11.2	-	-	-
Active power (kW)	508	508	575	319	239	275
Total power (kW)		1591			833	
Demand covering (%)		65.65			34.35	

* Actual measurements from the San Cristobal hybrid wind-diesel power system on 08/08/2015 at 19:10:00 (UTC—6:00).

Table 4. SCHPS power consumption *.

Primary Feeder Number	1	2	3	TOTAL
Active power (kW)	1072	1045	307	2424.00

* Actual measurements from the San Cristobal hybrid wind-diesel power system on 08/08/2015 at 19:10:00 (UTC—6:00).

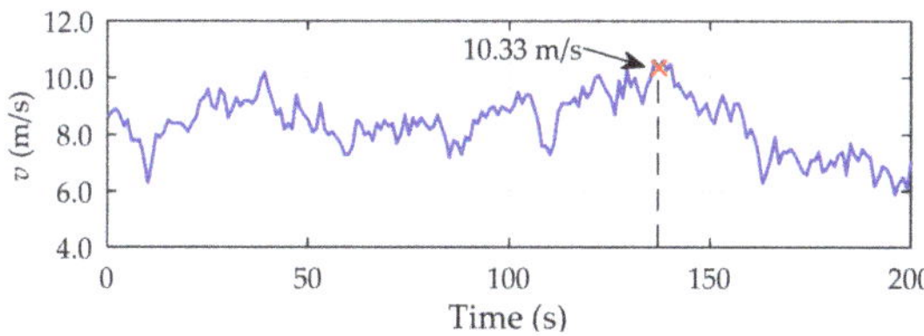

Figure 8. Actual wind speed profile used in the time domain simulations.

For the reproduction of this case of study in MATLAB/Simulink®, the numerical value of all the parameters of both the conventional power plant and the wind farm have been assigned by using the same criteria than in Section 2.1. Nevertheless, in this case, the three synchronous generation units are online, leading to have an equivalent constant of inertia of $H_T = 3 \times 0.4208$ s. Additionally, the geographic distribution of the wind turbines, wake effect and wind transportation delay are neglected in the simulation, so, the wind profile of Figure 8 is applied simultaneously to the three wind turbines. According to the procedure described in Section 2.2., the set-point value P^*_{WF} for this specific case, which corresponds to a wind curtailment situation, is established as follows:

- First control criterion:

$$P_L - n_{GD} \times P_{m,\min} = 2424 - 3 \times 162.5 = 1936.5 \text{ kW}$$

- Second control criterion:

$$\text{SR} = (P_{nom} - 0.25P_{nom}) = 487.5 \text{ kW} \approx 500 \text{ kW}$$

$$n_{DG} \times \text{SR} = 3 \times 500 = 1500 \text{ kW}$$

- WFC active power set-point: $P^*_{WF} = 1500 \text{ kW}$

Figure 9a shows the active power dispatched by each wind turbine obtained by simulation. The total power injected by the wind farm into the grid is illustrated in Figure 9b, along with the theoretical optimal wind power which is not dispatched due to the activation of the curtailment operation mode in the WFC.

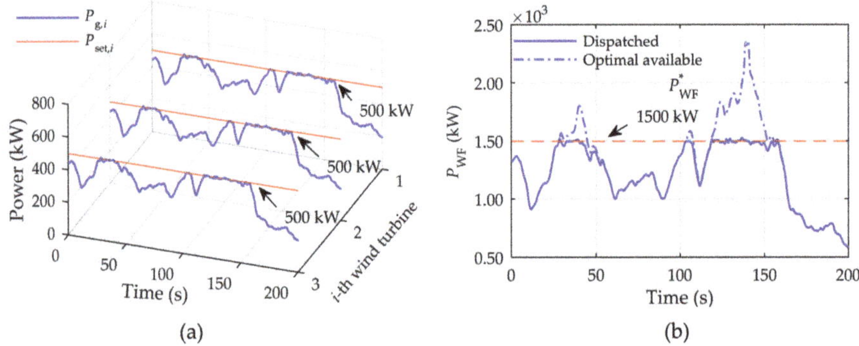

Figure 9. Dispatched active power: (**a**) Wind turbines; (**b**) Wind farm.

Finally, the simulated dynamics of the active power delivered by the SCHPS generation agents and the grid frequency are shown in Figures 10 and 11, respectively.

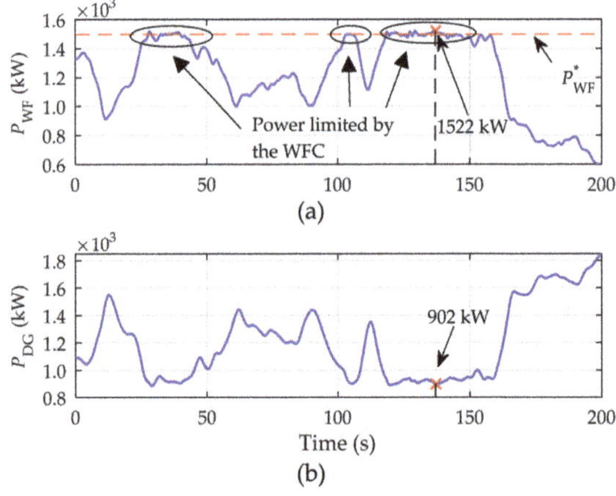

Figure 10. Delivered active power: (**a**) Wind farm; (**b**) Diesel power plant.

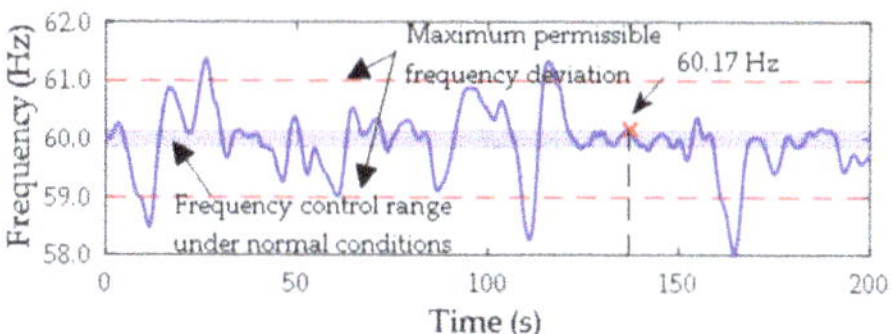

Figure 11. Time domain behavior of grid frequency.

2.4. Analysis of the Results and Problem Statement

The time domain behavior of the output power delivered by the wind turbine shown in Figure 10a reveals a correct performance in power curtailment around the predefined set point, $P^*_{WF} = 1500$ kW. When the available active power is below the set point value, the wind energy is dispatched according to the optimal wind power conversion criterion, as expected. Complementarily, in Figure 10b, it can be seen that the fluctuating wind power injection is effectively absorbed by diesel generators in order to match the power generation and consumption. It should be noted the similarity between the values highlighted in Figure 10 and those provided in Table 3. This fact allows to validate the implemented SCHPS-model in terms of accuracy.

In reference to Figure 11, the grid frequency presents an inadmissible evolution in terms of power quality. This figure shows the permissible frequency ranges established by the Ecuadorian Grid Code (EGC) [25], which are mandatory for any generation unit integrated to the National Bulk Power System. The EGC requires that the nominal frequency be set to 60 Hz, with a control range of $\pm$ 0.15 Hz (under normal conditions). In order to avoid unnecessary trip of frequency relays, the EGC recommends to maintain the grid frequency excursion between the values 59–61 Hz. However, since the particular characteristics of the SCHPS (isolated operation, small-capacity of synchronous generators, high wind energy penetration levels, and others) make the frequency control tasks difficult, some solutions aimed at improving the power quality indexes have to be proposed. These solutions are discussed in the following sections and are applied to both the diesel power plant and the wind farm.

3. Proposals to Enhance the Power System Frequency Control under Normal Conditions

In this section, some proposals intended to enhance frequency control actions in a weak and isolated power system with high wind energy penetration are presented. These solutions consider the adjustment of certain parameters of the diesel-engine speed governors and the incorporation of additional control strategies to the wind turbine controllers.

3.1. Adjustment of the Control Gains of the Diesel-Engine Speed Governor

The first improvement proposal, and the only one applied to the conventional power plant, consists of readjusting the controller gains associated to each diesel-engine speed governor (Figure 2) based on a stability analysis. By considering identical characteristics of the diesel generator units and their speed governors (a real situation in the SCHPS), it is possible to obtain an aggregate representation of the thermal power plant and incorporate it in the LFC-scheme as shown Figure 12. In this scheme, w_T denotes the equivalent energy share factor that, in this case, can be calculated as a sum of the individual energy share weights, w_i (see the definition of w_i in Section 2.1).

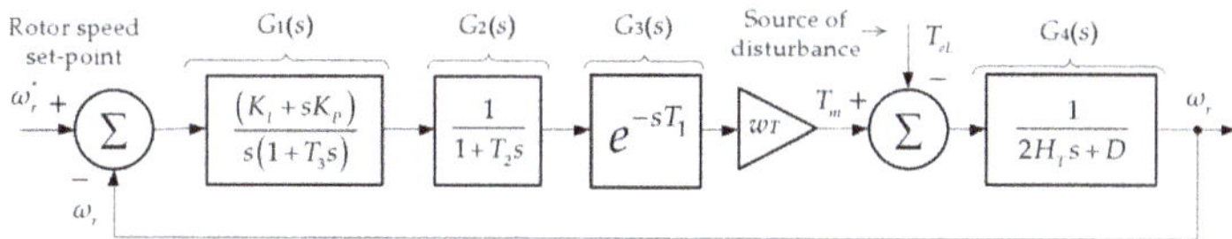

Figure 12. Aggregate representation of the diesel power plant in the LFC-scheme.

The linear system schematized above corresponds to an alternative representation of the LFC-scheme shown in Figure 1b, which is derived by considering the torque difference in the swing Equation (1) instead of small power deviations, as it is expressed in (4). In this equation, T_m and T_{eL} are the mechanical and the electromagnetic-load torque expressed in per unit. Since the variable-speed wind turbine lacks a natural response to the grid frequency deviations, its model is not considered in the stability analysis presented in this section. However, its potential contribution to the grid frequency support will be studied in detail in Section 3.3.

$$T_{m[\text{pu}]} - T_{eL[\text{pu}]} = 2H_{T[\text{s}]}\frac{d\omega_{r[\text{pu}]}}{dt} + D\omega_{r[\text{pu}]} \tag{4}$$

In order to evaluate the disturbance rejection ability of the speed governor while maintaining the rotor speed around its set-point (nominal grid frequency), the closed-loop transfer function that relates the output variable, ω_r, and the input variable, T_{eL}, in Figure 12, can be formulated as (5). The transfer functions $G_1(s)$–$G_4(s)$ are graphically defined in this figure.

$$\frac{\omega_r(s)}{T_{eL}(s)} = -\underbrace{\frac{G_4(s)}{1 + w_T G_1(s)G_2(s)G_3(s)G_4(s)}}_{\text{Characteristic equation}} \tag{5}$$

For obtaining a rational transfer function form from (5), the time-delay characteristic introduced by $G_3(s)$ is approximated by using the first-order Páde approximant [26], as follows:

$$e^{-T_1 s} \approx \frac{2/T_1 - s}{2/T_1 + s} \tag{6}$$

By substituting (6) in (5), the following symbolic equation is attained:

$$\frac{\omega_r(s)}{T_{eL}(s)} = -\frac{s(T_1 s + 2)(T_2 s + 1)(T_3 s + 1)}{s(T_1 s + 2)(T_2 s + 1)(T_3 s + 1)(2H_T s + D) - w_T(T_1 s - 2)(K_I + K_P s)} \tag{7}$$

For a numerical evaluation of the previous equation, the real case of study depicted in Section 2.1. will be considered (where: $H_T = 2 \times 0.4208$ s, $D = 0.5$, and $w_T = 2/3$, and the information summarized in the Appendix A). The resulting expression is:

$$\frac{\omega_r(s)}{T_{eL}(s)} = -\frac{0.001s^4 + 0.1933s^3 + 10.17s^2 + 83.33s}{0.001683s^5 + 0.3259s^4 + 17.21s^3 + 143.8s^2 + 168.1s + 81} \tag{8}$$

Now, the stability margin of the SCHPS is determined by analyzing the trajectory described by the eigenvalues on the s-plane as a consequence of a gradual increase of a system gain, K. According to the classical control theory, such gain is introduced in the characteristic equation as follows:

$$1 + K[w_T G_1(s)G_2(s)G_3(s)G_4(s)] = 0 \tag{9}$$

Note in (9) that, for the particular linear system studied in this work, the K gain affects to the entire set of parameters of the speed governor as well as the inertial characteristics of the power system. Given the difficulty of modifying the value of some intrinsic parameters of the LFC-scheme (such as time constants, constant of inertia and damping effects), the only parameters that can be adjusted to produce a change in K are the control gains of the speed governor of the diesel generator: K_P and K_I.

Figure 13 shows the root-locus plot generated from (8). It reveals that the SCHPS is stable when $K = 1$ (that is, by using the default values of the controller gains specified in the Appendix A) and it will remain under this condition as long as the gain K is less than 33.6. This critical value gives the maximum multiplying factor of the controller gains, K_P and K_I, that ensures a stable response of the system under contingency and guarantees a proper load sharing in normal operating conditions.

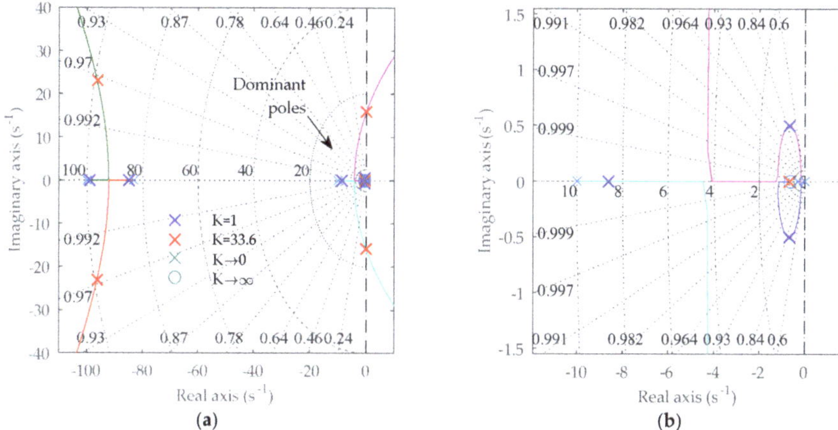

Figure 13. Root-locus stability analysis of the SCHPS: (**a**) All poles; (**b**) Detailed zone of dominant poles.

Complementarily, Figure 14 shows the effect of increasing K on the dynamics of the frequency of the SCHPS model implemented in MATLAB/Simulink®. As expected, larger values of K improve the frequency recovery process in terms of the maximum instantaneous frequency deviation (absolute peak value), the recovery time, and the settling time. However, increasing values of K give place to complex conjugated poles (See dominant poles trajectory in Figure 13b), and an oscillation will appear after the frequency rebound, the higher value of K, the more sustained the oscillation. When K reaches its critical value, the power system becomes unstable.

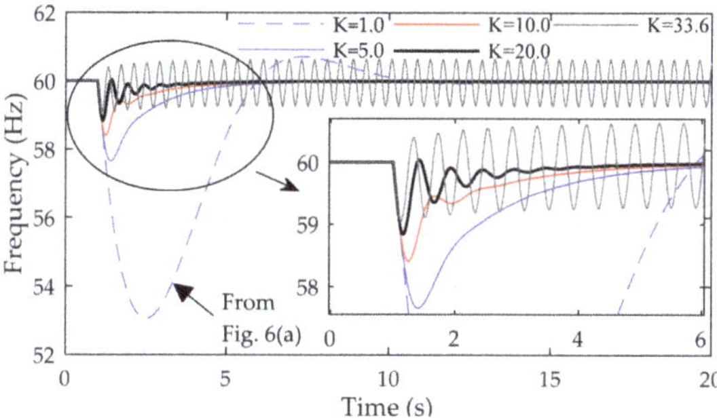

Figure 14. Effect of increasing K on the frequency recovery process of the SCHPS.

From the results presented in Figure 14, it can be concluded that there is a specific margin for increasing the K gain that could enhance the frequency response of the power system in amplitude and time. However, due to the subjectivity of the term "most suitable" to define the desired frequency response, the selection of the multiplying factor K has to be made by using a purely technical criterion based on three specific objectives: (1) attain a faster response from the controller in the frequency recovery process; (2) reduce the frequency drop and; (3) after the frequency rebound, damp the remaining frequency oscillations in a relative short time. Objectives 1 and 2 are designed to maintain an adequate power quality index and the latter is aimed to avoid causing additional mechanical stress in the diesel engine drives during the frequency control process. In this work, it has been verified that a factor of multiplication of $K = 10$ allows to fulfill the objectives described before. Figure 15 shows the results obtained by adjusting the diesel-engine controller gains and applying such modifications to the SCHPS model implementation particularized to the case of study presented in Section 2.3. In this figure, it can be seen that the power delivered by diesel generators improves the absorption of the fluctuating wind power injection in order to match the generated and demanded power in the power system. As a consequence, the excursions of the frequency are appreciably reduced, being great part of these within the band of control required by the applicable grid code. Similar benefits can be observed in the evolution of the rate of change of frequency (ROCOF). With the implementation of the corrective actions on the conventional generation, the average value of this variable is reduced by approximately 89%, which leads to substantially reduce the risk of triggering of frequency relays installed at different points in the SCHPS.

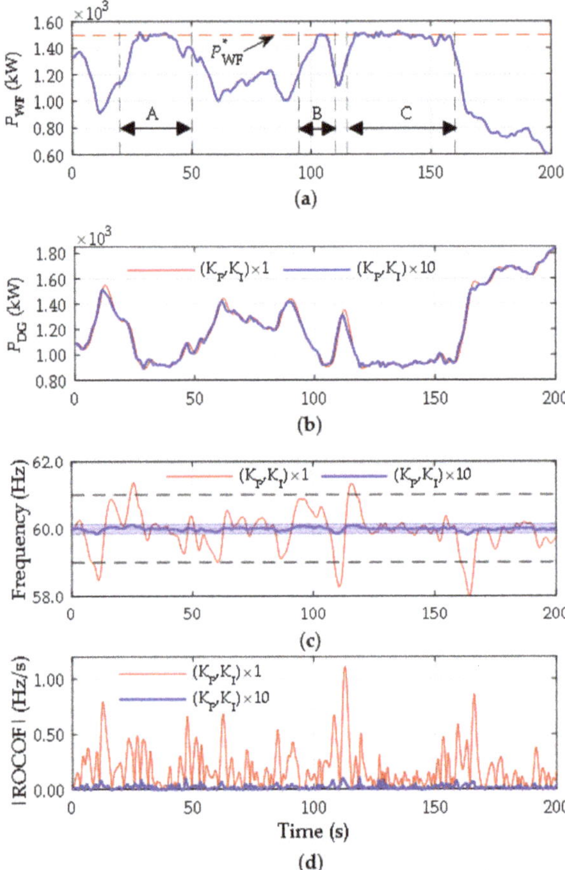

Figure 15. Adjustment of the control gains of the diesel-engine speed governor and its effect on the power system dynamics: (**a**) Wind farm generated power; (**b**) Diesel power plant production; (**c**) Grid frequency and; (**d**) ROCOF.

3.2. Improvement of the Wind Power Curtailment Strategy

The wind power curtailment strategy implemented in the SCHPS, and presented in Section 2.2., acts directly on the blade pitch angle to limit the power delivered by each wind turbine to a maximum value established by the centralized wind farm control, WFC. Although the results obtained in the simulation reveal that these control actions are performed properly, the handling of mechanical variables might introduce transient errors in the set-point tracking P_{set}, inducing small grid frequency deviations as a consequence of the non-continuous wind power injection under curtailment operation mode (Figures 10a and 11). At first glance, it would seem that an increase of the gains of the PI controller implemented in the WTC (Figure 7b) could improve the effectiveness of the existing control strategy. However, it is not a suitable solution since it would subject the turbine blades to an additional mechanical stress caused by greater variations of the pitch angle for wind power limitation. The control strategy proposed in this section aims to improve the performance of the control philosophy implemented in the real wind farm and is based on the idea of combining the use of electromagnetic and mechanical variables for wind power curtailment.

The method consists of modifying the MPPT curve, implemented in the wind turbine speed controller, by adjusting the upper limit of the active power to be generated, P_{max}, to the curtailment level required by the WFC, P_{set}, as illustrated in Figure 16a. If the incoming wind speed causes the mechanical power transmitted to the shaft to be below P_{set}, the wind turbine will dispatch its active power under the default optimal conversion criterion. On the contrary, if the mechanical power delivered by the turbine exceeds P_{set}, the wind turbine will inject at most the set-point power P_{set}. Since, in this situation, the inner wind turbine power mismatch would cause an undesirable rotor over-speeding, it is necessary to compensate such imbalance by acting on the blade pitch control system. In order to address this, a new maximum rotor speed reference, ω'_{max}, has to be specified. A new operating zone (points C′ and D′ in Figure 16b) is defined by displacing the quasi-constant speed zone of the MPPT curve along the optimal power locus until its maximum power value coincides with the specified P_{set}. Such operating zone is characterized by the power values P_{set} and P_1, and the rotor speeds ω'_1 and ω'_{max}.

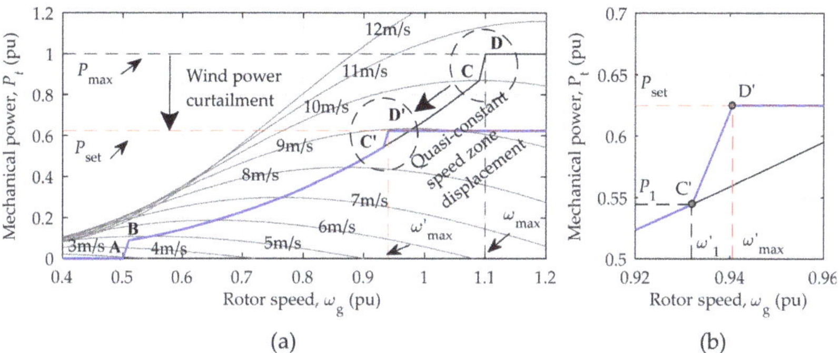

Figure 16. Active power curtailment procedure: (**a**) MPPT curve clipping; (**b**) Quasi-constant speed zone under curtailed condition.

The displacement of the quasi-constant speed zone is achieved by substituting the values P_{set}, P_1, ω'_1 and ω'_{max}, by P_{max}, $P_{v(base)}$, ω_1 and ω_{max}, in Equation (3), respectively. The values associated to the new operating points, C′ and D′, can be calculated by considering the same power and speed ratios than in the quasi-constant speed zone of the default MPPT curve, as follows:

$$\frac{P_1}{P_{set}} = \frac{P_{v(base)}}{P_{max}} \tag{10}$$

$$\frac{\omega'_{max}}{\omega'_1} = \frac{\omega_{max}}{\omega_1} \tag{11}$$

Since point C′ is still located in the optimization zone, ω'_1 can be calculated by using (3) as follows:

$$\omega'_1 = \left(\frac{P_1}{K_{opt}}\right)^{\frac{1}{3}} = \left(\frac{P_{v(base)}}{P_{max}}\frac{P_{set}}{K_{opt}}\right)^{\frac{1}{3}} \qquad (12)$$

Figure 17 shows the schematic implementation of the proposed strategy in the wind turbine. The values P_{set}, P_1, ω'_1 and ω'_{max}, are applied to the WTC, in order to evaluate the MPPT curve and, ω'_{max} to the blade pitch angle controller for a proper rotor speed limitation. Since the proposal is designed to be implemented locally in the WTC associated with the i-th wind turbine, the external configuration of the wind farm controller will maintain its original topology (Figure 7a).

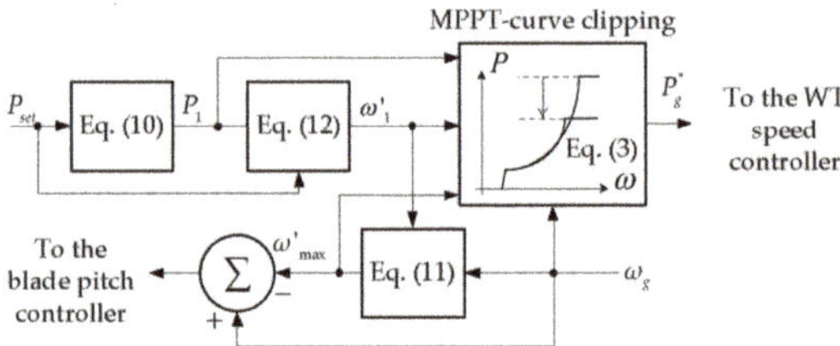

Figure 17. Implementation of the wind power curtailment proposal on the i-th WTC.

Figures 18 and 19 show the results obtained by implementing the proposal in the SCHPS model for the case of study described in the final part of previous subsection (with the diesel-engine speed governor gains readjusted). These results are compared with those generated by considering the original wind farm controller (blue line curves in Figure 15). In order to better show the time evolution of the variables of interest, only those intervals where the wind power is curtailed (defined as A, B, and C in Figure 15a) are shown in Figure 18. This is because outside these temporal ranges, the dynamics obtained in the two cases compared are practically the same.

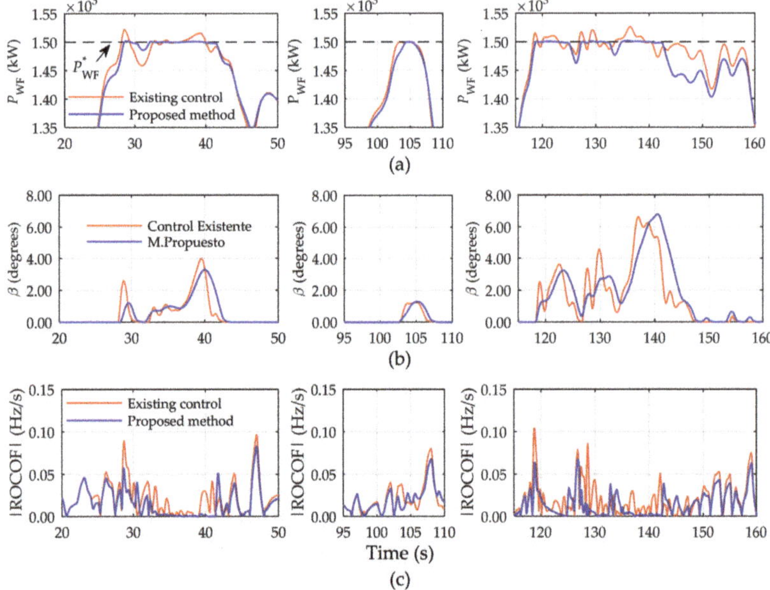

Figure 18. Time-domain simulation results by implementing the wind curtailment proposal: (**a**) Wind farm output power; (**b**) Blade pitch angle and; (**c**) ROCOF.

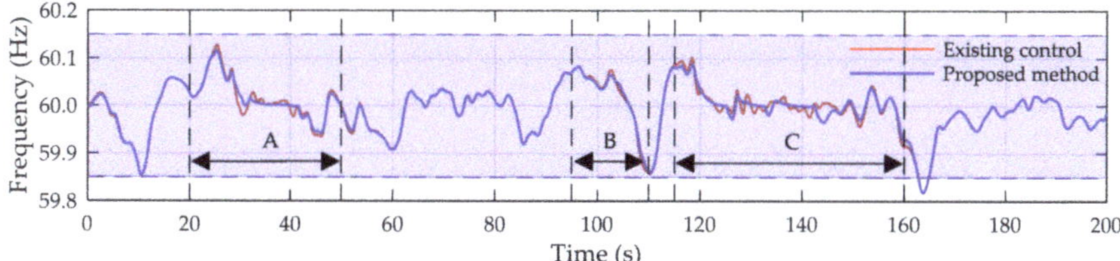

Figure 19. Effects of the proposal on the grid frequency.

In Figure 18a, it can be seen the notorious improvement in the wind curtailment performance achieved with the proposed method, as expected. Since the control strategy is based on a combined action of electromagnetic and mechanical variables to perform a finer tracking of the set-point signal P_{set}, the evolution of the blade pitch angle is much smoother as it can be seen in Figure 18b. Regarding the ROCOF, the implementation of the proposal allows to achieve an average reduction of this variable by 21.66% as shown in Figure 18c. Finally, Figure 19 shows the reduction of frequency fluctuations obtained by applying the proposal, specifically in those intervals corresponding to the actions of wind power curtailment. In order to better show the grid frequency excursions in this figure, the scale of the y-axis has been adjusted around the control frequency range specified by the grid-code described in Section 2.4., and it will be kept that way for depicting the rest of results.

3.3. Inertial Control Strategy Applied to the Wind Turbines

This section presents a third improvement proposal aimed to reduce the fluctuating wind power injection to the grid under normal operating conditions and, to provide grid frequency support in case of contingency. The inertial control strategy presented below is based on [12] and has been named Extended Optimal Power Point Tracking (EOPPT). It has been designed to be implemented in the speed controller of the wind turbine (Figure 3) and its principle of operation is based on the extraction of kinetic energy stored in the rotating masses to produce controlled variations of the output active power depending on two input variables: the deviation of the grid frequency, Δf, and the time derivative of frequency, df/dt. Unless this kind of control strategy is implemented, a full converter based wind turbine would not be able to reveal its inertial resources to the grid in a natural manner, because the DC-link decouples the frequencies of both sides of the converter (MSC and GSC), so that frequency variations in the grid do not reflect in frequency variations in the electrical generator terminals. To briefly explain the mechanics of the method, let us assume that the wind turbine is operating at point A in Figure 20 under a fixed wind speed condition. In case the power system experiences an under-frequency event, the wind turbine is required to increase its active power injection to the grid. This action can be done by moving the operation point to B immediately by means of a controlled shifting of the MPPT curve to the left. At this operating point, the imbalance between the mechanical power developed by the turbine and the electrical power required by the generator causes the rotor angular deceleration until a new equilibrium condition is reached at point C. Through the path A→B→C, the wind turbine will have to deliver part of its kinetic energy to provide frequency support to the grid. After that, the method will force the wind turbine to recover its pre-disturbance operating point, A. On the contrary, if an over-frequency event takes place, the active power set-point applied to the speed controller of the wind turbine will have to follow the path A→D→E→A.

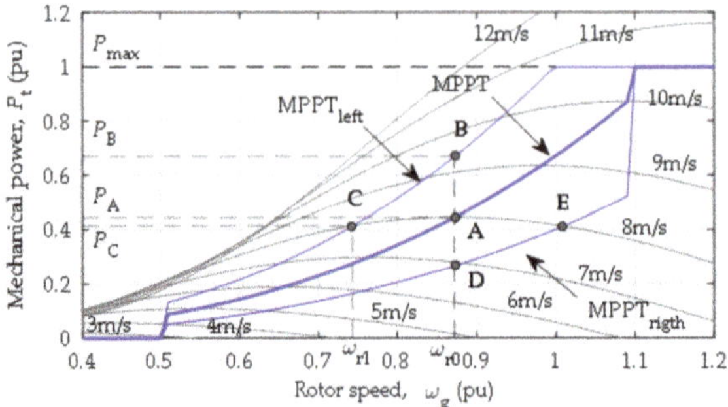

Figure 20. MPPT curve shifting procedure.

Figure 21 shows the general scheme for the implementation of the EOPPT-method in the wind turbine. This scheme is made up of two subsystems: the frequency deviation dependent loop and the time derivative of frequency loop. The first loop receives the rotor speed, ω_g, and the deviation of the grid frequency, Δf, as input variables. The latter variable is passed through a first-order low pass filter (with time constant, T_{wo}) in order to remove the steady-state error of the grid frequency introduced by the primary frequency regulation tasks. The control actions provided by this subsystem are driven by (13), which has been derived in [12]. In this equation, f_{Kopt}, is the virtual inertia factor, $\omega r0$ is the pre-disturbance rotor speed, $kvir$ is the virtual inertia coefficient, $\Delta f'$ is the filtered frequency deviation, and p is the number of pole pairs of the electrical generator. The shifting of the MPPT curve, either to the left or to the right, is achieved by multiplying f_{Kopt} by the optimization constant K_{opt} in (3). *CTRL* is a binary control signal that is activated whenever a non-zero frequency deviation is detected, which is necessary for allowing a continuous updating of ω_{r0}. Finally, in order to restrict the shifting of the MPPT curve within safe and stable operating ranges, the limits $f_{Kopt,max}$ and $f_{Kopt,min}$, have been included in the control scheme.

$$f_{Kopt} = \frac{\omega^3_{r0[\text{rad}/s]}}{\left(\omega_{r0[\text{rad}/s]} + 2\pi k_{vir}\Delta f'_{[\text{Hz}]}/p\right)^3} \tag{13}$$

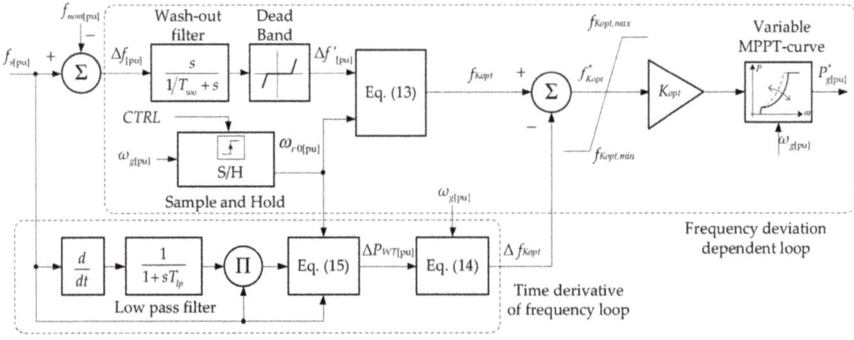

Figure 21. Extended Optimal Power Point Tracking method scheme.

On the other hand, the time derivative of frequency loop is conceived to provide a complementary frequency support and is focused on the reduction of the ROCOF. The control actions resulting from this subsystem are given by (14), which have been also derived in [12], where: Δf_{kopt} is the virtual inertia factor deviation, f_s is the measurement of the grid frequency ($f_{s[pu]} = \omega_{s[pu]}$), H_{eq} is the equivalent constant of inertia of the wind turbine, ω_{max} and ω_{min} are the rotor speed limits, graphically defined in Figure 4, and T_{lp} is the time constant of the low pass filter designed to eliminate the measurement noise. The values assigned to the parameters of the EOPPT-method are summarized in Appendix A.

$$\Delta f_{kopt} = \frac{\Delta P_{WT[pu]}}{K_{opt}\omega_{g[pu]}^3}\left(\frac{\omega_g^2 - \omega_{min}^2}{\omega_{max}^2 - \omega_{min}^2}\right) \tag{14}$$

where:

$$\Delta P_{WT[pu]} = 2\left(\frac{\omega_{r0[pu]}}{\omega_{s[pu]}}k_{vir}H_{eq[s]}\right)f_{s[pu]}\frac{df_{s[pu]}}{dt} \tag{15}$$

Figure 22 depicts the dynamics obtained by implementing the EOPPT-method in the wind turbines WT_1, WT_2 and WT_3 under operating conditions similar to those presented in Section 3.2. Note in Figure 22a that the generated wind power is slightly smoothed in those intervals outside the power curtailment condition. This produces a noticeably reduction of the ROCOF (Figure 22b), whose average value decreases in an additional 15% compared to the results achieved in the previous subsection. In addition, the application of the EOPPT-method in the wind turbines allows a milder evolution of the grid frequency as shown in Figure 22c.

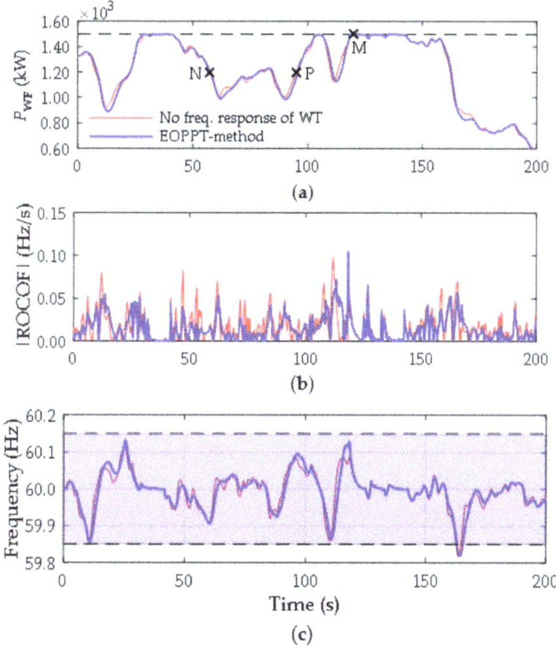

Figure 22. Time domain simulation with the application of the EOPPT-method: (**a**) Dispatched wind power; (**b**) ROCOF and; (**c**) Grid frequency.

4. Assessment of the Improvement Proposals under Contingency Conditions

The proposals aimed to improve the frequency control actions of the SCHPS, in the previous section, have been discussed and tested by considering normal operating conditions, characterized by a fluctuating behavior of the grid frequency due to the injection of a realistic wind power profile. Although the previous results reveal a significant reduction of frequency deviation and an important decrease of the average ROCOF by applying these solutions, it is essential to consider more severe operating scenarios in which the loss of generation or demand may involve a risk to the power supply continuity. The most probable scenario for the occurrence of a contingency in the actual power system corresponds to the sudden loss of wind generation, a fact that will be considered in the following study. For the test power system, similar operating conditions to those described in Section 3.3. are taken into account, and a new base case is defined. It considers all the improvement proposals introduced throughout the previous section, except the EOPPT-method (red line curves in Figure 22). The disturbance is generated by disconnecting the wind turbine WT$_3$ (Figure 1) at three different instants within the time frame considered in the simulations: points P, N, and M, which have been highlighted in Figure 22a. These points correspond to three operational status of the wind farm that are interesting for the assessment of the EOPPT-method under contingency conditions:

Case P: Tripping of WT$_3$ when the incoming wind speed is increasing ($dv/dt > 0$)
Case N: Tripping of WT$_3$ when the incoming wind speed is decreasing ($dv/dt < 0$)
Case M: Tripping of WT$_3$ under highly favorable wind conditions (Curtailment)

Figure 23 summarizes the results obtained by simulating each case of study.The performance of the power system for the first contingency scenario (Case P) is depicted in Figure 23a. The upper subfigure illustrates the time-domain behavior of the power delivered by each machine installed in the wind farm along with the maximum output power set-point imposed by the wind farm controller, P_{set}. Note that, at the moment of the disconnection of WT$_3$, the value of P_{set} increases from 500 to 750 kW. This is because the wind farm output power reference P^*_{WF} (Computed as 1500 kW in Section 2.3.) has to be fulfilled only by the remaining two online wind turbines. Since a non-variable load demand profile is considered in the simulation, the value of P^*_{WF} will remain unchanged throughout the simulated time frame. The subsequent subfigures show the dynamics of the power dispatched by the wind farm, the behavior of the power delivered by the diesel power plant, the response of the grid frequency, and the evolution of the ROCOF. These illustrations reveal that a sudden loss of wind power of 400 kW causes a grid frequency drop which is effectively mitigated by the wind turbines by implementing the EOPPT-method: the maximum instantaneous frequency deviation (MIFD) is reduced in 42.88% in comparison with the base case. The inertial response from wind turbines helps to the conventional generation in the frequency controls tasks, a fact that can be seen in the diesel power generation subplot, where a milder dynamic of the mechanical power is achieved. Since the increase of active power required for frequency support is backed up by an increase in the available mechanical power, due to $dv/dt > 0$, a reduction of the maximum rate of change of frequency, $|ROCOF|_{max}$, is achieved.

For the case N (Figure 23b), similar conclusions can be drawn: the wind turbines contribute to the reduction of the MIFD in 39.15% and facilitate the frequency regulation performed by synchronous generation. However, in this particular situation, where $dv/dt < 0$, the kinetic energy stored in the wind turbine will be spent faster during the frequency support stage which leads to have a minimum effect on the reduction of $|ROCOF|_{max}$. In the third case, the contingency occurs when the wind turbines are operating under curtailment conditions (Figure 23c). According to the curtailment philosophy discussed in Section 3.2., the wind turbine will be operating at power and rotor speed maximum conditions, P_{set} and ω'_{max}, and, unless these values undergo a change imposed by the WFC, the wind turbine will not be able to provide frequency support to the grid. Nevertheless, as it has been explained before, the disconnection of one wind turbine forces the others to cover the missed power by increasing the individual reference signal P_{set} until the total wind farm power reaches its set-point value, P^*_{WF}. This enables the online wind turbines to play an important role in the frequency recovery process, attaining a significant reduction of MIFD of 47.53% and a slightly improvement of $|ROCOF|_{max}$.

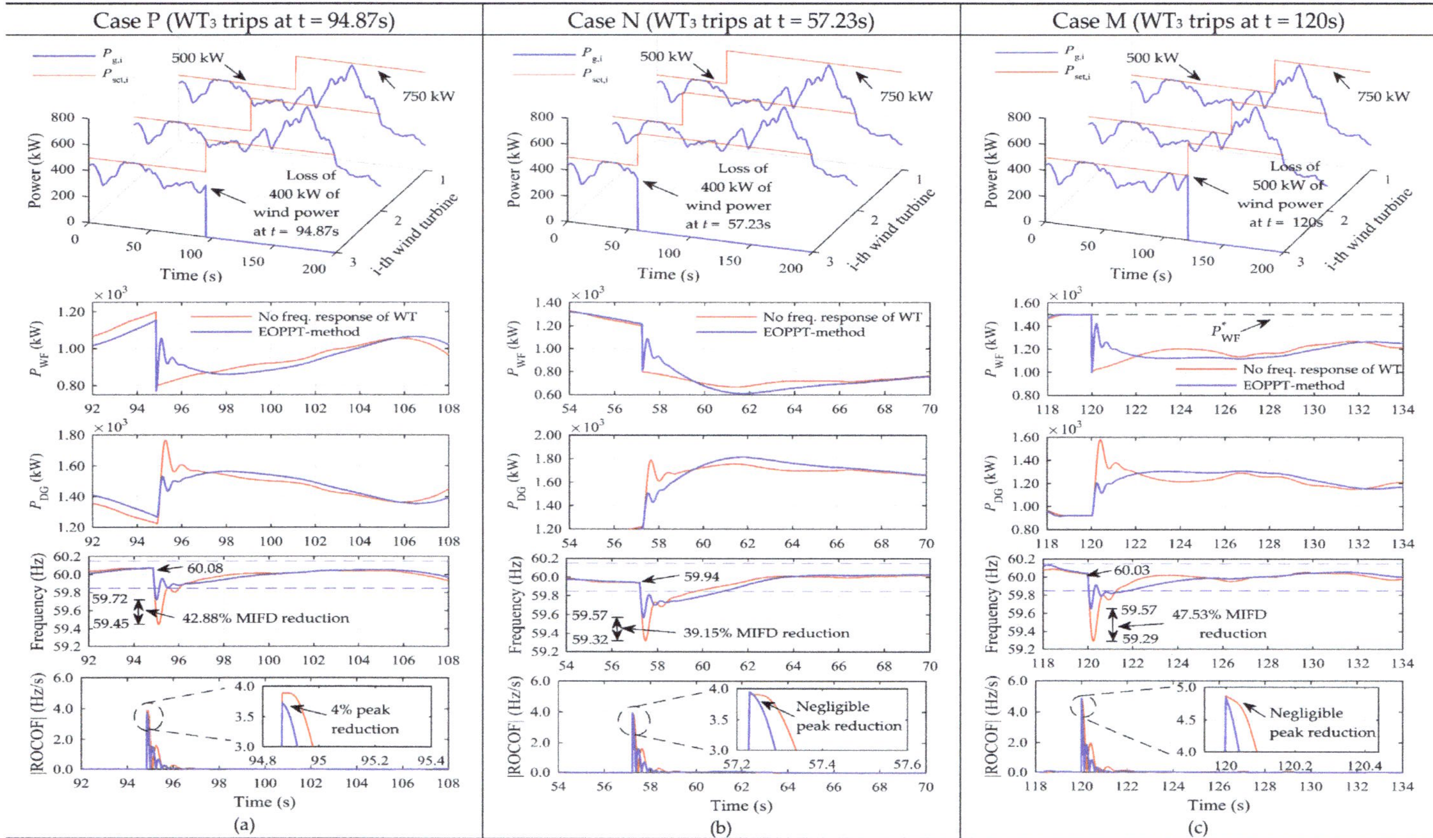

Figure 23. Power system response under different contingency scenarios: (**a**) Case P; (**b**) Case N and (**c**) Case M.

5. Conclusions

This paper presents some proposals aimed at enhancing frequency control actions in a weak and isolated power system with an important penetration of wind energy. As a case of study, the San Cristobal hybrid power system (Galapagos Islands-Ecuador) has been considered, whose time-domain behavior has been studied by using the Load Frequency Control approach described in the first part of this work. The implemented model has been validated after recreating certain actual operative situations of the power system and comparing the results with the available records acquired in field tests. The preliminary results show the technical challenge experienced by this type of systems in the frequency control tasks, specifically under scenarios characterized by high share of intermittent wind energy in the power demand covering. This results in prominent frequency fluctuations under normal operating conditions, which might cause under/over frequency relays tripping at different points of the power system, putting at risk the power supply continuity.

A first action intended to solve this problem, consists of a procedure to quantify the stability margin of the power system, whose frequency control is completely governed by the conventional synchronous generation. It has been found that a readjustment of the speed governor gains, by respecting the stability margin previously computed, allows to substantially improve the frequency quality indicators. The final selection of the controller gains has been done by applying a purely technical approach based on three objectives: fast frequency recovery, frequency drop reduction, and frequency rebound stage with small oscillations. However, better results could be obtained if an optimization process for tuning the controller gains is performed. Unfortunately, the applicable grid code does not provide enough criteria for defining an objective function (according to the objectives described above) and even for the constraints of the optimization problem, so, further studies on this topic is recommended as future work.

Another two improvement proposals are devised to be implemented in the wind turbine, as causing agent of the main issue addressed in this work. In this sense, a second corrective action, which constitutes the main contribution of this paper, consists in proposing an improved wind power curtailment strategy. This strategy combines the use of electromagnetic and mechanical variables to carry out such limitation in a more effective way, in response to the requirements established by the wind farm controller, either for technical or environmental reasons. As positive aspects of the implementation of this proposal, the results obtained by simulation show that the injection of a practically constant wind power under highly favorable wind conditions considerably reduces the associated frequency fluctuations, providing an additional margin of improvement in this regard. Also, the proposal produces a smaller and milder variation of the pitch angle during the power limitation, which could positively impact to the wear and tear of the wind turbine due to the reduction of the mechanical stress exerted on the blades. Nevertheless, it must be kept in mind that the actual effects of the proposal on the wind turbine aging can be only evaluated by long-term field research on real prototypes, therefore, a specialized study in this regard is suggested.

In addition, in order to cover those intervals where the wind energy is dispatched under the criterion of optimal conversion (non-curtailment condition), the implementation of an inertial control based strategy is presented as a third corrective action. Such strategy is designed to be incorporated in the wind turbine controllers to enable its participation in the grid frequency support tasks in cooperation with the conventional synchronous generation. After subjecting the test system to normal operating conditions and to contingency situations, it was found that this solution allows to enhance the overall performance of the system during the frequency recovery stage. This improvement would allow the system operator to define less restrictive wind power curtailment levels for a greater participation of the wind power in the electrical demand covering without compromising the security of the power system. This would lead to a smaller wind energy waste while increasing the economic and ecological benefits derived from electrical exploitation in the San Cristobal Island.

Acknowledgments: This work was supported by the Secretaria de Educacion Superior, Ciencia, Tecnologia e Innovacion (SENESCYT), Government of the Republic of Ecuador (grant number: 2015-AR6C5141). The authors would like to thank the companies ELECGALAPAGOS S.A. and EOLICSA for providing information regarding the operation and control of the San Cristobal power system (Galapagos Islands-Ecuador).

Author Contributions: Danny Ochoa conceived and designed the paper; Danny Ochoa carried out the time-domain simulations; Danny Ochoa and Sergio Martinez analyzed the results and wrote the paper; Sergio Martinez guided the whole work and the direction of the research.

Conflicts of Interest: The authors declare no conflict of interest.

Appendix

A. *Diesel-engine speed governor parameters:* $T_1 = 0.024$ s, $T_2 = 0.1$ s, $T_3 = 0.01$ s, $T_{max} = 1.1$ pu, $T_{min} = 0$ pu, $K_P = 2.294$, $K_I = 1.458$.

B. *Variable-speed wind turbine parameters:*

Table A1. Wind turbine parameters.

Parameter	Symbol	Value
Rated power (base power)	P_{base}	800 kW
Max./Min. active power	$P_{g,max}/P_{g,min}$	1/0.04 pu
Max./Min. electromagnetic torque	$T_{em.max}/T_{em.min}$	0.91/0.08 pu
Base wind speed	v_{base}	10 m/s
Active power at v_{base}	$P_{v(base)}$	697 kW
Generator pole pairs number	p	2
Generator nominal frequency	$f_{g,nom}$	50 Hz
Base speed of the turbine	$\omega_{t,base}$	2.37 rad/s
Base speed of the generator	$\omega_{g,base}$	157.08 rad/s
Gearbox speed ratio	n_t	66.185
Air density	ρ	1.225 kg/m^3
Radius of the rotor	R	29.5 m
Equivalent constant of inertia	H_{eq}	4.18 s
Optimization constant	K_{opt}	0.6728
Min./Max. blade pitch angle	β_{min}/β_{max}	0/88°
Maximum blade pitch angle rate	$(d\beta/dt)max$	10°/s
Generator and converter time constant	τ_C	20 ms
Blade pitch servo time constant	τ_P	0.3 s
Pitch controller gains	K_{Ppc}/K_{Ipc}	150/25
Speed controller gains	K_{Psc}/K_{Isc}	0.3/8

C. *MPPT curve parameters:* $\omega_{min} = 0.5$ pu, $\omega_0 = 0.51$ pu, $\omega_1 = 1.09$ pu, $\omega_{max} = 1.1$ pu.

Power coefficient equation [27]:

$$C_p(\lambda, \beta) = 0.5176(116/\lambda_i - 0.4\beta - 5)e^{-21/\lambda_i} + 0.0068\lambda \tag{A1}$$

where:

$$1/\lambda_i = (\lambda + 0.08\beta)^{-1} - 0.035\left(\beta^3 + 1\right)^{-1}$$

D. *Wind farm controller parameters* (Figure 7b): $K_{PWTC} = 300$, $K_{IWTC} = 150$.

E. *Extended OPPT method parameters:* Dead band = 1 mHz, $k_{vir} = 4$, $f_{Kopt,max} = 1.6$, $f_{Kopt,min} = 0.4$, $T_{wo} = 10$ s, $T_{lp} = 0.8$ s.

References

1. Lu, J.; Wang, W.; Zhang, Y.; Cheng, S. Multi-Objective Optimal Design of Stand-Alone Hybrid Energy System Using Entropy Weight Method Based on HOMER. *Energies* **2017**, *10*, 1664. [CrossRef]
2. Martínez-Lucas, G.; Sarasúa, J.; Sánchez-Fernández, J. Frequency Regulation of a Hybrid Wind–Hydro Power Plant in an Isolated Power System. *Energies* **2018**, *11*, 239. [CrossRef]
3. Kies, A.; Schyska, B.; von Bremen, L. Curtailment in a Highly Renewable Power System and Its Effect on Capacity Factors. *Energies* **2016**, *9*, 510. [CrossRef]
4. Bird, L.; Lew, D.; Milligan, M.; Carlini, E.M.; Estanqueiro, A.; Flynn, D.; Gomez-Lazaro, E.; Holttinen, H.; Menemenlis, N.; Orths, A.; et al. Wind and solar energy curtailment: A review of international experience. *Renew. Sustain. Energy Rev.* **2016**, *65*, 577–586. [CrossRef]
5. Al-Sarray, M.; McCann, R.A. Control of an SSSC for oscillation damping of power systems with wind turbine generators. In Proceedings of the 2017 IEEE Power Energy Society Innovative Smart Grid Technologies Conference (ISGT), Washington, DC, USA, 23–26 April 2017; pp. 1–5.
6. Kim, C.; Muljadi, E.; Chung, C. Coordinated Control of Wind Turbine and Energy Storage System for Reducing Wind Power Fluctuation. *Energies* **2017**, *11*, 52. [CrossRef]
7. Dou, X.; Quan, X.; Wu, Z.; Hu, M.; Sun, J.; Yang, K.; Xu, M. Improved Control Strategy for Microgrid Ultracapacitor Energy Storage Systems. *Energies* **2014**, *7*, 8095–8115. [CrossRef]
8. Diaz-Gonzalez, F.; Bianchi, F.D.; Sumper, A.; Gomis-Bellmunt, O. Control of a Flywheel Energy Storage System for Power Smoothing in Wind Power Plants. *IEEE Trans. Energy Convers.* **2014**, *29*, 204–214. [CrossRef]
9. Lotfy, M.E.; Senjyu, T.; Farahat, M.A.-F.; Abdel-Gawad, A.F.; Yona, A. A Frequency Control Approach for Hybrid Power System Using Multi-Objective Optimization. *Energies* **2017**, *10*, 80. [CrossRef]
10. Ren, G.; Liu, J.; Wan, J.; Guo, Y.; Yu, D. Overview of wind power intermittency: Impacts, measurements, and mitigation solutions. *Appl. Energy* **2017**, *204*, 47–65. [CrossRef]
11. Ma, H.; Wang, B.; Gao, W.; Liu, D.; Sun, Y.; Liu, Z. Optimal Scheduling of an Regional Integrated Energy System with Energy Storage Systems for Service Regulation. *Energies* **2018**, *11*, 195. [CrossRef]
12. Ochoa, D.; Martinez, S. Fast-Frequency Response Provided by DFIG-Wind Turbines and its Impact on the Grid. *IEEE Trans. Power Syst.* **2017**, *32*, 4002–4011. [CrossRef]
13. Naik, K.A.; Gupta, C.P. Output Power Smoothing and Voltage Regulation of a Fixed Speed Wind Generator in the Partial Load Region Using STATCOM and a Pitch Angle Controller. *Energies* **2017**, *11*, 58. [CrossRef]
14. Kerdphol, T.; Rahman, F.S.; Mitani, Y.; Hongesombut, K.; Küfeoğlu, S. Virtual Inertia Control-Based Model Predictive Control for Microgrid Frequency Stabilization Considering High Renewable Energy Integration. *Sustainability* **2017**, *9*, 773. [CrossRef]
15. Lee, H.; Hwang, M.; Muljadi, E.; Sørensen, P.; Kang, Y.C. Power-Smoothing Scheme of a DFIG Using the Adaptive Gain Depending on the Rotor Speed and Frequency Deviation. *Energies* **2017**, *10*, 555. [CrossRef]
16. Ma, S.; Geng, H.; Yang, G.; Pal, B.C. Clustering based Coordinated Control of Large Scale Wind Farm for Power System Frequency Support. *IEEE Trans. Sustain. Energy* **2018**, 1. [CrossRef]
17. Raoofsheibani, D.; Abbasi, E.; Pfeiffer, K. Provision of primary control reserve by DFIG-based wind farms in compliance with ENTSO-E frequency grid codes. In Proceedings of the IEEE PES Innovative Smart Grid Technologies, Istanbul, Turkey, 12–15 October 2014; pp. 1–6.
18. Bignucolo, F.; Cerretti, A.; Coppo, M.; Savio, A.; Turri, R. Effects of Energy Storage Systems Grid Code Requirements on Interface Protection Performances in Low Voltage Networks. *Energies* **2017**, *10*, 387. [CrossRef]
19. Yuan, C.; Xie, P.; Yang, D.; Xiao, X. Transient Stability Analysis of Islanded AC Microgrids with a Significant Share of Virtual Synchronous Generators. *Energies* **2018**, *11*, 44. [CrossRef]
20. Theubou, T.; Wamkeue, R.; Kamwa, I. Dynamic model of diesel generator set for hybrid wind-diesel small grids applications. In Proceedings of the 2012 25th IEEE Canadian Conference on Electrical Computer Engineering (CCECE), Montreal, QC, Canada, 29 April–2 May 2012; pp. 1–4.
21. Yoo, J.I.; Kim, J.; Park, J.W. Converter control of PMSG wind turbine system for inertia-free stand-alone microgrid. In Proceedings of the 2016 IEEE Industry Applications Society Annual Meeting, Portland, OR, USA, 2–6 October 2016; pp. 1–8.

22. Gonzalez-Longatt, F.; Bonfiglio, A.; Procopio, R.; Bogdanov, D. Practical limit of synthetic inertia in full converter wind turbine generators: Simulation approach. In Proceedings of the 2016 19th International Symposium on Electrical Apparatus and Technologies (SIELA), Bourgas, Bulgaria, 29 May–1 June 2016; pp. 1–5.
23. Ochoa, D.; Martinez, S. A Simplified Electro-Mechanical Model of a DFIG-based Wind Turbine for Primary Frequency Control Studies. *IEEE Lat. Am. Trans.* **2016**, *14*, 3614–3620. [CrossRef]
24. Proyecto Eólico Isla San Cristóbal—Galápagos 2003–2016. Available online: https://www.globalelectricity. org/content/uploads/Galapagos-Report-2016-Spanish.pdf (accessed on 11 April 2018).
25. Agencia de Regulación y Control de Electricidad. Procedimientos de Despacho y Operación. Available online: http://www.regulacionelectrica.gob.ec/wp-content/uploads/downloads/2015/10/ ProcedimientosDespacho.pdf (accessed on 13 February 2018).
26. Natori, K. A design method of time-delay systems with communication disturbance observer by using Pade approximation. In Proceedings of the 2012 12th IEEE International Workshop on Advanced Motion Control (AMC), Sarajevo, Bosnia and Herzegovina, 25–27 March 2012; pp. 1–6.
27. Junior, C.R.S.; Lima, F.K.A. Wind Turbine and PMSG Dynamic Modelling in PSIM. *IEEE Lat. Am. Trans.* **2016**, *14*, 4115–4120. [CrossRef]

Article

Hosting Capacity of the Power Grid for Renewable Electricity Production and New Large Consumption Equipment

Math H. J. Bollen * and Sarah K. Rönnberg *

Electric Power Engineering, Luleå University of Technology, 931 87 Skellefteå, Sweden
* Correspondence: math.bollen@ltu.se (M.H.J.B.); sarah.ronnberg@ltu.se (S.K.R.)

Academic Editor: Birgitte Bak-Jensen
Received: 27 July 2017; Accepted: 29 August 2017; Published: 2 September 2017

Abstract: After a brief historical introduction to the hosting-capacity approach, the hosting capacity is presented in this paper as a tool for distribution-system planning under uncertainty. This tool is illustrated by evaluating the readiness of two low-voltage networks for increasing amounts of customers with PV panels or with EV chargers. Both undervoltage and overvoltage are considered in the studies presented here. Probability distribution functions are calculated for the worst-case overvoltage and undervoltage as a function of the number of customers with PV or EV chargers. These distributions are used to obtain 90th percentile values that act as a performance index. This index is compared with an overvoltage or undervoltage limit to get the hosting capacity. General aspects of the hosting-capacity calculations (performance indices, limits, and calculation methods) are discussed for a number of other phenomena: overcurrent; fast voltage magnitude variations; voltage unbalance; harmonics and supraharmonics. The need for gathering data and further development of models for existing demand is emphasised in the discussion and conclusions.

Keywords: hosting capacity; power quality; solar power integration; electric vehicle integration; electricity distribution; distribution-system planning

---

1. Introduction

Changes in society are typically initiated and mainly take place outside of the power grid, however, such changes might still affect the electric power grid. Three specific types of changes are currently taking place at the same time, causing a range of serious challenges to the design, planning and operation of the grid:

i) *New types of electricity production.* There is a shift from large production units connected to the transmission grid to small units connected to the distribution grid, sometimes even connected at low voltage on the customer side of the electricity meter. Driven by the need for a more sustainable energy system, this new production is often of the renewable type and connected through a power-electronics interface. This shift in production, including the expected future developments, is rather well documented in papers, books, and government reports [1–4].

ii) *Changes in electricity consumption.* The electricity consumption is where societal changes often have the first impact. There is since many years a general increase in the number of electric devices used. The transition to a sustainable energy system is driving a shift towards more energy-efficient equipment, equipment with a power-electronics interface, and the introduction of new types of equipment. Examples of the latter are electric vehicles [5–7] and heat pumps (as a replacement of either gas heating or as a replacement for resistive electric heating) [8]. The replacement of incandescent lamps by compact fluorescent and LED lamps should be mentioned specifically [9]. In developing countries, an increase in wealth is strongly correlated to a growth in electricity consumption.

iii) *Changes in the grid.* A combination of technical developments on the one hand and societal, environmental and regulatory requirements on the other hand is leading to new types of solutions being used as part of the power grid. The shift from overhead lines to underground cables is one example, but also the slow but on-going introduction of a range of technology that comes under the term "smart grids" [10–12]. This new technology is obviously intended to offer better and/or more cost-effective solutions, but unintended consequences of the new technology may still pose a challenge. An overview of some of the unintended consequences for power quality of smart-grid technology is presented in [13].

All these changes create a range of challenges and a lot of research and practical studies have been done and are ongoing to address those changes and the possible challenges. Many studies have been done on the impact of local generation on the distribution grid, where overcurrent and overvoltage are typically the risks being studied [2,14–18]. An overview of the power-quality related challenges for some of the major changes is presented in [19,20]. Some of the specific conclusions from those studies are:

i) The large-scale introduction of active power electronics, in production as well as consumption equipment, results in additional phenomena becoming of importance: interharmonics; DC components and low-frequency subharmonics ("quasi-DC"); and components above 2 kHz ("supraharmonics" [21]).

ii) The amount of capacitance connected to the grid is expected to increase at all voltage levels. This will result in a shift of resonances to lower frequencies. The increased emission at higher frequencies may be (partly) compensated by the shift in resonance to lower frequencies. However, at the same time, the transfer of disturbances will become less predictable.

An international working group [22] recently presented a number of recommendations, key findings and open issues to be addressed because of the integration of solar power. Recommendations were given for a number of power-quality phenomena, including harmonics, supraharmonics, fast voltage variations (faster than 10-min time scale), slow voltage variations (slower than 10 min), overvoltage, flicker, and voltage unbalance. The hosting capacity approach, as was proposed by an international cooperation project in 2004 (see Section 2) is put forward in the report as an important tool for quantifying the impact of solar power on the power system. Additional work is needed, according to the report, for establishing suitable performance indices and corresponding limits.

This paper consists of three main parts. It starts in Section 2 with an overview, partly from a historical insider perspective, of the hosting-capacity approach. This part also includes a discussion on different kinds of uncertainly. A hosting-capacity-based planning approach is presented in Section 3, applied in Section 4.1 through Section 4.6 and discussed in Section 4.7. The description and the application are for overvoltage and undervoltage as these are the phenomena, with the possible exception of overcurrent, for which the best calculation tools, performance indices and limits are available. The third part of the paper contains a detailed discussion about including other phenomena in hosting-capacity studies. For each phenomenon, performance indices, limits and calculation methods are discussed.

2. Hosting Capacity

The term "hosting capacity" was coined in the context of distributed generation by André Even in March 2004 during discussions within the integrated European EU-DEEP project [23,24]. An approach for quantifying the hosting capacity was developed further by others within that same project [25]. The term "hosting capacity" was already in use before 2004, but in rather different context, for example for internet servers [26], for watermarking of images [27] and for the settlement of refugees [28]. Hosting capacity is now widely used as a term and as a methodology by network operators, by energy regulators and by researchers. Many studies have been done where the hosting capacity was calculated, especially for distribution networks and especially for new production, for example [29]. The hosting capacity has also been used to quantify the impact of electric vehicle charging on the grid, for example [30]. The basic hosting-capacity approach was later extended, among others to cover solutions like curtailment [24]. Studies towards estimating hosting capacity are common part of studies by large network operators as part of their strategies towards allowing the integration of more renewable electricity production into their electricity networks [31–33].

An important recent development is the introduction of a stochastic approach to hosting capacity [34–39]. The stochastic element introduced concerns especially the unknown location of future PV installations in the distribution grid, but it can be extended to include other uncertainties (see Section 2.2).

2.1. Definition and Aim of Hosting Capacity

The hosting-capacity approach, for distributed generation, has been introduced as a transparent communication tool between stakeholders concerning the connection of distributed generation to the grid. The hosting capacity is defined as the amount of new production or consumption that can be connected to the grid without endangering the reliability or voltage quality for other customers [2,25]. Essential to the approach are the selection of performance indices for the grid and acceptability limits for those indices. The hosting capacity is the amount of new production or consumption where the first performance index reaches its limit. This is illustrated in Figures 1 and 2, for two cases involving new production. For example, the risk of overvoltage already increases with very small amounts of new production connected to a part of the grid with only consumption (like in Figure 1); the risk of overcurrent will initially decrease (like in Figure 2). In cases like in Figure 2, it sometimes makes sense to introduce a second hosting capacity value (HC1) at which the performance is the same as for the system without any new generation.

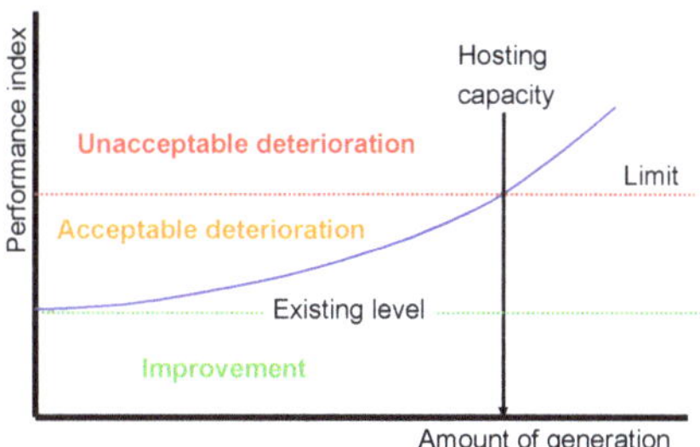

Figure 1. Hosting capacity approach, where the performance deteriorates already with small amounts of local generation.

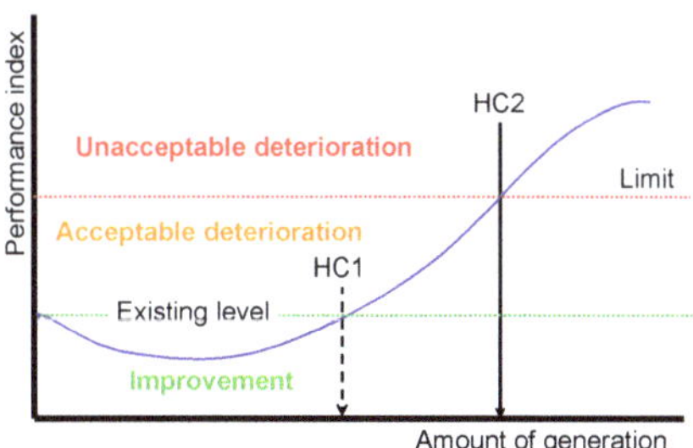

Figure 2. Hosting capacity approach where the performance initially improves and only deteriorates with larger amounts of local generation.

2.2. Uncertainties

When determining the hosting capacity of the grid at a certain location or for a certain part of the grid, several uncertainties play a role. The uncertainty that is most discussed and studied is the variation of wind or solar power production with time. This is somewhat incorrectly referred to as "intermittent" and regularly (but somewhat unwarranted) mentioned as the main concern with integration of renewable electricity production. It is admittedly not possible to predict the solar or wind-power production accurately, more than a few days ahead (for most countries). Even prediction a few hours ahead is often difficult. This limits the usefulness of wind and solar power as a dispatchable source of electricity.

It is however possible, with a reasonable accuracy, to obtain probability distributions for the amount if solar or wind-power production at a certain location, from some basic information about the size of the installation. Weather data, often obtained over several decades with high accuracy, is the basis for the calculation of such distribution functions. Long-term trends are most likely present in the weather, but their impact on the stochastic properties of wind and solar-power production is still expected to be smaller than the impact of other uncertainties.

In a similar way, probability distributions and time series can be obtained for the consumption of individual customers or for groups of customers (from the load of a single distribution transformer, up to the consumption of a whole country). Time series over many years are rare, but the introduction of smart metering makes that data for a few years becomes increasingly available. Consumption patterns are prone to change, and likely faster than weather patterns, but not to the extent that the measurements become useless.

Not all time variations in (renewable-electricity) production and consumption are uncertain. The daily and seasonal variations in solar-power production are very predictable, as an example. When studying solar-power integration for a limited part of the year (e.g., around noon during Summer) the appropriate probability distribution function should be used.

Next to the above-mentioned uncertainties, there exists another level of uncertainty. That kind of uncertainty cannot be obtained from past measurements. Some examples (all related to small-scale solar power), are:

i)   Which customers will have a PV installation and how big will these installations be?
ii)   Will these installations be three-phase or single-phase connected?
iii)   With single-phase connection: to which phase will it be connected?
iv)   What will be the direction and tilt of the panels?
v)   Will any of the panels follow the sun through single-axis or double-axis mounts?
vi)   What type of inverter will the installation have? Will it be one large inverter or a number of smaller inverters?
vii)   Will the installation have on-site storage or not? When it has on-site storage, what will its size be, which control algorithms will it use, and will the owner use the storage to participate in day-ahead and balancing markets?
viii)   Will the inverter be equipped with voltage and reactive power control?

The answers to all of these questions need to be known before an accurate hosting capacity can be calculated. However, the answers will most likely not be known. Either an educated guess will have to be made or a stochastic approach is needed. The latter is the approach that will be illustrated in Section 4.

2.3. Impacts on the Hosting Capacity

The hosting capacity approach has been developed as a transparent approach aimed at enabling a more open discussion between the different stakeholders. That does however not imply that there is a unique value of the hosting capacity, as resulting from the calculations.

The before-mentioned uncertainties all have their impact on the results of the calculations. In addition, the calculations themselves affect the results. Certain assumptions will have to be made and certain parameter values are needed. The choice of these assumptions and parameter values can have a big impact. Data, especially on the consumption patterns, is not always available when the studies are made and assumptions can have a serious impact. A sensitivity analysis is needed to evaluate if additional data collection and model development are needed. Alternatively, additional stochastic variables can accommodate for the uncertainty.

What has an even bigger potential impact, and where no amount of data collection or model development may help, is the choice of the performance index and the limit (as shown in Figures 1 and 2). These choices depend strongly on the amount of risk that the stakeholders (especially the network operators and their customers) are willing to take. The lower the risk the network operators are willing to take, the lower the hosting capacity.

3. A Hosting-Capacity-Based Planning Approach

A range of hosting-capacity studies are presented in the literature and more are being used in various studies without being published. The hosting-capacity-based planning approach proposed here, consists of the following steps:

i)   Estimate the no-load voltage variations in the low-voltage distribution network during those hours of the year that the production from solar power may be high. These are the voltage variations originating from the medium-voltage network.

ii)   Estimate the range of the lowest consumption during those hours of the year that the production from solar power may be high.

iii)   Estimate the production per installation, during the 10-min period with the highest impact from all installations together. This is not the same as the maximum production per panel, but it can be referred to as an "after diversity maximum production", next to an "after diversity minimum consumption".

iv)   Add solar power installations in a random way and calculate the distribution of worst-case voltage with increasing amount of solar power.

v)   Define a performance index for the network, an appropriate limit for this index, and determine the hosting capacity.

This approach for distribution-system planning can be seen as an adapted version of the classical approach, where an "after diversity maximum consumption" is compared with the capacity of lines, cables and transformers.

The difference between the planning approach used here, and the time-series approach often used in other studies, is that the proposed planning approach immediately considers the worst-case during a longer period, like several years. Distribution-network planning is largely about making sure that the network can cope with for example the highest current through a transformer. It is this worst case that matters. However, also the worst-case value can be treated as a random variable, which is the base for the planning approach presented here. Probability distributions are used, for example for the consumption. These are not the same as the probability distributions obtained from time series. Instead, they are the range of values during those hours of the year that the worst-case situation can occur.

The result of the calculations is thus not the probability distribution of the voltage magnitude, but instead the probability distribution of the worst-case voltage magnitude. Alternatively, one may consider this as the probability distribution obtained over a range of possible futures, all with a different worst-case value. Important advantages of the approach proposed here and applied in Section 4, are:

i)   The approach fits closely to existing planning approaches used by distribution companies.

ii)   A limited amount of input data is needed.

iii)   The results are such that they can be interpreted relatively easy by distribution companies.
iv)   Different kinds of uncertainties can be added without changing the basic approach.
v)   Any power-system analysis tool can be used to perform the actual calculations.

The limitation of the approach is that it requires certain assumptions, like in step (i) and (ii) above. In practical planning studies, these assumptions might become more based on expert opinions than on actual reproducible studies. Different persons may obtain different results. Such is however common in practical planning studies for distribution grids. With the proposed approach, the assumptions become clearly visible and might trigger further studies.

4. Overvoltage and Undervoltage

4.1. Two Swedish Low-Voltage Networks

To illustrate the calculation of the hosting capacity in a stochastic way for planning purposes, two low-voltage networks in the North of Sweden have been used: a 6-customer rural network; and a 28-customer suburban network. Both networks were used in [39], where the details of the networks can be found. Both networks contain only domestic customers and both networks are three-phase down to the customer. The customer has the possibility of connecting three-phase equipment and of spreading the equipment over the three phases. Note that this is not possible in distribution networks with single-phase laterals, as are common in North America.

4.2. Risks Due to Single-Phase Equipment

In this section (Section 4), the risk of overvoltage and undervoltage due to single-phase equipment will be quantified. Both new production (domestic PV installations) and new consumption (domestic EV chargers) are included in the study. The connection of single-phase equipment is more severe for the grid than the connection of three-phase equipment with the same power rating. Not only is the voltage rise or drop higher for single-phase equipment (three times the current and about twice the impedance), but the other phases are also impacted. Connection of a single-phase EV charger in one phase results in a voltage drop in that phase and a voltage rise in the other phases.

Many hosting capacity studies consider detailed time series of both consumption and production to find at which amount of production voltage or current limits are reached. At higher voltage levels, the data is more likely available than at lower voltage levels. The additional time needed to obtain the data (often several years of data collection are needed) will probably not result in a more accurate result because of remaining uncertainties for example in the details of the installations. This holds especially when the hosting-capacity study concerns the introduction of many devices or installations. In this paper, the authors have therefore chosen for a probability distribution instead of time series. Both methods should be further developed and compared. Some aspects of such comparison are discussed in Section 4.7.3.

4.3. Single-Phase PV in a Rural Network—Overvoltage

The risk of overvoltage due to the introduction of single-phase-connected PV has been studied for the six-customer rural network (see Section 4.1). The probability distribution for the highest voltage is shown in Figure 3 for one through six customers with PV. The following input data has been used for obtaining these distributions:

i)   Each installation is connected single-phase and each installation produces 6 kW.
ii)   The consumption per customer per phase, during the worst case, is uniformly distributed between 0 and 250 W. This range was obtained from measurements of 10-min values of two customers (one in the rural network and one in the suburban network) and the study of hourly consumption from other customers. The worst case, when considering the risk of overvoltage, is when consumption is low and production is high. High production will likely last a period of one or two hours

around noon. The regulation in most European countries sets limits to 10-min values of rms voltage. The lowest 10-min consumption values during one or two-hour periods were considered to obtain the range from 0 to 250 W. Only measurement values around noon during the summer months were used here.

iii) The no-load voltage in the low-voltage network, during the worst case, is uniformly distributed between 238 V and 242 V. The highest 10-min values during one or two-hour periods around noon in the summer months (as obtained from the same measurements) were used to obtain this distribution.

All the random variables in the model are non-correlated. It is further assumed that the PV installations are randomly distributed over the customers and over the phases. The network model used is a simple, linear one, consisting of the node-impedance matrix calculated from the positive and zero-sequence series impedance of the different branches. Positive-sequence and negative-sequence impedances are considered equal. The difference between positive and zero-sequence impedance is considered in the calculations. From the injected and consumed power for each customer and phase, a current vector is calculated. This vector is multiplied with the node-impedance matrix to obtain the vector with phase-to-neutral voltages for all customers and phases. All calculations are performed in Matlab.

A Monte-Carlo simulation is used to generate a large number of combinations of customers and phases with PV. For each combination, the highest phase-to-ground voltage is calculated for each customer terminals. The probability distribution for this voltage is shown in Figure 3. In this case, each distribution was obtained from 10,000 samples.

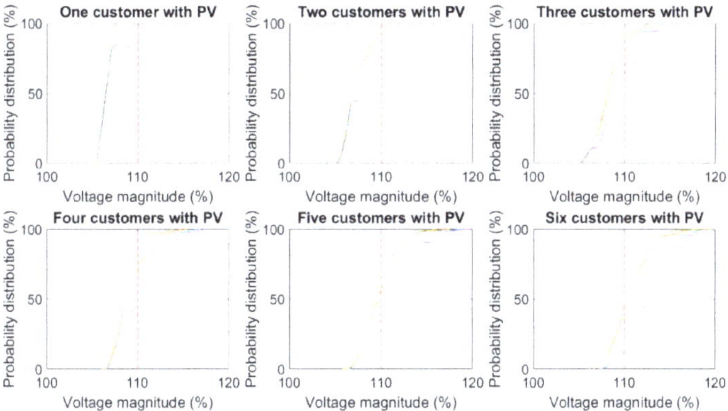

Figure 3. Probability distribution (cumulative distribution function) for the highest voltage (worst-case voltage) for increasing amount of single-phase connected solar power in the 6-customer rural network. The different colours refer to the six different customers. The red dashed vertical line is the overvoltage limit at 110% of the nominal voltage.

The plots show how the distribution shifts towards higher voltage magnitudes with increasing amount of solar power. The probability that the overvoltage limit (110% of nominal voltage) is exceeded increases because of this. Already for one customer with PV, there is a small probability that the voltage with one of the customers exceeds the overvoltage limit.

An important advantage of the hosting-capacity approach is its transparency: a well-defined performance index is compared with a well-defined limit. The same should hold when the hosting capacity is used as a planning tool. In this example, the 90th percentile of the distribution shown in Figure 3 has been used as an index and 110% of nominal voltage as a limit. The index value with increasing amount of solar power is shown in Figure 4, where each point is the result of 100,000 simulations. The results are shows for each of the six customers. When one of the 90th percentiles exceeds the 110% limit for one of the customers, the hosting capacity has been exceeded. The hosting capacity is in this case equal to only one customer with PV.

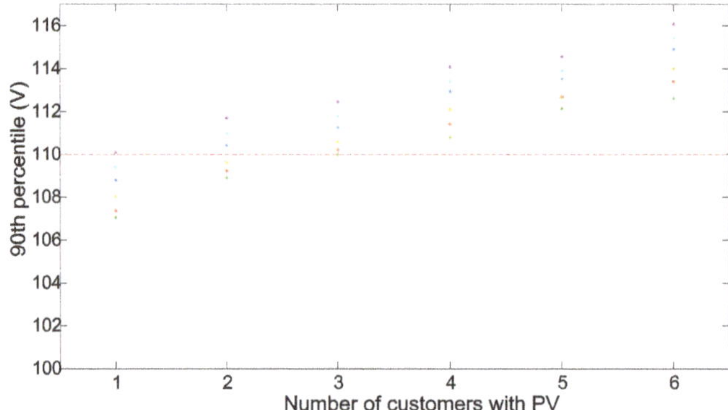

Figure 4. 90th percentile of the worst-case overvoltage as a function of the number of customers with PV in the 6-customer rural network.

It has been mentioned before (in Section 2.3) that the hosting capacity is not a unique value. Instead, it depends strongly on the model and data used to calculate the index, on the choice of index and on the choice of limit. An example of the latter, with reference to Figure 4, is that the hosting capacity increases to two customers when 112% instead of 110% of nominal is used for the limit.

Figure 5 shows the performance index when it is assumed that each PV installation injects 4 kW instead of 6 kW during the worst case. The resulting hosting capacity is five customers. An example of the impact of the choice of performance index is shown in Figure 6, where the 75th percentile is used instead of the 90th percentile. The production per installation is again assumed 6 kW. The resulting hosting capacity is two customers with PV. In the latter example (Figure 6), it is clear that the increase in hosting capacity goes together with an increase in risk. However, even for the example in Figure 5, the risk of overvoltage increases, as installations may contribute more than 4 kW to the worst case. The choice of model, data, etc. with hosting-capacity calculations is strongly related to the risks that the different stakeholders are willing to take. A brief discussion on this is part of Section 4.7.7.

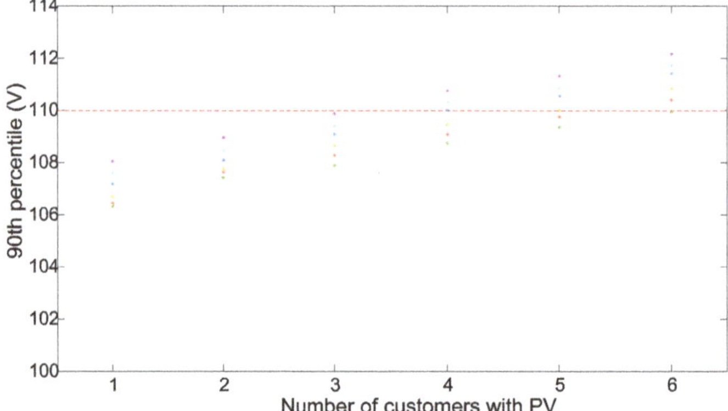

Figure 5. 90th percentile of the worst-case overvoltage as a function of the number of customers with PV in the 6-customer rural network; 4 kW production per PV installation.

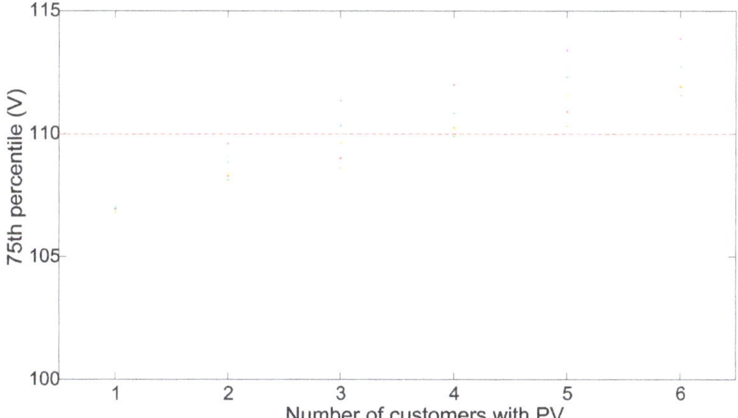

Figure 6. 75th percentile of the worst-case overvoltage as a function of the number of customers with PV in the 6-customer rural network.

4.4. Single-Phase PV in a Rural Network—Undervoltage

When considering the risk of undervoltage, other distributions have to be considered for the no-load and no-PV voltages than when considering the risk of overvoltage (Section 4.3). Undervoltage occurs in the phases without PV during periods with high production and at the same time high consumption and low no-load voltages. The high production can occur, like before, around noon during the summer months. Instead of the lowest consumption and the highest no-load voltage, the highest consumption and the lowest no-load voltage should be used as input to the calculation. From the same data as in the previous section, the following input data to the hosting-capacity calculation has been used:

i)   Consumption: 1000 W–2500 W per customer per phase. Note again that this is not a typical consumption but an estimation of the amount of consumption that may occur during a worst case for undervoltage due to PV.

ii) No-load voltage: 232 V–236 V.

Like before, 6 kW production per PV installation has been assumed. The resulting probability distributions are shown in Figure 7. Compared to the previous figures, the lowest of the three voltages for each customer has been used as input to the probability distribution function (also known as "cumulative distribution function" or CDF). The results in the figure show that a voltage drop may occur due to solar power, but values less than 95% of nominal during sunny hours are very unlikely. Undervoltage does not set the hosting capacity in this network. However, when the highest consumption occurs during periods with high solar power production, undervoltage may be a bigger concern than overvoltage.

A further look at the results (not presented here), including simulations with 1,000,000 samples, showed that voltages as low as 92% of nominal are actually possible, but with extremely small probabilities: of the order of one per million. This is a good example showing that considering a stochastic approach is the way to go. Such low probabilities do not need to be considered in distribution-system planning.

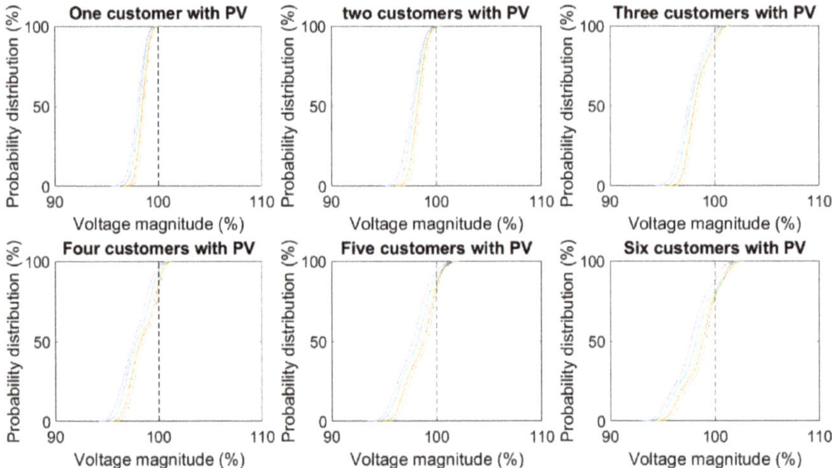

Figure 7. Probability distribution (cumulative distribution function) for the lowest voltage (worst-case voltage) for increasing amount of single-phase connected solar power in the 6-customer rural network. The different colours refer to the six different customers. The black dashed vertical line is the nominal voltage.

4.5. Single-Phase PV in a Suburban Network—Overvoltage

The calculations presented in the previous two sections have been repeated for a suburban network with 28 customers (see Section 4.1). The distributions for no-load voltage and consumption are the same as used for the rural network. The results are shown in Figures 8 and 9. The hosting capacity, using the same values as before, equals three customers (see Figure 9).

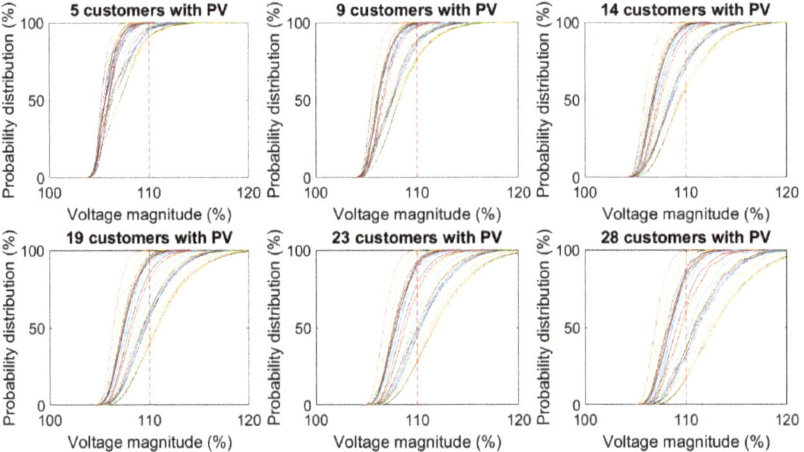

Figure 8. Probability distribution (cumulative distribution function) for the highest voltage (worst-case voltage) for increasing amount of single-phase connected solar power in the 28-customer suburban network.

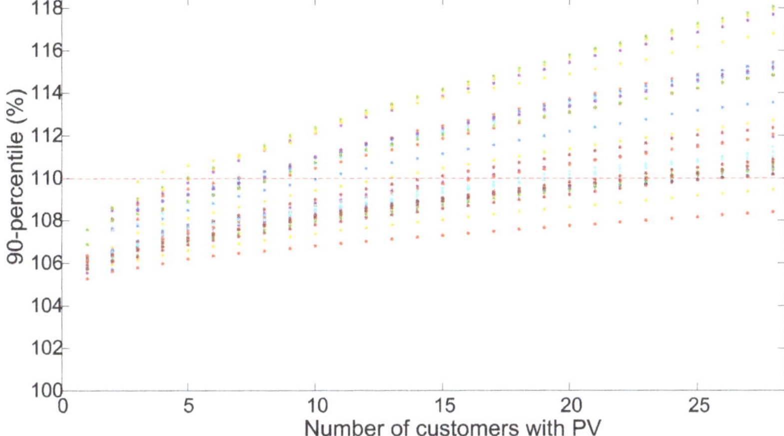

Figure 9. 90th percentile of the worst-case overvoltage as a function of the number of customers with PV in the 28-customer suburban network.

To illustrate how different parameters affect the outcome of the calculation, a sensitivity analysis has been done. The results of this are shown in Table 1. The hosting capacity turns out to be most sensitive to the percentile used and to the produced power per installation. The percentile used is a matter of how much risk the network operator is willing or even allowed to take in the planning stage. This remains largely an un-explored but very important area. The value of the produced power used in the calculation depends on the size of the individual installations and on the spread in tilt direction and angle between the installations. Some specific studies, with more detailed models including this, are needed for a more accurate estimation of the hosting capacity. See Section 4.7.6.

Table 1. Sensitivity Analysis of the Hosting Capacity.

Case	Parameter	Default Value	New Value	Hosting Capacity
0				3 customers
1	Produced power per installation	6 kW	7 kW	2 customers
2			5 kW	6 customers
3			4 kW	11 customers
4	Percentile	90th	95th	1 customer
5			85th	5 customers
6			75th	8 customers
7	load per customer per phase	[0, 250 W]	[0, 150 W]	3 customers
8			[0, 350 W]	3 customers
9	No-load voltage	[238 V, 242 V]	[240 V, 244 V]	2 customers
10			[239 V, 243 V]	2 customers
11			[237 V, 241 V]	4 customers
12			[236 V, 240 V]	6 customers

The hosting capacity turns out to be most sensitive to the percentile used and to the produced power per installation. The percentile used is a matter of how much risk the network operator is willing or even allowed to take in the planning stage. This remains largely an un-explored but very important area. The value of the produced power used in the calculation depends on the size of the individual installations and on the spread in tilt direction and angle between the installations. Some specific studies, with more detailed models including these parameters, are needed for a more accurate estimation of the hosting capacity. See also Section 4.7.6.

4.6. Single-Phase Electric Vehicle Chargers in the Suburban Grid

The same methodology as before has been used to study the risk of undervoltage due to large numbers of single-phase EV chargers. Charging may take place any time of the day and any time of the year. Therefore, lower no-load voltages and higher consumption should be considered than when studying the risk of undervoltage due to PV. The following values have been used for the calculations:

i)   Active-power consumption: uniformly distributed between 1500 W and 3000 W.
ii) No-load voltage: uniformly distributed between 230 V and 234 V.
iii) Charging power: 2300 W, 3680 W, 4600 W and 5750 W (corresponding to 10 A, 16 A, 20 A and 25 A).

The resulting performance index is shown in Figure 10 as function of the number of EV chargers that are drawing current at the same time, for 5760 W per charger. The hosting capacity equals nine EV chargers. Assuming 4600 W per charger results in a hosting capacity of 14 chargers (not shown here). For 3680 W and 2300 W per charger, 22 and 28 chargers, respectively can be operating at the same time without the undervoltage limit being exceeded.

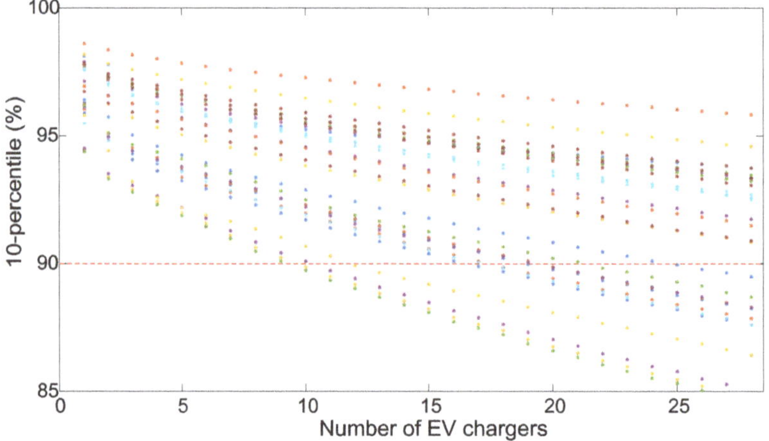

Figure 10. 10th percentile of the lowest voltage (worst-case voltage) as a function of the number of EV chargers in a 28-customer suburban network. A consumption of 5750 W (25 A) per charger has been used in the calculations.

4.7. Discussion

The subsections below present some points of discussion, including the need for further work resulting from the presented study. The discussion text refers only to the impact of new production, but a similar discussion is possible for new consumption.

4.7.1. Reactive Power

In this study, only the active power has been included. The reason for this is that the reactive-power consumption of modern domestic customers is small and that the X/R ratio with the supply terminals in a low-voltage network is small. For connection of installations to medium-voltage networks, e.g., solar parks and wind turbines, the reactive power needs to be considered.

4.7.2. Probability Distributions

The study presented here used uniform distributions for the no-load voltage and the consumption during critical hours. More studies, based on measurements as well as on simulations, are needed to find out what are reasonable distributions for these input variables. More advanced stochastic load models exist (for example [40]) and further development and application of those is important. There is, however, a risk of over-sophistication here, where the other uncertainties dominate and make that such advanced models have no higher accuracy than simplified models. With advanced models, there is also the risk that the study loses generality. The development of appropriate load models is an essential part of hosting-capacity studies, next to the development of appropriate models for new production and consumption.

In the study, it was assumed that the no-load voltage was independent on the amount of customers with PV. It requires further study to find out to which extent this no-load voltage is affected by solar power elsewhere in the same medium-voltage network. Some positive correlation can be expected here: when the number of customers with PV increases in a low-voltage network, it is also likely to increase in neighbouring low-voltage networks supplied from the same medium-voltage feeder.

4.7.3. Time Series

The study presented here used probability distribution functions together with a fixed value of the solar-power production. An alternative, commonly used in other studies, is to use time series for both production and consumption. A discussion is needed on the advantages and disadvantages of these two methods. In that discussion, the main comparison should be between time series based on measurements and probability distributions or time series obtained from stochastic models. With the former approach, all the time series should be obtained over the same or an equivalent measurement period. The time series should be sufficiently long to cover all combinations (maximum and minimum production, maximum and minimum consumption, but even medium values that could combine towards a worst case). This requires data collection over several years, to include year-to-year variations.

The main advantage of using measured time series is that these are closest to reality and therefore include all phenomena and correlations as they occur in reality. An example is the relation between solar power production and electric heating or cooling [41].

Comparison studies are needed, between measured time series, probability distributions, and time series obtained from more advanced stochastic models. Differences between the results from the different approaches do however not always imply that one method would be superior compared to another.

The approach presented here can be combined with the use of (measured or generated) time series. Instead of taking one value for consumption and no-load voltage for each spread of the PV panels over the customers and phases, time series are used for each of such spreads. This will bypass the discussion on when the production is expected to be highest (Section 4.7.6), but it will require much more data and calculations.

The voltage-quality regulations in many countries set limits to the 95th percentile of the rms voltage calculated over a 10-min window. This value shall not exceed the overvoltage limit, typically at 110% of nominal voltage. During the planning, the distribution company may still decide to use the highest voltage as a criterion. Alternatively, the 95th percentiles can be used for planning. The plots in Figure 3 and similar would in that case have to show the distribution of the 95th percentile of the voltage for random customers and phases with PV. The use of time series would be a possible choice; but not necessarily, time series for the whole year would be needed. Those time series can be limited to periods with high or low production and/or consumption; for example around noon during summer. Alternatively, the estimated values (i, ii and iii) in Section 3 can be adjusted (see Section 4.7.2).

4.7.4. Need for Data Collection

Much of the work towards further developing the hosting-capacity approach requires measurement data. The use of measured time series obviously requires large amounts of data. The choice of probability distribution for the consumption and the development of more advanced load models is also not possible without significant amounts of data available.

Such data might be available to the network operator, for example in the form of hourly metering. However, the data is in most cases not available to universities or research institutes that work on model development. There is no direct solution to this dilemma as this involves consumption patterns from individual domestic customers with important privacy implications.

Data collection on consumption and consumption patterns for many domestic customers should however still be pursuit. One way or the other, such data should become available for research and education. The set of test feeders, some with load data, collected by [42] is an appropriate platform for making such data available.

4.7.5. Generality of the Results and of the Method

The results of the study presented here are not valid beyond the specific two networks studied, and even for those only under the assumptions presented. The example's only aim is to illustrate the method. Network operators should use data from their own networks to estimate the hosting capacity as part of their distribution-system planning.

The calculations used here were performed using Matlab only, but any other power-system analysis package can be used for this. In fact, several network operators are developing tools using such packages to estimate the hosting capacity and to create hosting-capacity maps.

Although the specific results are only valid for the studied networks, the method itself has a broader applicability. The method should be applied to different low and medium-voltage networks, to find out how general the results are, but also to evaluate the broader applicability of the method.

The two example networks used here are typical Swedish low-voltage networks, dimensioned to accommodate electric heating. This will result in a higher hosting capacity than for grids only dimensioned for non-electric domestic consumption. The approach should be applied to networks in countries without electric heating, like in the more southern parts of Europe. Both networks, like all Swedish distribution grids, are three phase all the way down to the domestic customer. Two-phase and three-phase equipment is common with domestic customers in Sweden. The approach should be applied to networks dominated by single-phase laterals, like in North America.

The consumption of domestic customers in Sweden is rather constant, with mainly a small downward trend due to increased energy efficiency. This allows the use of a "background consumption" based on measurements with the new production or consumption added to this. This approach is not possible for grids in developing countries where load growth could be significant.

4.7.6. Production per Installation during the Worst Case

In the study presented here, a fixed production per PV installation during the worst-case was used. It was shown from a sensitivity analysis (Table 1) that this value has a big impact on the obtained value for the hosting capacity.

Studies are needed to determine how to get a good model for the production per installation during those hours of the year that the highest production can be expected. One of the questions that has to be answered is, "what are the hours of the year during which the production from solar may be high?" The answer to this question depends for example on the spread of roof top directions. When all rooftops face south, the panels will produce the same amount and the period of peak production will be relatively short. When there is a spread in rooftop direction, the period of expected peak production will be longer but the peak will be smaller. Both measurements and simulations are needed to answer this question.

Instead of a fixed value for the production during the worst case, a stochastic model could also be considered here. The stochastic element is again not in the variation of production with time, but in the uncertainty of the contribution of an individual panel to the worst-case situation.

4.7.7. Choice of Performance Indices and Limits

An essential discussion in hosting capacity studies concerns the choice of performance indices. This discussion started soon after the introduction of the term for the integration of distributed generation. A subject heavily discussed initially was the choice of the appropriate index to quantify the performance of the supply, for example the highest 10-min rms voltage or the 95th percentile of the 3-s rms voltages. This is largely a regulatory issue. More recently, and during the study presented here, the discussion is moving towards the choice of indices for planning under uncertainty. The sensitivity study (Table 1) showed that the percentile value has a big impact on the obtained hosting capacity.

The choice of percentile value is strongly related to the amount of risk carried by the different stakeholders. Important input to the choice of percentile values is who carries this risk and what the possible consequences are. The smaller the consequences, the higher the risk that can be taken.

The selection of the percentile values (and risk management in general) is not a sole power-engineering issue, but it is still very important. This is where economics, social aspects, legal and regulatory aspects, politics and even psychological aspects have to be included in the studies. A new line of interdisciplinary research is needed to address this.

5. Other Phenomena

Section 4 illustrates several aspects of the hosting-capacity approach, through overvoltage and undervoltage. These are however not the only phenomena that can set the hosting capacity (i.e., can set limits to the amount of new production or new consumption that can be connected). Several other phenomena are discussed in the forthcoming sections. As of yet, there is no complete list of phenomena and indices for hosting-capacity studies. An attempt at creating such a list is presented in [43]. With reference to Figures 1 and 2, calculating the hosting capacity requires the following tools and information:

i)   A generally accepted performance index.
ii)   A limit for the performance index that forms a border between "acceptable performance" and "unacceptable performance".
iii) A method for calculating the value of this index, either deterministically or in a stochastic way, as a function of the amount of new production or consumption.

These three requirements will be used in the forthcoming sections to discuss the hosting-capacity approach for different phenomena. In those sections, like in Section 4, the aim is to strike a balance between the need for accuracy on one hand and the availability of data and models on the other hand.

5.1. Overcurrent

5.1.1. Performance Indices

Excessive currents (overcurrent) lead to excessive heating with too high temperatures as a result. As this is a thermal issue, the rms current is the obvious choice as base for performance indices. As voltage variations are normally small, the apparent power can be used instead. If reactive power is small, active power can be used instead of apparent power.

Methods for calculating currents through the grid are very similar to methods for calculating voltages. In fact, a load-flow calculation will result in both voltages and currents. When considering them in hosting-capacity studies, some differences need to be considered. The first difference concerns the importance of the phenomenon for the performance of the grid. In [2] and [44], a distinction is made between primary and secondary aims of the power system. In the same way, a distinction can be made between "primary performance indices" and "secondary performance indices" for the hosting-capacity approach. Voltage magnitude would be a primary performance index: something that matters directly to the network users. The current through for example a transformer is however not of direct concern for the network users; it is a secondary performance index. The overloading of a component will quickly result in an interruption for all network users downstream of the overloaded component (either because that component fails or because the protection removes the component to avoid its failure). Avoiding overloading is thus also in the interest of the network user. The probability of an interruption due to transformer overloading would be a primary performance index.

The second difference is in the limit values to be used in the hosting capacity studies. The actual values for current can be obtained in similar ways as before, through "after diversity maximum production", "after diversity minimum consumption", etc. The challenge lies in the choice of maximum-permitted-value of, for example, the current through the transformer. For voltage-magnitude variations, the 10-min rms value is defined in standards and regulation, so that the choice becomes easy.

The window over which the value should be calculated depends on the thermal time constant of the cable, line or transformer under study. Information on this can be obtained, for example, from the increasing volume of works on dynamic rating [45]. Here it should be considered that much of the work on thermal time constants is directed towards short circuits where adiabatic temperature rise can be considered. For the lesser overloads that are part of hosting capacity studies, this assumption would likely lead to an underestimation of the hosting capacity. From the literature [46–48], time constants between minutes and hours were found. A 10-min value would be a good compromise, given the above range of thermal time constants. Nevertheless, when only 1-hour values are available (e.g., from hourly metering) those also seem to be reasonable.

5.1.2. Limits

In case apparent power or active power is used for the performance index, a margin is appropriate to compensate for the incompleteness in the model. For example, a 10% reduction in limit could be used to compensate for the possible 10% reduction in voltage when apparent power is used instead of rms current. This reduction in limit corresponds mathematically to the assumption that voltage, active power and reactive power are stochastically independent. A more complete model, supported by more complete measurement data, will typically result in an increase in hosting capacity, as was illustrated for a simple case in [49]. With many hosting capacity studies, the more data is available, the less safety margins are needed, and the higher the hosting capacity.

5.1.3. Calculation Methods

The calculation of the currents through a cable, line or transformer, with increasing levels of new production or consumption, is straightforward. The new current is added, in the complex plane, to the existing current. Stochastic methods can be used in the same way as before: a fixed limit for the maximum-permissible current has to be decided, corresponding to the 110% and 90% limits used for the studies in Section 4. An example of such a study is given in Section 5.1.4. Alternatively, a detailed thermal model of the line, cable or transformer can be used together with time series of production and consumption. Many studies use such time series and compare the resulting currents with a maximum current value, e.g., [50].

5.1.4. Planning Example

To illustrate the use of the hosting-capacity approach in distribution, assessing the risk of transformer overload, a similar example as in Section 4 is presented here. This example concerns a 500-kVA transformer supplying 83 customers. When all customers would have a single-phase-connected PV installation (6 kW injected power), they would all together produce 500 kW. When equally distributed over the three phases, this would exactly correspond to the transformer rating. Note that reactive power has not been considered here, partly to simplify the calculations, but also because its impact is limited and because reactive-power consumption from domestic customers is small. For an actual planning study, it is recommended to consider the reactive power and even its possible changes in the future. The results are shown in Figures 11 and 12. Figure 11 gives the probability distribution of the highest single-phase current through the transformer. The consumption per phase has again been assumed uniformly distributed between 0 and 250 W. The current increases when more customers install solar power. However, even when 40 (out of 83) customers have solar power, single-phase connected, the probability that the transformer will be overloaded is still very small.

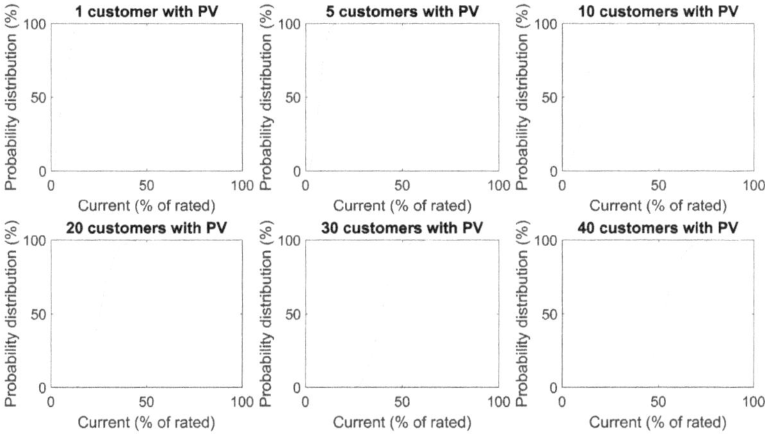

Figure 11. Probability distribution (cumulative distribution function) of the highest single-phase current through the distribution transformer, with increasing number of customers with single-phase connected PV.

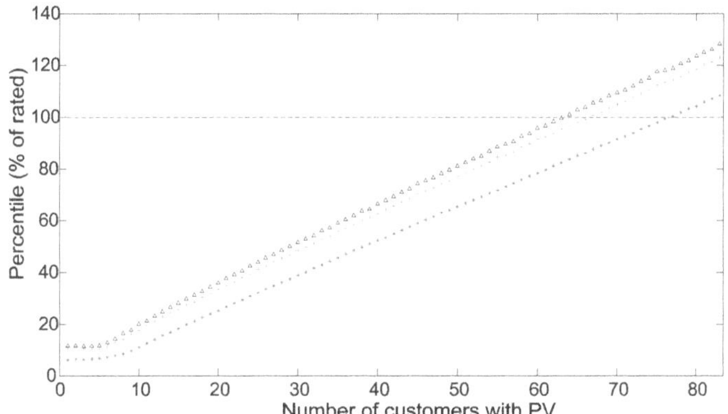

Figure 12. Four different percentiles (50th: red star; 75th: green circle; 90th: blue plus; 95th: black triangle) for the current through the transformer, as a function of the number of customers (out of 83) with single-phase connected PV.

For the application of the hosting-capacity approach, it is again necessary to define a performance index and a corresponding limit. For this example, four different percentile values (50th, 75th, 90th and 95th) have been used as indices, with the (single-phase) transformer rating as a limit. The result is shown in Figure 12. For up to about five customers with PV, the current index is not affected. For low amounts of PV, the highest transformer loading occurs for maximum load and not for maximum production. Here it is important to realize that "maximum load" corresponds to the period of the day and year when highest solar production can be expected. This is not in any way related to the highest annual consumption.

From Figure 12, the hosting capacity is shown to depend on the percentile value, i.e., on the amount of risk that the network operator is willing and allowed to take. The hosting capacity ranges from 63 (when using the 95th percentile) up to 76 (50th percentile). The impact of the percentile value on the hosting capacity is less than for overvoltage (Table 1).

5.2. Fast Voltage Magnitude Variations

Fast voltage magnitude variations occur because of fast variations in production or consumption.

5.2.1. Performance Indices

Standard indices exist for the fastest voltage magnitude variations; at time scales less than one minute. The short-term and long-term flicker severity (Pst and Plt, respectively) are defined in IEC 61000-4-15; IEC 61000-4-30 gives a definition of (individual) rapid voltage changes. Both definitions are however strictly measurement based, which especially for flicker severity requires detailed models.

Incandescent lamps, which have been the base for the flicker-severity indices, are replaced by other types of lamps in most countries. The flicker-severity indices can no longer be considered as primary performance indices. Their future role as secondary performance indices (of importance to the network operator but not of direct importance to the network user) remains unclear.

For the time scale between 1 s and 10 min (the lower limit for standards on slow voltage magnitude variations in most countries) no standardized or generally accepted indices exist. An index referred to as "very-short variations", was proposed in 2005 [51,52], but received only limited following [53–59]. However, the need for performance indices in this range of time scale remains big: this is where the main impact of fast variations in production with solar-power installations will be, as was quantified among others by [60].

5.2.2. Limits

Limits exist for flicker severity and for the number of rapid voltage changes. No limits exist for voltage magnitude variations in the time scale between 1 s and 10 min.

5.2.3. Calculation Methods

Three different phenomena were mentioned in Section 5.2.1, all of which require different methods for calculation. Flicker severity requires either time series with very high time resolution (for example 20 ms) or the use of simplified aggregation rules for example as proposed in IEC/TR 61000-3-7.

For rapid voltage changes, simple voltage calculations (using only the source impedance at the point of connection or at point of common coupling) are sufficient to obtain the step in voltage due to a step in current. The latter can be the start of charging of an electric vehicle but also the unnecessary tripping of a PV installation due to its anti-islanding protection.

5.3. Voltage Unbalance

Voltage unbalance occurs due to the connection of large single-phase devices; both single-phase-connected PV and electric car chargers fall in this category.

5.3.1. Performance Indices

The ratio between negative-sequence and positive-sequence voltage is used as performance index in both IEC and IEEE standards. As the magnitude of the positive-sequence voltage shows limited variation, the negative-sequence voltage can also be used as an index. Next to that, IEEE and NEMA define some additional indices including the difference between highest and lowest line voltage. The different IEC, IEEE and NEMA definitions, all referred to as "unbalance", are compared in [61–63] where it for example is shown that different definitions can give rather different values for the unbalance.

The negative-sequence voltage is a useful index to quantify the impact of unbalance on three-phase rotating machines directly connected to the grid. The impact of unbalance on such devices is a thermal issue, so that 10-min values are appropriate.

For machines connected through a three-phase rectifier, like adjustable-speed drives, and for other equipment connected through a three-phase converter, the difference between the line voltages is a better index. This difference results in current unbalance for three-phase converters that in turn can result in unwanted tripping of the converter [64]. As shown in [62] this value is 87% to 101% of the absolute value of negative-sequence voltage, depending of the phase angle of the latter. As this is a protection issue, a value over a much shorter time scale is needed, like a one-second or three-second value.

5.3.2. Limits

For the negative-sequence unbalance (the IEC definition), a limit of 2% holds for low-voltage networks in many countries. See [65] for an overview of regulatory limits in European countries. Strictly speaking, the indictor is defined as the ratio between negative and positive-sequence voltage. When the calculations give the negative-sequence voltage (or when nominal positive-sequence voltage is assumed) a correction might have to be made for the limit. Assuming that the positive-sequence voltage can be as low as 90% of nominal, and assuming that negative and positive-sequence voltage are stochastically independent, the limit has to be reduced from 2% to 1.8%. There are no limits based on the IEEE and NEMA definitions of unbalance. When 3-s values are used instead of 10-min values, a higher limit than the IEC limit of 2% seems appropriate.

5.3.3. Calculation Methods

The negative-sequence transfer impedance matrix is used in [39] to calculate the increase in negative-sequence voltage in a low-voltage network with increasing number of single-phase-connected PV installations. The negative-sequence impedance of lines, cables and transformers is equal to their positive-sequence value. That can also be generally assumed for the source impedance of public medium-voltage networks. For industrial medium-voltage networks, large rotating machines might result in a lower value for the negative-sequence than for the positive-sequence source impedance.

The presence of three-phase equipment in the low-voltage network may have to be considered as well in the transfer-impedance matrix. The presence of three-phase equipment, like variable-speed heat pumps (common in Sweden), results in a reduction of the negative-sequence transfer impedance and thus in an increase of the hosting capacity. For a planning study, a value or range of values has to be estimated for three-phase equipment that is connected during hours with high amount of production from solar power. In [39] it was assumed that this could be a very small amount, so that its impact on negative-sequence impedance could be neglected. This assumption may result in an underestimation of the hosting capacity.

In [39] the same networks were used as in Section 4. That allows a comparison of the results. The hosting capacity for unbalance, assuming a 1.5% limit for the negative-sequence voltage and the 90th percentile as performance index, was estimated as 3 customers for the 6-customer network and 26 customers for the 28-customer network. The results from the overvoltage studies were: two customers for the 6-customer network (from Figure 4) and up to 22 customers for the 28-customer network (from Table 1).

Important for a complete picture of the hosting capacity is to consider the background voltage unbalance, i.e., the voltage unbalance without connection of any new production or consumption. Here it is important to consider that negative-sequence voltage is a complex quantity, consisting of magnitude and phase angle. During power-quality measurements, the phase angle of the negative-sequence voltage is normally not recorded. This is partly because there is no measurement definition of this angle available.

When the NEMA/IEEE definition of unbalance is used, the negative-sequence transfer impedance is not sufficient for the calculations. Either the complex phase voltages should be used; or separate impedances for positive, negative and zero-sequence components. The latter allows the incorporation of the impact of three-phase-connected equipment on the unbalance. In addition, information on background unbalance may be hard to obtain and require dedicated measurement campaigns.

5.4. Harmonic Voltage Distortion

Harmonic voltage distortion occurs because the current injected by the device is not sinusoidal. The bigger the device, the bigger the impact is of even a small amount of waveform distortion. Even when the relative emission (in percent of rated current) is small, the total impact on the grid can still be significant. With the connection of multiple devices, both the transfer impedance and aggregation rules should be considered.

5.4.1. Performance Indices

Standard indices are defined for each of the individual harmonics and interharmonics, and for total harmonic distortion in IEEE 519, IEC 61000-4-7 and IEC 61000-4-30. Normally, only 10-min values are considered as the impact of harmonics is considered a thermal phenomenon. However, for power-electronic converters and for certain types of protection of other equipment, unwanted protection operation can result from high harmonic or interharmonic distortion. Values obtained over shorter periods, like 1 or 3 seconds, should be used to study such impacts.

In IEC 61000-4-7, both groups and subgroups are defined for harmonics up to order 40 (i.e., 2 kHz in a 50 Hz system) and interharmonics. However, in IEC 61000-4-30 it is stated that the harmonic and interharmonic subgroups should be used when quantifying power quality. Therefore, those are also most appropriate for hosting-capacity studies.

5.4.2. Limits

The selection of limits for use in hosting-capacity studies is not obvious, despite the presence of a range of limits ("objective values") in national and international standards. Voltage characteristics are given in EN 50160; compatibility levels in IEC 61000-2-2; planning levels can be selected by each network operator themselves, but most of them follow the indicative planning levels from IEC/TR 61000-3-6. Especially the latter deviate a lot from the other ones and it is not obvious which limits should be used.

One approach is to consider a hosting-capacity study as a planning study, in which the planning levels would be appropriate. However, in more and more countries are the planning levels replaced by the limits set in local regulations. These are in turn, at least in Europe, strongly based on the voltage characteristics in EN 50160.

The final decision on the choice of limits, as many other choices in a hosting-capacity study, is part of the risk management between the different stakeholders. A complete hosting capacity study should therefore also consider a mapping of the risks as carried by the different stakeholders. Such a mapping is beyond the scope of this paper.

5.4.3. Calculation Methods

The choice of performance indices and limits is relatively straightforward for harmonics. However, the main barrier is the lack of appropriate calculation models, especially when considering low-voltage and medium-voltage networks.

The basic approach used in the literature is the source impedance or the transfer impedance matrix [66–70]; for harmonic frequencies instead of for the negative-sequence voltage as in the case of unbalance [39] or the equivalent impedances for single-phase loads as in the case of overvoltage due to single-phase PV. Once the injected harmonic currents and the impedance values are known, the harmonic voltage distortion can be calculated as a function of the amount of new production or consumption. Obtaining those values is however not easy and with the current state of the art no generally accepted method is available. This was illustrated by several studies on the impact of compact fluorescent lamps and LED lamps on the waveform distortion. Simulation studies concluded that the voltage and current distortion would increase [71,72]; measurements showed however that this was not the case and that the distortion was simply not impacted [73–76]. Acceptable harmonic models exist for the components of the power system (cables, lines and transformers) but several barriers remain before a complete model is available for use in hosting-capacity studies:

i)   The emitted harmonic current is impacted by the voltage distortion at the terminals of the equipment. This phenomenon has been observed early for diode rectifiers like the ones used in televisions [77] and was explained and quantified by a model in [78]. That impact was shown to be limited and the emission for a clean supply voltage was in most cases the worst case and the one reasonably useable for harmonic studies. With modern power-electronic equipment, with an actively controlled interface, there are observations showing the contrary, the emission for a distorted supply voltage can be much higher than for a sinusoidal supply [79–82].

ii) No suitable models exist for the customers connected to a low-voltage network. Some recent studies [83–85] show that the customer model is the main determining factor for the resonant frequency and thus for the harmonic transfer impedance.

iii) The connecting of new production or consumption will change the impedance of the low-voltage customer and therewith the harmonic transfer impedance. There exist no acceptable models for new devices like PV inverters or EV chargers. Some measurements are presented in [86–89] but there are big variations between manufacturers and for future equipment guesses have to be made.

iv) The statistical aggregation between different sources of harmonics is not known. This holds for the aggregation between different new devices (e.g., between different PV inverters) but also for the aggregation between the new devices and the background distortion. The aggregation between individual wind turbines has been studied by some authors [90–92] but it is not known if similar conclusions hold for PV inverters, EV chargers or other large low-voltage equipment like heat pumps [22,71,93].

5.5. Supraharmonics

Supraharmonics (waveform distortion in the frequency range between 2 kHz and 150 kHz) are injected by an increasing amount of devices connected to the grid. Supraharmonics are mainly due to active switching in the grid-interface of the devices. The transfer and aggregation of supraharmonics is significantly different from the transfer and aggregation of harmonics. To do a complete hosting capacity study of supraharmonics is to date not feasible, simply because there are no established limits or indices for distortion in this frequency range. This section will instead give a general description of supraharmonics, what levels can be expected and how they are transferred through the grid.

Supraharmonics commonly originate from two sources: power-electronic converters and transmitters of power line communication [21], the former being the focus here. Magnitude, frequency and duration of the emission vary between different devices but come in three general types: constant in magnitude and/or frequency over one cycle of the power system frequency; varying in magnitude and/or frequency over one cycle of the power system frequency or having a transient character [94]. For household devices the emission is in most cases of the varying type, two examples can be seen in Figure 13, where the Short Time Fourier Transform of a fluorescent lamp and a heat pump is shown. The emission from the fluorescent lamp varies in frequency and in magnitude during 20 ms whereas the emission from the heat pump varies only in amplitude as it appears and disappears four times per cycle.

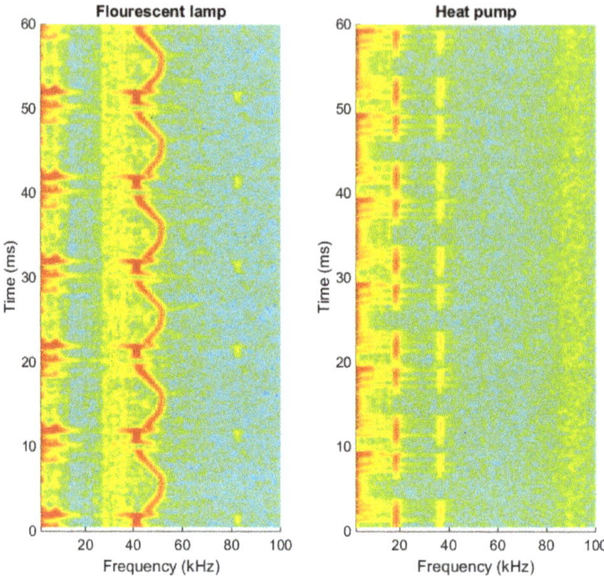

Figure 13. Suraharmonics from a fluorescent lamp (**left**) and from a heat pump (**right**).

In most cases, suraharmonics do not propagate over long distances. Damping in cables is one of the causes of this. More importantly, other devices connected to the low voltage network offer a low impedance path for these currents so that they tend to flow mainly between devices [95,96]. However, it is still feasible that suraharmonics can transfer through the grid. In [97] it is shown, through both measurements and simulations, that a resonance between the distribution transformer and an 800 m long cable amplified a 35 kHz voltage component on the low-voltage side of the transformer five times. To predict how the emission from an installation or device will transfer through the grid is difficult because of the dominating impact from connected. To correctly predict what is connected at a certain grid at any given moment in time would not be possible.

Many modern household devices emit suraharmonics to some extent. In [94] it is concluded that more than half of the household devices on the market have identifiable suraharmonic emission. There is a large diversity in magnitude and frequency between different devices, even if they are of the same type. For instance, measurements of about 80% of the types of EV chargers on the German market show that the switching frequency is between 9 and 100 kHz with a magnitude between 8 mA and 1.8 A [98].

The emission from a PV inverter and a heat pump are shown in Figure 14. The dominating emission is seen at 16 kHz from the inverter and at 18 kHz from the heat pump (note that these are just two examples, other PV inverters and heat pumps on the market will typically show a completely different behavior). For the measurements seen on the left in Figure 14, the devices are connected alone at the test site. For the measurements seen on the right, other devices are connected nearby; a TV close to the PV inverter and an induction stove close by the heat pump. When other devices are connected to the same phase, the emission at 16 kHz and 18 kHz originating from the inverter and the heat pump is increased, two and five times respectively. This indicates a resonant frequency close to 16 kHz or 18 kHz. The impact of neighbouring equipment on the emission level of suraharmonics from an EV charger is discussed in detail in [99].

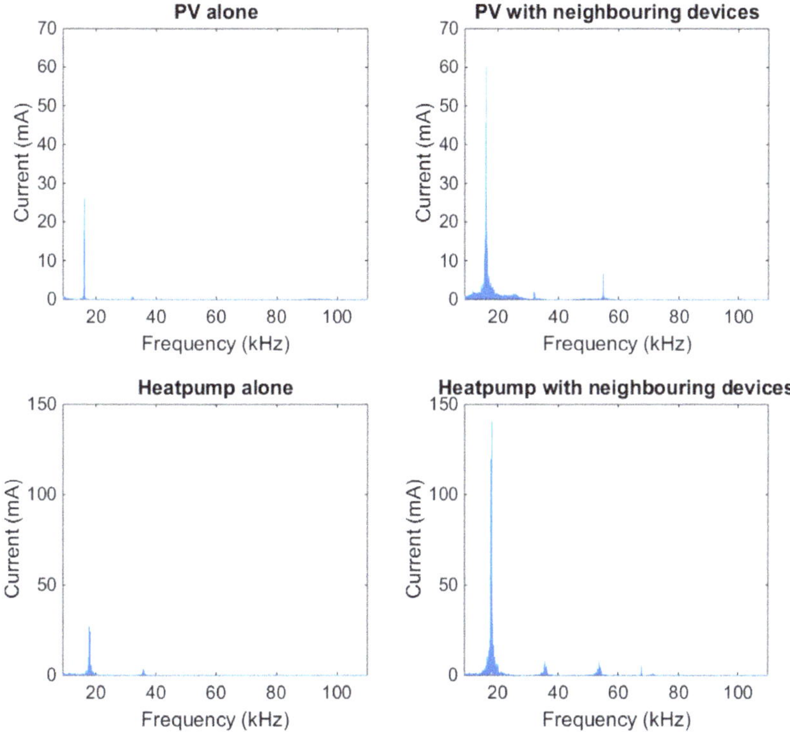

Figure 14. Supraharmonics from a PV inverter connected alone (upper **left**), the same inverter while neighbouring devices are connected (upper **right**). A heat pump connected alone (lower **left**), the same heat pump while neighbouring devices are connected (lower **right**).

6. Conclusions

The concept of hosting capacity has been introduced as a transparent tool allowing an open discussion between different stakeholders with the integration of distributed generation. This transparency remains an important characteristic of the hosting-capacity approach also when it is used as a planning tool for both new production and new consumption, as illustrated in this paper. Any hosting-capacity study requires (directly or indirectly) three parts: a performance index; a corresponding limit; and a method to calculate the value of the performance index as a function of the amount of new production or consumption. Further work is needed towards almost all of these before complete hosting-capacity studies can be done.

A hosting-capacity-based planning approach has been presented in this paper. The approach requires a network model, limited input data and a Monte-Carlo simulation to address the uncertainties. The approach has been applied to overvoltage and undervoltage due to increasing amounts of solar power and electric-vehicle chargers in low-voltage networks. A sensitivity analysis shows that the size of the PV inverters and the performance index have the main impact on the hosting capacity.

It is shown that, with single-phase connection of PV, not only overvoltage but also undervoltage limits may be exceeded. The hosting-capacity values obtained from the case studies cannot be generalized and applied to other low-voltage networks.

For the main impacts of new production and consumption, overvoltage, undervoltage and overcurrent, the calculation tools are available. However, suitable models to describe the existing situation in terms of voltage and current are missing. Different approaches are discussed in the paper, but as of yet no method is the dominating one. Data gathering is an important step in the development of such models.

For harmonics and interharmonics, acceptable calculation models remain missing and further work is needed towards those. Measurements are needed to develop component models so that harmonics and supraharmonics can be sufficiently accurately predicted for future grids. Generally, further work is needed towards selecting appropriate performance indices and the corresponding limits. This requires an interdisciplinary research effort.

Acknowledgments: This studies presented in Section 3 of this paper have been funded by a number of Swedish network operators through the smart-grid program of Energiforsk, by Skellefteå Kraft and by Umeå Energi. Some of the simulations shown in Section 3 were performed by Cecilia Karlsson as part of her final-year project. The measurements referred to in Section 3 were performed by Anders Larsson, Martin Lundmark and Mikael Byström.

Author Contributions: Both authors contributed to this publication; Math H. J. Bollen has been involved in the hosting-capacity approach since the term was first coined in 2004; Sarah K. Rönnberg has been a contributor to introductory parts of the paper; a discussion partner for the simulations presented in Section 3 and the main author for Sections 5.4 and 5.5.

Conflicts of Interest: The authors declare no conflict of interest.

References

1. Tagare, D.M. *Electricity Power Generation: The Changing Dimensions*; Wiley: Hoboken, NJ, USA, 2011.
2. Bollen, M.; Hassan, F. *Integration of Distributed Generation in the Power System*; Wiley-IEEE Press: Hoboken, NJ, USA, 2011.
3. Staffell, I.; Brett, D.J.L.; Brandon, N.P.; Hawkes, A.D. *Domestic Microgeneration: Renewable and Distributed Energy Technologies, Policies and Economics*; Routledge: Abingdon, UK, 2015.
4. International Energy Agency. *Medium-Term Renewable Energy Market Report 2016—Market Analysis and Forecasts to 2021*; International Energy Agency: Paris, France, 2016.
5. International Energy Agency. *Global EV Outlook 2016—Beyond One Million Electric Cars*; International Energy Agency: Paris, France, 2016.
6. Matulka, R. *The History of the Electric Car*; Department of Energy: Washington, DC, USA, 2014.
7. Wu, Q. *Grid Integration of Electric Vehicles in Open Electricity Markets*; Wiley: Hoboken, NJ, USA, 2013.
8. Chua, K.J.; Chou, S.K.; Yang, W.M. Advances in heat pump systems: A review. *Appl. Energy* **2010**, *87*, 13611–13624. [CrossRef]
9. Waide, P. *Phase out of Incandescent Lamps—Implications for International Supply and Demand for Regulatory Compliant Lamps*; International Energy Agency: Paris, France, 2010.
10. Bollen, M. *The Smart Grid—Adapting the Power System to New Challenges*; Morgan & Claypool: San Rafael, CA, USA, 2011.
11. Hatziargyriou, N. *Microgrids: Architectures and Control*; Wiley-IEEE Press: Hoboken, NJ, USA, 2014.
12. Liu, C.-C.; McArthur, S.; Lee, S.-J. *Smart Grid Handbook*; Wiley: Hoboken, NJ, USA, 2016.
13. Bollen, M.H.J.; Das, R.; Djokic, S.; Ciufo, P.; Meyer, J.; Rönnberg, S.K.; Zavoda, F. Power quality concerns in implementing smart distribution-grid applications. *IEEE Trans. Smart Grid* **2017**, *8*, 391–399. [CrossRef]
14. Dugan, R.C.; McGranaghan, M.F.; Santoso, S.; Beaty, H.W. *Electric Power Systems Quality*; McGraw-Hill: New York, NY, USA, 2012.
15. Jenkins, N.; Allan, R.; Crossley, P.; Kirschen, D.; Strbac, G. *Embedded Generation*; Institution of Engineering and Technology: Stevenage, UK, 2000.
16. Lopes, J.A.P.; Hatziargyriou, N.; Mutale, J.; Djapic, P.; Jenkins, N. Integrating distributed generation into electric power systems: A review of drivers, challenges and opportunities. *Electr. Power Syst. Res.* **2007**, *77*, 1189–1203. [CrossRef]
17. Freris, L.; Infield, D. *Renewable Energy in Power Systems*; Wiley: Hoboken, NJ, USA, 2008.
18. Masters, C.L. Voltage rise: The big issue when connecting embedded generation to long 11 kV overhead lines. *IET Power Eng. J.* **2002**, *16*, 5–12. [CrossRef]
19. Rönnberg, S.K.; Bollen, M.H.J. Power quality issues in the future electric power system. *Electr. J.* **2016**, *29*, 49–61. [CrossRef]

20. Rönnberg, S.K.; Bollen, M.H.J.; Langella, R.; Zavoda, F.; Hasler, J.-P.; Ciufo, P.; Cuk, V.; Meyer, J. The expected impact of four major changes in the grid on the power quality—A review. *CIGRE Sci. Eng.* **2017**, *8*, 5.

21. Rönnberg, S.K.; Bollen, M.H.J.; Amaris, H.; Chang, G.W.; Gu, I.Y.H.; Kocewiak, Ł.H.; Meyer, J.; Olofsson, M.; Ribeiro, P.F.; Desmet, J. On waveform distortion in the frequency range of 2 kHz–150 kHz—Review and research challenges. *Electr. Power Syst. Res.* **2017**, *150*, 1–10. [CrossRef]

22. CIGRE JWG C4/C4.29, Power Quality Aspects of Solar Power, CIGRE Technical Brochure 672. 2016. Available online: www.e-cigre.org (accessed on 24 July 2017).

23. Bourgain, G. *Integrating Distributed Energy Resources in Today's Electrical Energy System*; Book Presenting the Results of the EU-DEEP Project; ExpandDER: Saint Denis le Plaine, France, 2009.

24. Etherden, N. Increasing the Hosting Capacity of Distributed Energy Resources Using Storage and Communication. Ph.D. Thesis, Luleå University of Technology, Luleå, Sweden, 2014.

25. Bollen, M.H.J.; Häger, M. Power quality: Interactions between distributed energy resources, the grid and other customers. In Proceedings of the 1st International Conference on Renewable Energy Sources and Distributed Energy Resources, Brussels, Belgium, 1–3 December 2004.

26. Carnelley, P.; Kasica, C. Choosing a Web Hosting Provider. *Computer Weekly*, September 2001. Available online: www.computerweekly.com (accessed on 31 August 2017).

27. Bastug, A.; Sankur, B. Improving the payload of watermarking channels via LDPC coding. *IEEE Signal Process. Lett.* **2014**, *11*, 90–91. [CrossRef]

28. Ahrens, J.D. *Evacuees from Border Towns in Tigray Setting up Makeshift Camps*; UNDP Emergencies Unit for Ethiopia: Addis Ababa, Ethiopia, 1999.

29. Etherden, N.; Bollen, M.H.J. Increasing the hosting capacity of distribution networks by curtailment of renewable energy resources. In Proceedings of the IEEE Trondheim PowerTech, Trondheim, Norway, 19–23 June 2011.

30. Leemput, N.; Geth, F.; van Roy, J.; Olivella-Rosell, P.; Sumper, J.D.A. MV and LV Residential grid impact of combined slow and fast charging of electric vehicles. *Energies* **2015**, *8*, 1760–1783. [CrossRef]

31. It's All about the Value of the Network: ComEd Gears up for a Distributed Energy Boom. *Green Tech Media*, 1 July 2016. Available online: https://www.greentechmedia.com (accessed on 24 July 2017).

32. Hosting Capacity Map. Available online: http://www.pepco.com/Hosting-Capacity-Map.aspx (accessed on 24 July 2017).

33. Currie, B.; Abbey, C.; Ault, G.; Ballard, J.; Conroy, B.; Sims, R.; Williams, C. Flexibility is key in New York: New tools and operational solutions for managing distributed energy resources. *IEEE Power Energy Mag.* **2017**, *15*, 20–29. [CrossRef]

34. Smith, J. *Stochastic Analysis to Determine Feeder Hosting Capacity for Distributed Solar PV*; EPRI Technical Update: Knoxville, TN, USA, 2012.

35. Dubey, D.; Santoso, S.; Maitra, A. Understanding photovoltaic hosting capacity of distribution circuits. In Proceedings of the IEEE Power & Energy Society General Meeting, Denver, CO, USA, 26–30 July 2015.

36. Widén, J.; Wäckelgård, E.; Paatero, J.; Lund, P. Impacts of distributed photovoltaics on network voltages: Stochastic simulations of three Swedish low-voltage distribution grids. *Electr. Power Syst. Res.* **2010**, *80*, 1562–1571. [CrossRef]

37. McGranaghan, M.F. Grid modernization challenges for the integrated grid. In Proceedings of the IEEE PowerTech Manchester, Manchester, UK, 18–22 June 2017.

38. Bahramirad, S. Design and planning of the grid of the future. In Proceedings of the IEEE PowerTech Manchester, Manchester, UK, 18–22 June 2017.

39. Schwanz, D.; Möller, F.; Rönnberg, S.K.; Meyer, J.; Bollen, M.H.J. Stochastic assessment of voltage unbalance due to single-phase-connected solar power. *IEEE Trans. Power Deliv.* **2017**, *32*, 852–861. [CrossRef]

40. Richardson, I.; Thomson, M.; Infield, D.; Clifford, C. Domestic electricity use: A high-resolution energy demand model. *Energy Build.* **2010**, *42*, 1878–1887. [CrossRef]

41. Paisios, A.; Ferguson, A.; Djokic, S.Z. Solar analemma for assessing variations in electricity demands at MV buses. In Proceedings of the Mediterranean Conference on Power Generation, Transmission, Distribution and Energy Conversion (MedPower), Belgrade, Serbia, 6–9 November 2016.

42. Distribution Test Feeders, IEEE PES Distribution System Analysis Subcommittee's Distribution Test Feeder Working Group. Available online: https://ewh.ieee.org/soc/pes/dsacom/testfeeders/ (accessed on 24 July 2017).

43. Lennerhag, G.; Pinares, M.; Bollen, G.; Foskolos, O.; Gafurov, T. Performance indicators for quantifying the ability of the grid to host renewable electricity production. In Proceedings of the 24th International Conference on Electricity Distribution (CIRED), Glasgow, UK, 12–15 June 2017.

44. Pukhrem, S. *Investigation into Photovoltaic Distributed Generation Penetration in the Low Voltage Distribution Network*; School of Electrical and Electronic Engineering, Dublin Institute of Technology: Dublin, Ireland, 2017.

45. Fernandez, E.; Albizu, I.; Bedialauneta, M.T.; Mazon, A.J.; Leite, P.T. Review of dynamic line rating systems for wind power integration. *Renew. Sustain. Energy Rev.* **2016**, *53*, 80–92. [CrossRef]

46. Garniwa, I.; Burhani, A. Thermal incremental and time constant analysis on 20 kV XLPE cable with current vary. In Proceedings of the 8th International Conference on Properties and Applications of Dielectric Materials, Bali, Indonesia, 26–30 June 2006.

47. Douglass, D.A.; Edris, A.-A. Real-time monitoring and dynamic thermal rating of power transmission circuits. *IEEE Trans. Power Deliv.* **1996**, *11*, 1407–1418. [CrossRef]

48. Swift, G.; Molinski, T.S.; Bray, R.; Menzies, R. A fundamental approach to transformer thermal modeling. II. Field verification. *IEEE Trans. Power Deliv.* **2001**, *16*, 176–180. [CrossRef]

49. Lennerhag, O.; Ackeby, S.; Bollen, M.H.J.; Foskolos, G.; Gafurov, T. Using measurements to increase the accuracy of hosting capacity calculations. In Proceedings of the 24th International Conference on Electricity Distribution (CIRED), Glasgow, UK, 12–15 June 2017.

50. Etherden, N.; Bollen, M.H.J. Dimensioning of energy storage for increased integration of wind power. *IEEE Trans. Sustain. Energy* **2016**, *4*, 546–553. [CrossRef]

51. Bollen, M.H.J.; Gu, I.Y.H. Characterization of voltage variations in the very-short time-scale. *IEEE Trans. Power Deliv.* **2005**, *20*, 1198–1199. [CrossRef]

52. Bollen, M.H.J.; Häger, M.; Schwaegerl, C. Quantifying voltage variations on a time scale between 3 seconds and 10 minutes. In Proceedings of the International Conference on Electricity Distribution (CIRED), Turin, Italy, 6–9 June 2005.

53. Otomanski, P.; Wiczynski, G. Search for disturbing loads in power network with the use of voltage and current fluctuation. In Proceedings of the International School on Nonsinusoidal Currents and Compensation (ISNCC), Lagow, Poland, 15–18 June 2010.

54. Suwanapingkarl, P. Power Quality Analysis of Future Power Networks. Ph.D. Thesis, Northumbria University, Northumbria, UK, 2012.

55. Ciontea, C.I.; Sera, D.; Iov, F. Influence of resolution of the input data on distributed generation integration studies. In Proceedings of the International Conference on Optimization of Electrical and Electronic Equipment (OPTIM), Brasov, Romania, 22–24 May 2014.

56. Wiczynski, G. Voltage-fluctuation-based identification of noxious loads in power network. *IEEE Trans. Instrum. Meas.* **2009**, *58*, 2893–2898. [CrossRef]

57. Nambiar, A.J. Coordinated Control and Network Integration of Wave Power Farms. Ph.D. Thesis, The University of Edinburgh, Edinburgh, UK, 2012.

58. Wiczynski, G. Description of voltage fluctuations in LV power network with the use of PST indicator and voltage fluctuation indices. In Proceedings of the 13th International Conference on Harmonics and Quality of Power (ICHQP), Wollongong, Australia, 28 September–1 October 2008.

59. Pakonen, P.; Hilden, A.; Suntio, T.; Verho, P. Grid-connected PV power plant induced power quality problems—Experimental evidence. In Proceedings of the 18th European Conference on Power Electronics and Applications (EPE'16 ECCE Europe), Rheinstetten, Germany, 5–8 September 2016.

60. Lennerhag, O.; Bollen, M.H.J.; Ackeby, S.; Rönnberg, S.K. Very short variations in voltage (timescale less than 10 minutes) due to variations in wind and solar power. In Proceedings of the International Conference on Electricity Distribution (CIRED), Lyon, France, 15–18 June 2015.

61. Pillay, P.; Manyage, M. Definitions of Voltage Unbalance. *IEEE Power Eng. Rev.* **2001**, *21*, 50–51. [CrossRef]

62. Bollen, M.H.J. Definitions of Voltage Unbalance. *IEEE Power Eng. Rev.* **2002**, *22*, 49–50. [CrossRef]

63. Rodriguez, A.D.; Fuentes, F.M.; Matta, A.J. Comparative analysis between voltage unbalance definitions. In Proceedings of the Workshop on Engineering Applications—International Congress on Engineering (WEA), Bogota, Colombia, 28–30 October 2015.

64. Jouanne, V.; Banerjee, B. Assessment of voltage unbalance. *IEEE Trans. Power Deliv.* **2001**, *16*, 782–790. [CrossRef]

65. Council of European Energy Regulators. *6th CEER Benchmarking Report on the Quality of Electricity and Gas Supply*; Council of European Energy Regulators: Brussels, Belgium, 2016.

66. Sun, W.; Harrison, G.P.; Djokic, S.Z. Distribution network capacity assessment: Incorporating harmonic distortion limits. In Proceedings of the IEEE Power and Energy Society General Meeting, San Diego, CA, USA, 22–26 July 2012.

67. Pandi, V.R.; Zeineldin, H.H.; Xiao, W. Determining optimal location and size of distributed generation resources considering harmonic and protection coordination limits. *IEEE Trans. Power Syst.* **2013**, *28*, 1245–1254. [CrossRef]

68. Barutcu, I.C.; Karatepe, E. Influence of phasor adjustment of harmonic sources on the allowable penetration level of distributed generation. *Int. J. Electr. Power Energy Syst.* **2017**, *87*, 1–15. [CrossRef]

69. Sakar, S.; Balci, M.E.; Aleem, S.H.A.; Zobaa, A.F. Increasing PV hosting capacity in distorted distribution systems using passive harmonic filtering. *Electr. Power Syst. Res.* **2017**, *148*, 74–86. [CrossRef]

70. Santos, I.N.; Ćuk, V.; Almeida, P.M.; Bollen, M.H.J.; Ribeiro, P.F. Considerations on hosting capacity for harmonic distortions on transmission and distribution systems. *Electr. Power Syst. Res.* **2015**, *119*, 199–206. [CrossRef]

71. Wang, Y.; Yong, J.; Sun, Y.; Xu, W.; Wong, D. Characteristics of harmonic distortions in residential distribution systems. *IEEE Trans. Power Deliv.* **2017**, *32*, 1495–1504. [CrossRef]

72. Watson, N.R.; Scott, T.L.; Hirsch, S.J.J. Implications for distribution networks of high penetration of compact fluorescent lamps. *IEEE Trans. Power Deliv.* **2009**, *24*, 1521–1528. [CrossRef]

73. Blanco, A.M.; Stiegler, R.; Meyer, J. Power quality disturbances caused by modern lighting equipment (CFL and LED). In Proceedings of the IEEE PowerTech, Grenoble, France, 16–20 June 2013.

74. Rönnberg, S.K.; Wahlberg, M.; Bollen, M.H.J. Harmonic emission before and after changing to LED and CFL—Part II: Field measurements for a hotel. In Proceedings of the International Conference Harmonics and Quality of Power (ICHQP), Bergamo, Italy, 26–29 September 2010.

75. Rönnberg, S.; Wahlberg, M.; Bollen, M. Harmonic emission before and after changing to LED lamps—Field measurements for an urban area. In Proceedings of the International Conference Harmonics and Quality of Power (ICHQP), Hong Kong, China, 17–20 June 2012.

76. Gil-de-Castro, A.; Rönnberg, S.K.; Bollen, M.H.; Moreno-Muñoz, A. Harmonic phase angles for a domestic customer with different types of lighting. *Int. Trans. Electr. Energy Syst.* **2015**, *25*, 1281–1296. [CrossRef]

77. Arrillaga, J.; Watson, N.R. *Power System Harmonics*, 2nd ed.; Wiley: Hoboken, NJ, USA, 1997.

78. Mansoor, A.; Grady, W.M.; Staats, P.T.; Thallam, R.S.; Doyle, M.T.; Samotyj, M.J. Predicting the net harmonic currents produced by large numbers of distributed single-phase computer loads. *IEEE Trans. Power Deliv.* **1995**, *10*, 2001–2006. [CrossRef]

79. Yanchenko, S.; Meyer, J. Harmonic emission of household devices in presence of typical voltage distortions. In Proceedings of the IEEE Eindhoven PowerTech, Eindhoven, The Netherlands, 29 June–2 July 2015.

80. Bosman, A.J.A.; Cobben, J.F.G.; Myrzik, J.M.A.; Kling, W.L. Harmonic modelling of solar inverters and their interaction with the distribution grid. In Proceedings of the 41st International Universities Power Engineering Conference (UPEC), Northumbria, UK, 6–8 September 2006.

81. Müller, S.; Meyer, J.; Schegner, P. Characterization of small photovoltaic inverters for harmonic modeling. In Proceedings of the International Conference on Harmonics and Quality of Power (ICHQP), Bucharest, Romania, 25–28 May 2014.

82. Djokic, S.; Meyer, J.; Möller, F.; Langella, R.; Testa, A. Impact of operating conditions on harmonic and interharmonic emission of PV inverters. In Proceedings of the IEEE International Workshop on Applied Measurements for Power Systems (AMPS), Aachen, Germany, 23–25 September 2015.

83. Meyer, J.; Stiegler, R.; Schengner, P.; Röder, I.; Belger, A. Harmonic resonances in residential low voltage networks caused by consumer electronics. In Proceedings of the 24th International Conference on Electricity Distribution (CIRED), Glasgow, UK, 12–15 June 2017.

84. Pomilio, J.A.; Deckmann, S.M. Characterization and compensation of harmonics and reactive power of residential and commercial loads. *IEEE Trans. Power Deliv.* **2007**, *22*, 1049–1055. [CrossRef]

85. Chakravorty, D.; Meyer, J.; Schegner, P.; Yanchenko, S.; Schocke, M. Impact of modern electronic equipment on the assessment of network harmonic impedance. *IEEE Trans. Smart Grid* **2017**, *8*, 382–390. [CrossRef]

86. Collin, A.J.; Djokic, S.Z.; Thomas, H.F.; Meyer, J. Modelling of electric vehicle chargers for power system analysis. In Proceedings of the 11th International Conference on Electrical Power Quality and Utilisation, Lisbon, Portugal, 17–19 October 2011.

87. Dubey, A.; Santoso, S.; Cloud, M.P. Average-value model of electric vehicle chargers. *IEEE Trans. Smart Grid* **2013**, *4*, 1549–1557. [CrossRef]

88. Jiang, C.; Torquato, R.; Salles, D.; Xu, W. Method to assess the power-quality impact of plug-in electric vehicles. *IEEE Trans. Power Deliv.* **2014**, *29*, 958–965. [CrossRef]

89. Ackermann, F.; Moghadam, H.; Meyer, J.; Mueller, S.; Domagk, M.; Santjer, F.; Athamna, I.; Klosse, R. Characterization of harmonic emission of individual wind turbines and pv inverters—Part II: Photovoltaic inverters. In Proceedings of the 6th Solar Integration Workshop, Vienna, Austria, 14–15 November 2016.

90. Larose, C.; Gagnon, R.; Prud'Homme, P.; Fecteau, M.; Asmine, M. Type-III wind power plant harmonic emissions: Field measurements and aggregation guidelines for adequate representation of harmonics. *IEEE Trans. Sustain. Energy* **2013**, *4*, 797–804. [CrossRef]

91. Ghassemi, F.; Koo, K.L. Equivalent network for wind farm harmonic assessments. *IEEE Trans. Power Deliv.* **2010**, *25*, 1808–1815. [CrossRef]

92. Yang, K.; Bollen, M.H.J.; Larsson, E.O.A. Aggregation and amplification of wind-turbine harmonic emission in a wind park. *IEEE Trans. Power Deliv.* **2015**, *30*, 791–799. [CrossRef]

93. Jayasekara, N.; Wolfs, P. Analysis of power quality impact of high penetration PV in residential feeders. In Proceedings of the Australian Universities Power Engineering Conference (AUPEC), Christchurch, New Zealand, 5–8 December 2010.

94. Grevener, A.; Rönnberg, S.; Meyer, J.; Bollen, M.; Myrzik, J. Survey of Supraharmonic Emission of Household Appliances. In Proceedings of the International Conference and Exhibition on Electricity Distribution (CIRED), Glasgow, UK, 12–15 June 2017.

95. Rönnberg, S.; Wahlberg, M.; Bollen, M.; Lundmark, M. Equipment currents in the frequency range 9–95 kHz, measured in a realistic environment. In Proceedings of the 13th International Conference on Harmonics and Quality of Power (ICHQP), Wollongong, Australia, 28 September–1 October 2008.

96. Klatt, M.; Meyer, J.; Schegner, P.; Koch, A.; Myrzik, J.; Korner, C.; Darda, T.; Eberl, G. Emission levels above 2 kHz—Laboratory results and survey measurements in public low voltage grids. In Proceedings of the International Conference and Exhibition on Electricity Distribution (CIRED), Stockholm, Sweden, 10–13 June 2013.

97. Leroi, C.; De Jaeger, E. Conducted disturbances in the frequency range 2–150 kHz: Influence of the LV distribution grids. In Proceedings of the International Conference on Electricity Distribution (CIRED), Lyon, France, 15–18 June 2015.

98. Meyer, J.; Klatt, M.; Grevener, A. Supraharmonics—Future challenges in the frequency range 2–150 kHz, Keynote presentation. In Proceedings of the International Conference on Renewable Energy and Power Quality (ICREPQ), Malaga, Spain, 4–6 April 2017.

99. Gil-de-Castro, D.; Rönnberg, S.K.; Bollen, M.H.J. Harmonic interaction between an electric vehicle and different domestic equipment. In Proceedings of the International Symposium on Electromagnetic Compatibility (EMC Europe), Gothenburg, Sweden, 1–4 September 2014.

Article

A Piecewise Bound Constrained Optimization for Harmonic Responsibilities Assessment under Utility Harmonic Impedance Changes

Tianlei Zang [1,2,3] , Zhengyou He [1,*], Yan Wang [1], Ling Fu [1], Zhiyu Peng [2] and Qingquan Qian [1]

[1] School of Electrical Engineering, Southwest Jiaotong University, Chengdu 610031, China; zangtianlei@126.com (T.Z.); wangyanchn_cd@163.com (Y.W.); lingfu@swjtu.cn (L.F.); qqq@swjtu.cn (Q.Q.)

[2] Guizhou Key Laboratory of Electric Power Big Data, Guizhou Institute of Technology, Guiyang 550003, China; gitbigdata@163.com

[3] State Key Laboratory of Security Control and Simulation of Power Systems and Large Scale Generation Equipment, Department of Electrical Engineering, Tsinghua University, Beijing 100084, China

* Correspondence: hezy@swjtu.cn; Tel.: +86-28-8760-2445

Academic Editor: Frede Blaabjerg
Received: 22 April 2017; Accepted: 1 July 2017; Published: 6 July 2017

Abstract: Considering the effect of the utility harmonic impedance variations on harmonic responsibility, a method based on piecewise bound constrained optimization is proposed in this paper to evaluate the load harmonic responsibilities. The wavelet packet transform is employed to determine the change times of the utility harmonic impedances. The harmonic monitoring data is divided into several segments where the utility harmonic impedances are considered as constants. Then, the problem of harmonic responsibility assessment under utility harmonic impedance changes are settled by the piecewise bound constrained optimization model. Furthermore, the interior point, the sequential quadratic programming and the active set algorithm are respectively adopted to calculate all the instantaneous harmonic responsibilities of harmonic loads. Finally, the weighted summation is used to calculate the total harmonic responsibility. To demonstrate the validity, simulation tests are carried out on an experimental circuit and the IEEE 13-bus distribution system.

Keywords: harmonic responsibilities assessment; utility harmonic impedance changes; wavelet packet transform; piecewise approach; bound constrained optimization

---

1. Introduction

With the development of smart grids, increasing numbers of power electronic devices have been connected to distribution networks, which inject a large amount of harmonics [1–5]. Various electrical power equipment and electronic products have a strong sensitivity to the harmonics in the distribution network, making harmonic elimination of great importance [6]. To address the problem of harmonic pollution, appropriate punishment scheme should be executed according to the harmonic limits recommended by the IEEE or IEC standards. To ensure its implementation, it is necessary to quantitatively evaluate the harmonic responsibility of the major harmonic loads at the point of common coupling (PCC) in distribution networks [7–9].

In traditional methods, the key of harmonic responsibility evaluation is to determine the utility harmonic impedance. These works can be mainly classified as fluctuation quantity methods [10,11], linear regression methods [12–14] and independent component analysis (ICA) [15,16] methods. Fluctuation quantity methods rely on the fluctuation quantity proportion of harmonic voltage to current for calculating the harmonic impedance. The various regression analysis methods, such as the complex linear least squares [12], non-parametric regression [13] and multiple linear regression [14] methods, formulate an equation and solve the regression coefficient so as to get the utility harmonic impedance. The complex ICA [15] and FastICA [16] are usually used to estimate the utility harmonic impedance when the utility harmonic variations are neglected. Meanwhile, most of the above methods are based upon the supposition that the utility harmonic are invariant. In a real power grid, the utility harmonic voltage fluctuates due to the load fluctuation. The utility harmonic voltage has a certain influence on the amplitude as well as angle of the harmonic current which affects the harmonic voltage [17,18]. Under certain condition, the harmonic voltage and current all fluctuate simultaneously. The methods above cannot reflect the variation of harmonic voltage and current while the influence of utility harmonic voltage fluctuation is considered in [19–21]. In the previous study of the authors, an adaptive assessment approach [19] for harmonic responsibility under utility harmonic voltage variation was proposed. It has been proved that the utility harmonic voltage can be segmented by hierarchical K-means clustering under the condition of the same utility harmonic impedance. Then, regression methods can be effectively used to calculate the harmonic responsibilities. In the study of utility harmonic voltage fluctuation, the utility harmonic impedance is supposed to be invariant, but the switching of the equipment [22], changes in the reactive power compensation, the state of the distributed generators and the adjustment of the interruptible loads [23] can all result in variations in utility harmonic impedance. Under such an unrealistic assumption, a series of errors may be introduced in the assessment results. Therefore, it is of great significance to evaluate the harmonic responsibility in the presence of utility harmonic impedance changes.

Based on the analysis above, and considering the utility harmonic impedance changes, this paper firstly adopts the wavelet packet transform to detect the change points of the utility harmonic impedance. Then, the harmonic measurement data are segmented. Besides, in order to more accurately evaluate harmonic responsibility, the piecewise bound constrained optimization model and nonlinear optimization method are used to calculate the responsibility of each segment. Finally, the total harmonic responsibility of each harmonic load is obtained based on the data segment length. Section 2 describes the basic principles and conventional method of harmonic responsibility assessment. In Section 3, determination method of the change times of utility harmonic impedance is formulated. Section 4 introduces the piecewise bound constrained optimization method for harmonic responsibility assessment. The process of the novel method for harmonic responsibility assessment, numerical experiments and conclusion are provided in Sections 5–7, respectively.

2. Basic Principle and Conventional Method of Harmonic Responsibility Assessment

The Norton equivalent circuits can be applied for harmonic modelling of utility and loads [19,24]. Figure 1a shows a typical distribution system with two major harmonic loads, where h stands for the harmonic order; Z_s^h and $\dot{I}_s^h$ are the equivalent harmonic impedance and injected current at the utility side, respectively, while Z_k^h and $\dot{I}_k^h (k = 1,2)$ are the equivalent harmonic impedance and injected current of each nonlinear load; $\dot{I}_{bk}^h (k = 1,2)$ is the branch harmonic current; $\dot{V}_{pcc}^h$ and $\dot{I}_{pcc}^h$ are the h-th harmonic voltage and current at the PCC, respectively.

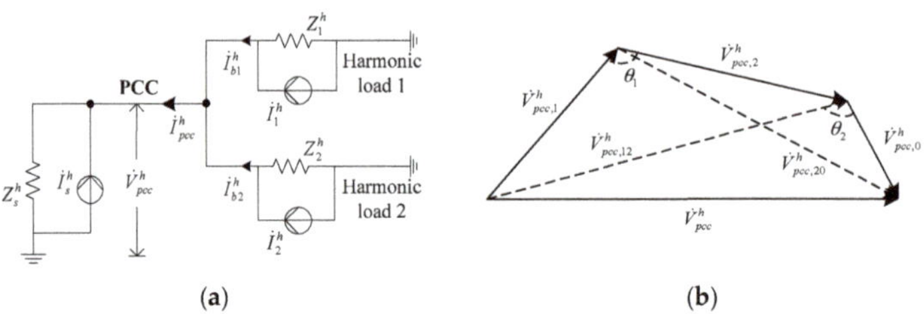

Figure 1. A typical distribution system with two major harmonic loads and its harmonic voltage phasors: (**a**) The Norton equivalent circuit; (**b**) Phasor diagram of the h-th harmonic voltages.

According to the superposition principle, the *h*-th harmonic voltage at the PCC is:

$$\dot{V}^h_{pcc} = \dot{V}^h_{pcc,1} + \dot{V}^h_{pcc,2} + \dot{V}^h_{pcc,0} = Z^h_{pcc,1}\dot{I}^h_{b1} + Z^h_{pcc,2}\dot{I}^h_{b2} + \dot{V}^h_{pcc,0} \tag{1}$$

where dots represent the phasors of the voltages or currents, $\dot{V}^h_{pcc,1}$ and $\dot{V}^h_{pcc,2}$ denote the harmonic voltage at harmonic load 1 and 2 at the PCC, respectively; $Z^h_{pcc,1}$ and $Z^h_{pcc,2}$ are the equivalent harmonic impedance but harmonic load 1 or 2, respectively; $\dot{V}^h_{pcc,0}$ is the harmonic voltage from the utility at the PCC, also known as the utility harmonic voltage. The phasor diagram of the *h*-th harmonic voltages is shown in Figure 1b.

The harmonic responsibility of harmonic load *i* (*i* = 1, 2) at the PCC can be calculated as:

$$\mu_{pcc,i} = \frac{\left|\dot{V}^h_{pcc,i}\right|}{\left|\dot{V}^h_{pcc}\right|} \cos \beta_i \times 100\% = \frac{\left|Z^h_{pcc,i}\right|\left|\dot{I}^h_{bi}\right|}{\left|\dot{V}^h_{pcc}\right|} \cos \beta_i \times 100\% \tag{2}$$

where β_i is the phase angle between $\dot{V}^h_{pcc,i}$ and $\dot{V}^h_{pcc}$.

Linear regression is a common assessment method for harmonic responsibilities [12,13], which is based on monitoring the harmonic voltage and current at the PCC.

The normalized *h*-th harmonic voltage and current at the PCC are related by:

$$\left|\dot{V}^h_{pcc}\right| = K\left|\dot{I}^h_{pcc}\right| + B \tag{3}$$

Figure 1 indicates the harmonic voltage at the PCC is:

$$\left|\dot{V}^h_{pcc}\right| = \left|\dot{V}^h_{pcc,0}\right| + \left|Z^h_{pcc,1}\right|\left|\dot{I}^h_{b1}\right| \cos \beta_1 + \left|Z^h_{pcc,2}\right|\left|\dot{I}^h_{b2}\right| \cos \beta_2 \tag{4}$$

Let $\dot{V}^h_{pcc,0} = \alpha_0$, $\left|Z^h_{pcc,1}\right| \cos \beta_1 = \alpha_1$, and $\left|Z^h_{pcc,2}\right| \cos \beta_2 = \alpha_2$ then $\left|\dot{V}^h_{pcc}\right|$ can be expressed as:

$$\left|\dot{V}^h_{pcc}\right| = \alpha_0 + \alpha_1\left|\dot{I}^h_{b1}\right| + \alpha_2\left|\dot{I}^h_{b2}\right| \tag{5}$$

It can be seen from Equations (3)–(5) that in the application of linear regression methods, the harmonic data should meet that the change of the utility harmonic voltage cannot influence the change of the harmonic current. Furthermore, if either the harmonic voltage, current or impedance changes, the accuracy of the regression analysis will be affected. Therefore, the variations of utility harmonics are the main error sources when the regression methods are employed.

3. Determination of the Change Time of Utility Harmonic Impedance Using Wavelet Packet Transform

In the distribution system, the changes of the operation mode, load or reactive compensation can all lead to changes of the utility harmonic impedance. To accurately calculate the harmonic responsibility, harmonic monitoring data must be properly segmented according to the identified utility harmonic impedance. In this article, the roughly estimates of utility harmonic impedance are used to segment the data.

Due to the complexity of the actual distribution system and the existence of transient processes, the actual utility harmonic impedance changes in a gradual manner. Therefore, it is necessary to choose an effective method to adaptively detect the change. In view of the good performance of wavelet package transform in signal singularity detection, this paper employs the wavelet package transform to detect the change points of the utility harmonic impedance.

In wavelet packet transform, the input signal can be decomposed into low frequency and high frequency components level by level to represent the approximations and details of signal respectively [25]. Figure 2 shows a wavelet packet transform tree with three decomposition levels. The wavelet packet coefficients at each level can be obtained by:

$$D_j^{2n}(t) = \sum_{\omega} G(\omega) D_{j-1}^n(2t - \omega)$$
$$D_j^{2n+1}(t) = \sum_{\omega} H(\omega) D_{j-1}^n(2t - \omega)$$

$$(6)$$

where G and H represent a low-pass filter and a high-pass filter, respectively; t is the sampling point; ω is the displacement factor; D_{j-1}^n represents the component at the level $j-1$, D_j^{2n} and D_j^{2n+1} represent the low frequency and high frequency components at the level j, respectively.

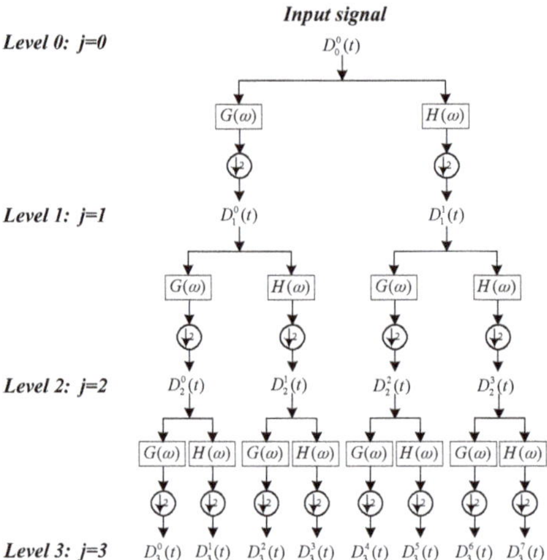

Figure 2. Wavelet packet decomposition tree with three decomposition levels.

The main idea of identifying the variation of utility harmonic impedance using wavelet packet transform is described in the following paragraphs. Since the boundary of the utility harmonic impedance change is not obvious, this paper transforms the identification of change point into the identification of change time window by adding windows, in order to reduce the identification error. The window length is denoted by L. According to Equation (3), if the values of the load harmonic impedances and injected harmonic currents can be considered as constants, the slope of the fitting curve is then a rough estimate of the utility harmonic impedance, and the slopes are approximately equal in this period. If mutation exists in the utility harmonic impedance, the slope of the fitting curve will change sharply compared with the adjacent window.

With regard to small samples, the window length $L = 3$ can be used to carry out the regression analysis. In the harmonic responsibility assessment, the data points in the time window, which correspond to the mutational utility harmonic impedance, are deleted. The sampling data points in the segments on both sides of the deleted time window can be considered as the data points under the same utility harmonic impedance.

For large samples, a long window length, such as $L = 30$, should be used to carry out the regression analysis, and the utility harmonic impedance change time can be directly determined through the wavelet packet decomposition curve.

Since high frequency components can reflect the mutational point of the signal, the high frequency band D_1^1, D_2^3 and D_3^7 and obtained by wavelet packet transform are used. Set the threshold value $T = \sigma\sqrt{2\ln(M)}$, where M is the sampling number, σ is the standard deviation of the high frequency band signal, $\sigma = \sqrt{\frac{1}{M}\sum_{i=1}^{M}(S_{wp}(i) - \mu)^2}$, S_{wp} represents the wavelet packet coefficients of a high frequency band and its mean is μ. The value below the threshold T is considered to be noise, and the value above T that is the mutation of the signal. Set A, B, and C as the sampling point sets of the change time window in the high frequency band D_1^1, D_2^3 and D_3^7 respectively. According to the importance of the three high frequency components, in this article, the change time window of the utility harmonic impedance determined by:

$$M = A \cup (B \cap C) \tag{7}$$

4. The Piecewise Bound Constrained Optimization Model of Harmonic Responsibility Assessment and Its Solution Algorithms

In order to accurately calculate the single sampling point harmonic responsibility and the total harmonic responsibility of harmonic loads, upon the segmentation of the harmonic monitoring data, the harmonic responsibility is then assessed by the piecewise bound constrained optimization in this paper. According to the law of cosines [26], for the triangle XYZ:

$$z^2 = x^2 + y^2 - 2xy\cos\varphi \tag{8}$$

where φ represents the angle contained between sides of lengths x and y and opposite the side of length z.

For Figure 1b, the following equations can be obtained for each measurement at time t_i:

$$\left|\dot{V}_{pcc}^h(t_i)\right|^2 = \left|\dot{V}_{pcc,12}^h(t_i)\right|^2 + \left|\dot{V}_{pcc,0}^h\right|^2 - 2\left|\dot{V}_{pcc,12}^h(t_i)\right|\left|\dot{V}_{pcc,0}^h\right|\cos\theta_2$$

$$\left|\dot{V}_{pcc,12}^h(t_i)\right|^2 = \left|Z_{pcc,1}^h\right|^2\left|\dot{I}_{b1}(t_i)\right|^2 + \left|Z_{pcc,2}^h\right|^2\left|\dot{I}_{b2}(t_i)\right|^2 - 2\left|Z_{pcc,1}^h\right|\left|\dot{I}_{b1}(t_i)\right|\left|Z_{pcc,2}^h\right|\left|\dot{I}_{b2}(t_i)\right|\cos\theta_1 \tag{9}$$

For simplicity, let $\left|\dot{V}_{pcc,12}^h(t_i)\right| = \gamma_0$, $\left|Z_{pcc,1}^h\right| = \gamma_1$, $\left|Z_{pcc,2}^h\right| = \gamma_2$, $\cos\theta_1 = \gamma_3$, $\left|\dot{V}_{pcc,0}^h\right| = \gamma_4$, and $\cos\theta_2 = \gamma_5$. In order to estimate the independent variables $\gamma = [\gamma_1, \gamma_2, \gamma_3, \gamma_4, \gamma_5]$, the absolute value of square error can be used as the objective function. The bound constrained optimization model is established as:

$$
\begin{aligned}
\min f(\gamma_1, \gamma_2, \gamma_3, \gamma_4, \gamma_5) &= \sum_{i=1}^{T}\left|\left|\dot{V}_{pcc}^h(\gamma_1, \gamma_2, \gamma_3, \gamma_4, \gamma_5, t_i)\right|^2 - \left|\dot{V}_{pcc}^h(t_i)\right|^2\right| \\
&= \sum_{i=1}^{T}\left|\left|\dot{V}_{pcc,12}^h(t_i)\right|^2 + \left|\dot{V}_{pcc,0}^h(t_i)\right|^2 - 2\left|\dot{V}_{pcc,12}^h(t_i)\right|\left|\dot{V}_{pcc,0}^h(t_i)\right|\cos\theta_2 - \left|\dot{V}_{pcc}^h(t_i)\right|^2\right| \\
&= \sum_{i=1}^{T}\left|\gamma_0^2 + \gamma_4^2 - 2\gamma_0\gamma_4\gamma_5 - \left|\dot{V}_{pcc}^h(t_i)\right|^2\right|
\end{aligned} \tag{10}
$$

$$
\begin{aligned}
\text{s.t.} \quad & \gamma_1 > 0 \\
& \gamma_2 > 0 \\
& -1 \leq \gamma_3 \leq 1 \\
& \gamma_4 > 0 \\
& -1 \leq \gamma_5 \leq 1
\end{aligned}
$$

where $\gamma_0^2 = \left|\dot{V}_{pcc,12}^h(t_i)\right|^2 = \gamma_1^2\left|\dot{I}_{b1}^h(t_i)\right|^2 + \gamma_2^2\left|\dot{I}_{b2}^h(t_i)\right|^2 - 2\gamma_1\left|\dot{I}_{b1}^h(t_i)\right|\gamma_2\left|\dot{I}_{b2}^h(t_i)\right|\gamma_3$; T is the number of sample points; $\dot{V}_{pcc}^h(\gamma_1,\gamma_2,\gamma_3,\gamma_4,\gamma_5,t_i)$ is the calculated value of harmonic voltage, $\dot{V}_{pcc}^h(t_i)$, $\dot{I}_{b1}^h(t_i)$ and $\dot{I}_{b2}^h(t_i)$ are the measured values of harmonic voltage and currents respectively.

Once the estimated $\gamma = [\gamma_1,\gamma_2,\gamma_3,\gamma_4,\gamma_5]$ is obtained, according to Equation (2), the harmonic responsibility from two major harmonic loads, for each sampling point at t_i can be calculated as:

$$
\begin{aligned}
u_{pcc,1}(t_i) &= \left(\frac{\left|\dot{V}_{pcc}^h(t_i)\right|^2 + \gamma_0^2 - \gamma_4^2}{2\left|\dot{V}_{pcc}^h(t_i)\right|^2}\right)\left(\frac{\gamma_0^2 + \gamma_1^2\left|\dot{I}_{b1}^h(t_i)\right|^2 - \gamma_2^2\left|\dot{I}_{b2}^h(t_i)\right|^2}{2\gamma_0^2}\right) \\[2mm]
u_{pcc,2}(t_i) &= \left(\frac{\left|\dot{V}_{pcc}^h(t_i)\right|^2 + \gamma_0^2 - \gamma_4^2}{2\left|\dot{V}_{pcc}^h(t_i)\right|^2}\right)\left(\frac{\gamma_0^2 + \gamma_2^2\left|\dot{I}_{b2}^h(t_i)\right|^2 - \gamma_1^2\left|\dot{I}_{b1}^h(t_i)\right|^2}{2\gamma_0^2}\right)
\end{aligned}
\tag{11}
$$

Assuming that the monitoring data is divided into N segments and each segment $S_j(j = 1,2,\cdots,N)$ corresponds to different utility harmonic impedance, the harmonic responsibility at each segment can be determined by:

$$
\mu_{pcc,i}^{S_j} = \sum_{i=1}^{T_j}\mu_{pcc,i}(i)\Big/T_j
\tag{12}
$$

where $T_j(j = 1,2,\cdots,N)$ is the number of data points in the segment S_j.

Total harmonic responsibility can be calculated by:

$$
\mu_{pcc,i} = \sum_{j=1}^{N}\omega_j\mu_{pcc,i}^{S_j}
\tag{13}
$$

$$
\omega_j = T_j\Big/\sum_{j=1}^{N}T_j
\tag{14}
$$

where $\mu_{pcc,i}^{S_j}$ is the harmonic responsibility of segment j; ω_j is the weight of each segment.

Numerous methods have been developed to figure out the bound constrained optimization problem. In this article, the interior-point (IP), the sequential quadratic programming (SQP) and the active set (AS) algorithm, which are all regarded as effective tools for solving nonlinear optimization problems, are selected.

A typical constrained programming problem can be expressed as:

$$
\begin{aligned}
&\min f(\gamma) \\
&\text{s.t.} \quad h_i(\gamma) = 0; \\
&\qquad\quad g_i(\gamma) \le 0; \\
&\qquad\quad \gamma = (\gamma_1,\gamma_2,\cdots,\gamma_n)^T
\end{aligned}
\tag{15}
$$

The interior-point [27] for constrained optimization is mainly used to solve a variety of approximate optimization problem. For each $\eta > 0$ the approximate model can be expressed as:

$$
\begin{aligned}
&\min_{\gamma,s} f_\eta(\gamma) - \eta\sum_i\ln(s_i) \\
&\text{s.t.} \quad h(\gamma) = 0 \\
&\qquad\quad g(\gamma) + s = 0
\end{aligned}
\tag{16}
$$

where s_i denotes the slack variable. As η decreases to 0, the minimum of f_η approaches to the minimum of f.

The SQP [28] is a kind of approximate Newton's method for solving constrained optimization problems. In each major iteration, the quasi-Newton updating method is firstly used to approach the Hessian of the Lagrangian function. The result is then employed to solve the QP (17) sub-problem:

$$\begin{aligned}
\min \nabla f(\gamma^k)^T q &+ \tfrac{1}{2} q^T H q \\
\text{s.t.} \quad h_i(\gamma^k) + \nabla h_i(\gamma^k)^T q &= 0; \\
g_i(\gamma^k) + \nabla g_i(\gamma^k)^T q &\leq 0; \\
\gamma &= (\gamma_1, \gamma_2, \cdots, \gamma_n)^T
\end{aligned} \tag{17}$$

where $H \in \Re^{n \times n}$ and $q = \gamma^{k+1} - \gamma^k, d \in \Re^n$.

For the constrained optimization problem, the following equation can be derived by Newton's method:

$$\begin{aligned}
\min \nabla f(\gamma)^T \Delta\gamma &+ \tfrac{1}{2} \Delta\gamma^T \nabla^2_{\gamma\gamma} L(\gamma, \lambda) \Delta\gamma \\
\text{s.t.} \quad h(\gamma) + \nabla h(\gamma)^T \Delta\gamma &= 0 \\
g(\gamma) + \nabla g(\gamma)^T \Delta\gamma &\leq 0
\end{aligned} \tag{18}$$

where λ is the multiplier; $L(\gamma, \lambda)$ denotes the Lagrangian expression for (17); and $\nabla^2_{\gamma\gamma} L(\gamma, \lambda)$ represents the Hessian of the Lagrangian.

The active set [29] solves the constrained optimization problem by determining the constraints that impact the results. Equation (15) can be rewritten as:

$$\begin{aligned}
\min f(\gamma) & \\
\text{s.t.} \quad A_i(\gamma) &= 0; i = 1, \cdots, n_e \\
A_i(\gamma) &\leq 0; i = n_e + 1, \cdots, n
\end{aligned} \tag{19}$$

where $A(\gamma)$ is a n-dimensional vector containing the evaluated values γ.

In the AS algorithm, the solutions of the Karush-Kuhn-Tucker (KKT) equations can be used to calculate the Lagrange multipliers (L_i, $i = 1, 2, ..., n$). For Equation (19), the KKT equations can be expressed as:

$$\begin{aligned}
\nabla f(\gamma^*) + \sum_{i=1}^{n} L_i \nabla A_i(\gamma^*) &= 0 \\
L_i A_i(\gamma^*) &= 0; i = 1, \cdots, n_e \\
L_i &\geq 0; i = n_e + 1, \cdots, n
\end{aligned} \tag{20}$$

Equation (20) demonstrates a cancelling of the gradients between the objective function and the active constraints at the solution point. Since the cancelling operation only involves active constraints, the Lagrange multipliers are therefore equal to 0.

5. The Proposed Harmonic Responsibility Assessment Approach

The working process of the proposed method for harmonic responsibility assessment is illustrated in Figure 3, where ε_c and ε_t represent the calculation error and the termination tolerance. Firstly, the utility harmonic impedance is roughly calculated by least squares linear regression. Second, the change time windows of the utility harmonic impedance are identified by the wavelet packet transform. Then, the harmonic responsibility of each segment is evaluated by the piecewise bound constrained optimization method. Finally, the total responsibilities of harmonic loads can be obtained by the weighted summation, based on the point numbers of segments.

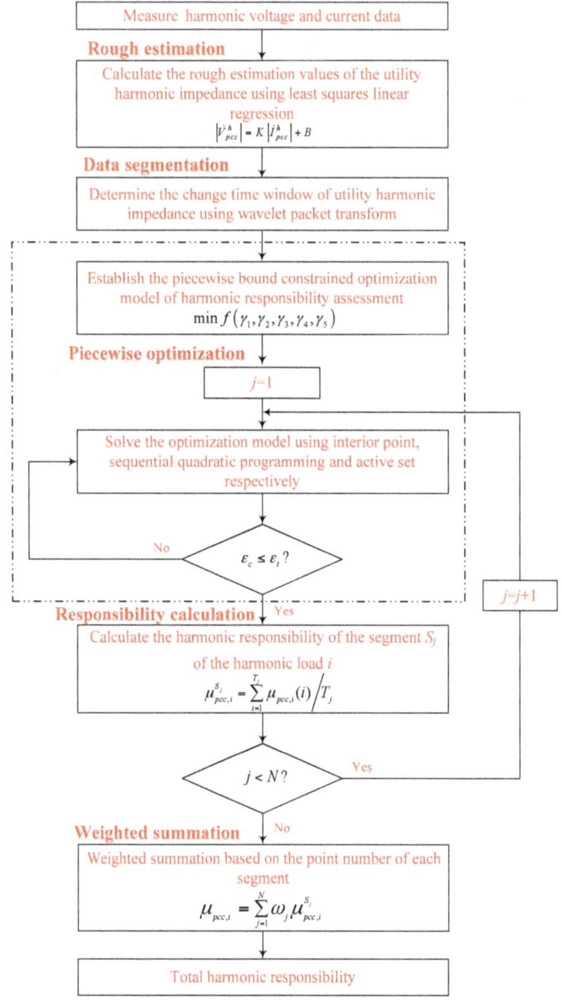

Figure 3. The process of the proposed approach.

6. Numerical Experiments

For the distribution system with two major harmonic loads as shown in Figure 1a, a Norton equivalent circuit model is established in MATLAB [30]. Taking the fifth harmonic as example, Table 1 presents the setting of parameter values. The harmonic impedance is modelled as the resistance R and reactance L in series. The system parameters and the injected harmonic currents are set and modified on the basis of [24]. In order to simulate practical engineering data, stochastic fluctuations are added to the harmonic data. For harmonic load 1 and 2, the means of the injected harmonic current data are $2.0788 + j0.1356$ (A) and $3.8849 - j0.8549$ (A) respectively, while the variances are 0.0018 and 0.0061 respectively. For the utility side, the mean of the injected utility harmonic current is $1.0243 - j0.3233$ (A). The variances of all the harmonic impedances are 0.001. A total of 1440 harmonic sampling points of harmonic voltage and current data are generated. The change time of utility harmonic impedance are set as 501 and 1001.

Table 1. Parameter values of the distribution system Norton equivalent circuit.

Parameters	Values (Ω)	The Numbers of Sampling Points
Z_s^5	$Z_{s,1}^5 = 0.5000 + j0.7854$	500
	$Z_{s,2}^5 = 0.2500 + j0.3927$	500
	$Z_{s,3}^5 = 1.0000 + j1.5708$	1440
Z_1^5	$15.0000 + j47.1239$	1440
Z_2^5	$9.0000 + j23.5619$	1440

For all the measured harmonic data, the least squares are used to carry out linear regression when the value of significance level is 0.05, and the regression coefficient is:

$$\alpha = \begin{bmatrix} \hat{\alpha}_0 \\ \hat{\alpha}_1 \\ \hat{\alpha}_2 \end{bmatrix} = \begin{bmatrix} 4.4903 \\ -0.1373 \\ 0.7400 \end{bmatrix}$$

where $\hat{\alpha}_0, \hat{\alpha}_1, \hat{\alpha}_2$ are the estimated values of $\alpha_0, \alpha_1, \alpha_2$ in Equation (5), that is, $\left| \dot{V}_{pcc}^h \right| \approx 4.4903 - 0.1373 \left| \dot{I}_{b1}^h \right| + 0.7400 \left| \dot{I}_{b2}^h \right|$.

The 95% confidence intervals for the coefficient estimations are:

$$C_{0.95} = \begin{bmatrix} -5.1372 & 14.1178 \\ -3.3517 & 3.0772 \\ -1.0726 & 2.5527 \end{bmatrix}$$

The R^2 statistic, the F statistic and its p value can be calculated as $R^2 = 4.44781 \times 10^{-4}$, $F = 0.3219$ and $p = 0.7248$, respectively. From the above results, the R^2 statistic and the F statistic are insignificant, and $p > 0.05$. It is indicated that the regression should be rejected, and the harmonic responsibilities cannot be accurately calculated by the linear regression method under the changes of utility harmonic impedance. Thus, identifying the change times of the utility harmonic impedances is considered in this paper. It is solved by the wavelet packet decomposition method.

The wavelet packet decomposition results of the rough estimation of utility harmonic impedance using the Haar wavelet bases function are shown in Figure 4a,b. In order to select the appropriate wavelet basis, the Haar wavelet 'haar (db1)', Daubechies wavelet 'db4', Symlet wavelets 'sym1' and 'sym4', Coiflet wavelet 'coif4', and Demy wavelet 'dmey' [31] are used respectively to perform analysis and compared in this paper. The identification results of the change time window of utility harmonic impedance under different wavelet bases are shown in Table 2. As the sampling window length is 3, the sampling points of utility harmonic impedance changes set preciously (501 and 1001) are all included in the change time windows determined by wavelet packet transform. It can be seen from Table 2, the various wavelet basis functions can all deliver good performances in identifying the change time window, especially for haar (db1) and sym1 wavelet.

According to the change time window of the utility harmonic impedance [167, 168, 333, 334, 335, 336] determined by wavelet packet transform with haar wavelet, the sampling points (499–504) and (997–1008) are deleted. Then, the harmonic data are divided into three segments, that is (1–498), (505–996) and (1009–1440).

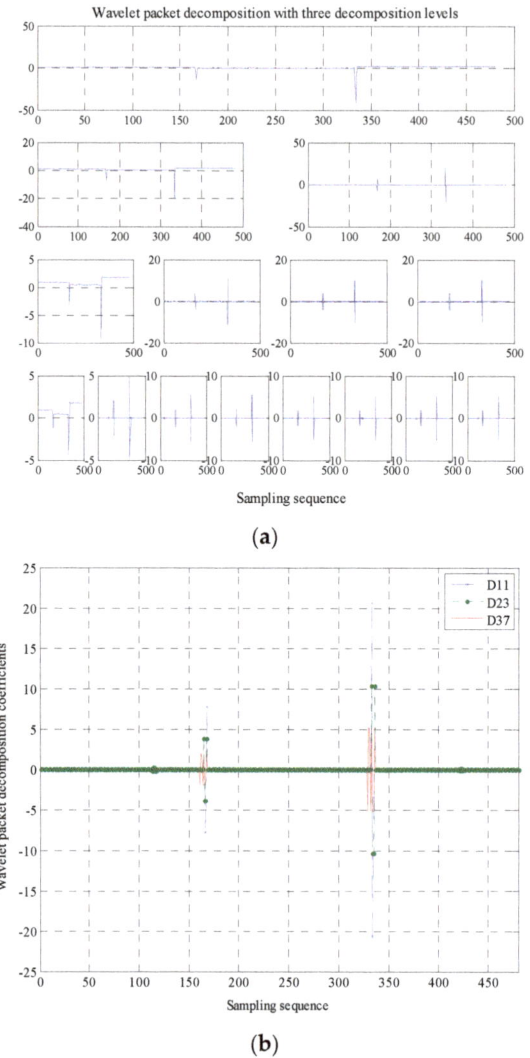

(a)

(b)

Figure 4. Wavelet packet decomposition curves: (**a**) Wavelet packet decomposition curves of the utility harmonic impedance rough estimation values; (**b**) The three concerned high frequency components.

Table 2. Identification results of the change time windows of utility harmonic impedance under different wavelet bases.

Wavelet Basis	Identified Change Time Window	Actual Change Time Window
haar (db1), sym1	167 168 333 334 335 336	
db4	166 167 333 334 335 336 337	
sym4	168 331 332 333 334 335 336 338	167
coif4	167 330 331 333 334 335 336 337 339	334
dmey	167 331 333 334 335 337	

Energies **2017**, 10, 936

On the basis of data segmentation, the piecewise bound constrained optimization methods with three algorithms are used respectively to assess the harmonic responsibility. To accelerate convergence, the least square is adopted to solve the regression coefficients, which are taken as the initial values of the load harmonic impedances $\left|\hat{Z}_{pcc,1}^{h}\right|$ and $\left|\hat{Z}_{pcc,2}^{h}\right|$. Then, the boundary constraints of the load harmonic impedances are set as $\left[0,2\left|\hat{Z}_{pcc,1}^{h}\right|\right]$ and $\left[0,2\left|\hat{Z}_{pcc,2}^{h}\right|\right]$. The initial values for the cosine of phase angle value are set to be 0. Since $\left|\dot{V}_{pcc,0}^{h}\right| \leq \left|\dot{V}_{pcc}^{h}\right|$ in general, the initial values and the boundary constraint of $\left|\dot{V}_{pcc,0}^{h}\right|$ are set to be $0.5\left|\dot{V}_{pcc}^{h}\right|$ and $\left[0,\left|\dot{V}_{pcc}^{h}\right|\right]$, which means the initial value of the independent variable is $\left(\left|\hat{Z}_{pcc,1}^{h}\right|,\left|\hat{Z}_{pcc,2}^{h}\right|,0,0.5\left|\dot{V}_{pcc}^{h}\right|,0\right)$, and the boundary constraint values are $(0,0,-1,0,-1)$ and $\left(2\left|\hat{Z}_{pcc,1}^{h}\right|,2\left|\hat{Z}_{pcc,2}^{h}\right|,1,\left|\dot{V}_{pcc}^{h}\right|,1\right)$. The implementation of IP, SQP and AS are based on the 'fmincon' function in MATLAB. In order to ensure the fairness of algorithm comparison, the parameter settings of the three algorithms in Matlab are modified as follows. The maximum number of iterations is 1000, the maximum number of function evaluations is 5000, the termination tolerance ε_t is 1×10^{-10}, and the values of other parameters are the default values of the 'fmincon' function [30].

For each segment, the theoretical harmonic responsibilities and calculated values obtained by the three algorithms, as well as the means and variances of the relative error between the calculated value and the theoretical value are presented in Table 3.

Table 3. The harmonic responsibilities obtained by the three algorithms and error statistics.

Segment No.	Algorithm	Harmonic Responsibility of Load 1				Harmonic Responsibility of Load 2			
		Theoretical Value (%)	Calculated Value (%)	The mean of the Relative Error	The Variance of the Relative Error	Theoretical Value (%)	Calculated Value (%)	The Mean of the Relative Error	The Variance of the Relative Error
1	IP		28.63	1.07×10^{-3}	6.45×10^{-7}		57.40	1.18×10^{-2}	9.21×10^{-7}
	SQP	28.61	28.61	9.12×10^{-4}	4.20×10^{-7}	56.73	57.36	1.11×10^{-2}	9.41×10^{-7}
	AS		28.96	1.22×10^{-2}	1.53×10^{-5}		58.05	2.32×10^{-2}	3.32×10^{-6}
2	IP		27.92	3.10×10^{-2}	3.90×10^{-4}		55.51	1.75×10^{-2}	1.28×10^{-4}
	SQP	28.72	2760	3.91×10^{-2}	1.16×10^{-4}	56.44	54.84	2.85×10^{-2}	3.90×10^{-5}
	AS		29.55	3.39×10^{-2}	5.63×10^{-4}		58.75	4.06×10^{-2}	2.34×10^{-4}
3	IP		28.88	1.78×10^{-2}	2.15×10^{-5}		58.75	2.56×10^{-2}	3.91×10^{-6}
	SQP	28.37	28.91	1.89×10^{-2}	2.36×10^{-5}	57.28	58.81	2.67×10^{-2}	4.33×10^{-6}
	AS		28.86	1.70×10^{-2}	1.89×10^{-5}		58.70	2.47×10^{-2}	3.39×10^{-6}
Total	IP		28.46	1.65×10^{-2}	2.98×10^{-4}		57.16	1.80×10^{-2}	7.67×10^{-5}
	SQP	28.58	28.35	1.96×10^{-2}	3.01×10^{-4}	56.80	56.93	2.19×10^{-2}	7.77×10^{-5}
	AS		29.13	2.12×10^{-2}	2.96×10^{-4}		58.49	2.97×10^{-2}	1.46×10^{-4}

For the two harmonic loads, the harmonic responsibilities for each sample point obtained by the three algorithms are shown in Figure 5.

From the tables and figure above, the calculated values of harmonic responsibility are basically consistent with the theoretical values. For the three algorithms, the mean and the variance of the relative error are all below 0.05 and 4×10^{-4} respectively. It is evidenced that the piecewise bound constrained optimization model with the three algorithms can assess the harmonic responsibility of harmonic load accurately. In addition, compared with the AS algorithm, the results of IP and SQP algorithms are closer to the theoretical values.

In order to examine how the fluctuation of the harmonic data can affect the three algorithms, the harmonic responsibilities are evaluated while the variances of the load harmonic impedances are set within the range of 0.005 to 0.1. The calculated values of the harmonic responsibilities obtained by the three algorithms are shown in Table 4. In comparison, the SQP algorithm can provide the most accurate and stable results.

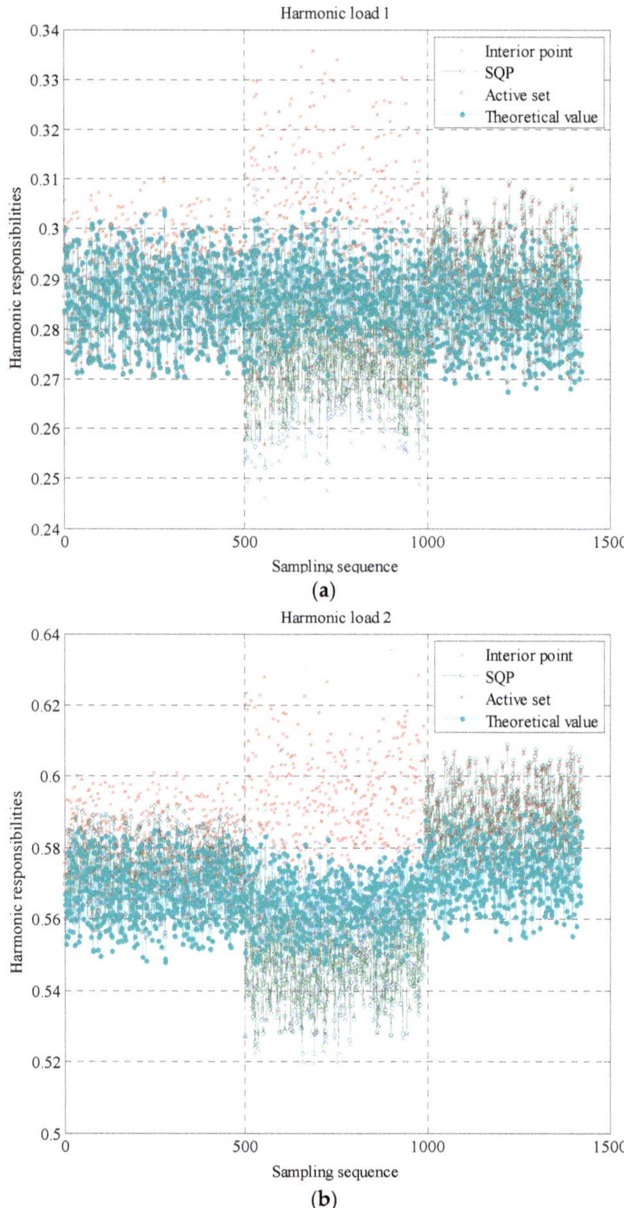

Figure 5. The harmonic responsibilities for each sample point of harmonic loads: (**a**) The harmonic responsibilities of load 1; (**b**) The harmonic responsibilities of load 2.

Table 4. The harmonic responsibilities under different variances of the load harmonic impedances.

Variance of Load Harmonic Impedances	Algorithm	Harmonic Responsibility of Load 1		Harmonic Responsibility of Load 1	
		Calculated Value (%)	Theoretical Value (%)	Calculated Value (%)	Theoretical Value (%)
0.005	IP	29.27		58.53	
	SQP	28.49	28.58	56.97	56.80
	AS	29.24		58.47	
0.010	IP	29.54		59.22	
	SQP	28.39	28.58	56.95	56.80
	AS	29.40		58.93	
0.050	IP	29.76		59.83	
	SQP	29.42	28.58	59.14	56.80
	AS	29.39		59.08	
0.100	IP	29.00		58.04	
	SQP	28.52	28.58	57.09	56.80
	AS	28.94		57.93	

To compare the calculation times of the three optimization algorithms, the statistic results of calculated time including the maximum values, minimum values, mean values and standard deviations of 100 consecutive runs are shown in Table 5. The results show that the AS algorithm is the fastest, while the IP algorithm is the slowest. Since the harmonic responsibility is usually assessed over a period of time, such as 24 h, the computation times of all three algorithms are acceptable.

Table 5. The calculated time statistic results of the three optimization algorithms.

Algorithms	Statistical Value of Calculation Times			
	Maximum Values (s)	Minimum Values (s)	Mean Values (s)	Standard Deviations
IP	38.6060	32.1420	33.9347	1.1632
SQP	4.3210	3.5920	3.8050	0.1344
AS	2.6990	2.2150	2.3595	0.0986

In order to further reflect the complexity of the actual distribution system, this article also carries out simulations on the IEEE 13-bus distribution system [19,32] as shown in Figure 6. The introduction of IEEE 13-bus distribution system can be seen in Appendix A.

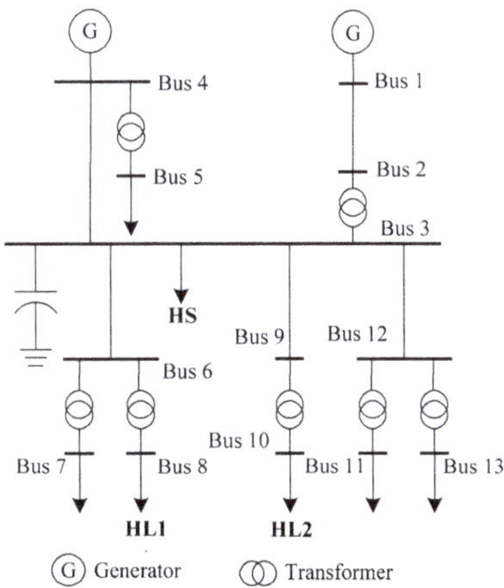

Figure 6. IEEE 13-bus distribution system.

The system parameters are referred to [32] and Table A1. In this work, the parameters of the IEEE 13-bus system are converted into the per unit values. In addition, all loads are modeled as the resistance R and reactance L in series. We take bus 3 as the bus of interest, and set load 8 and 10 as the harmonic load 1 (HL1) and harmonic load 2 (HL2), respectively. To simulate the harmonics at the utility side, the harmonic source (HS) is also injected into bus 3. In this work, the change of the reactive power compensation which can lead to the utility harmonic impedance change is analyzed.

The reactive power compensation of the system is set as 4500 kvar, 5500 kvar, and 6500 kvar. As mentioned above, the variances of the injected harmonic currents are set to be 0.005 and 0.1 so as to evaluate how different data fluctuations influence the harmonic responsibility assessment. The injected harmonic currents of each bus for the two cases are shown in Table 6. The simulations of harmonic responsibility assessment are performed in MATLAB. In the simulation process, the harmonic loads are regarded as known PQ constant loads. The Newton-Raphson method [33] is used to calculate the fundamental power flow, and the injected harmonic currents are calculated according to the typical harmonic current frequency spectrum in [32]. The fifth harmonic is taken as the example for simulation.

Table 6. The injected harmonic current for the two cases.

Case	Injected Bus	Injected Harmonic Current	
		Mean (p.u.)	Variance
Case 1	3	$I_3 = 0.6073 - j0.8896$	
	8	$I_8 = 1.1757 - j1.7222$	0.005
	10	$I_{10} = 2.2429 - j3.2855$	
Case 2	3	$I_3 = 0.6324 - j0.9263$	
	8	$I_8 = 1.1877 - j1.7398$	0.100
	10	$I_{10} = 2.2482 - j3.2932$	

A total of 14,400 sampling points are generated, and the reactive compensation quantity is changed for every 4800 points. The results of harmonic load flow are assumed as the measured harmonic data. In consideration of the data fluctuations in the actual system, the window length of wavelet packet transform is set to $L = 30$. Figure 7 presents the wavelet packet decomposition curves for the harmonic data of the two cases when a Haar wavelet is applied.

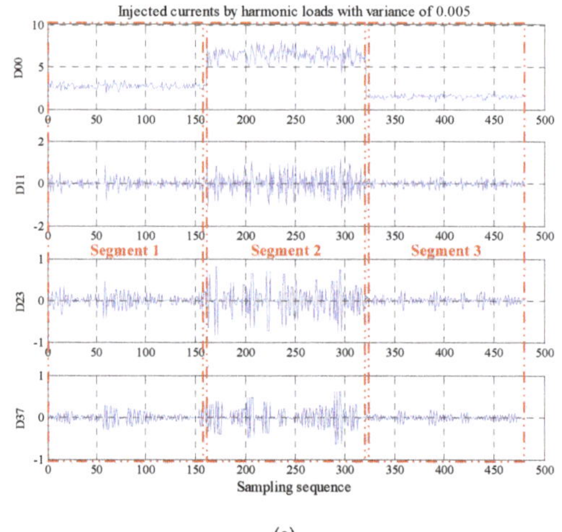

(a)

Figure 7. *Cont.*

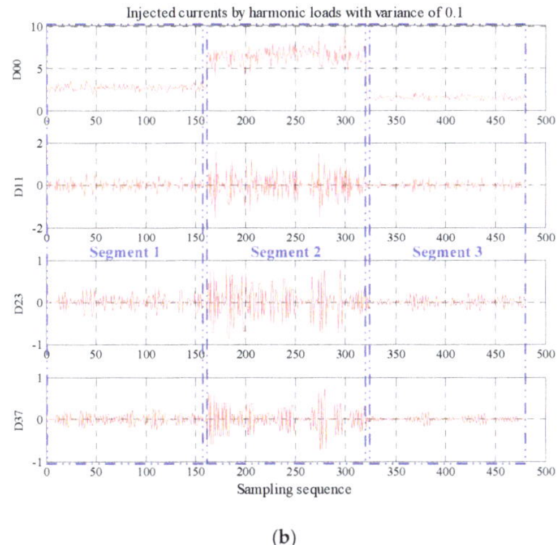

(**b**)

Figure 7. The wavelet packet decomposition curves for the harmonic data of the two cases: (**a**) The Case 1; (**b**) The Case 2.

Referring to Figure 7, the change time window can be approximatively identified as 160 and 320. As the sampling window length is 30, the sampling time of the utility harmonic impedance changes is approximatively to 4800 and 9600, which is consistent with the settings of change times. Thus, the harmonic data are divided into three segments (1–4800), (4801–9600) and (9601–14,400). Then, the piecewise bound constrained optimization model, the three algorithms and the weighted summation are also utilized to compute the total harmonic responsibility for the two cases. The setting of parameters for the three algorithms, as well as the initial values and the boundary constraints have been described previously. Tables 7 and 8 present the theoretical values and calculated values of the harmonic responsibilities obtained by the three algorithms for Case 1 and Case 2, as well as the corresponding means and variances of the relative error between the calculated value and the theoretical values.

Table 7. The harmonic responsibilities and error statistics (Case 1).

Segment No.	Algorithm	Harmonic Responsibility of Load 1				Harmonic Responsibility of Load 2			
		Theoretical Value (%)	Calculated Value (%)	The Mean of the Relative Error	The Variance of the Relative Error	Theoretical Value (%)	Calculated Value (%)	The Mean of the Relative Error	The Variance of the Relative Error
1	IP	29.84	30.78	3.12×10^{-2}	2.03×10^{-4}	55.51	55.71	7.09×10^{-3}	2.85×10^{-5}
	SQP		30.77	3.11×10^{-2}	2.04×10^{-4}		55.70	7.06×10^{-3}	2.83×10^{-5}
	AS		33.48	1.22×10^{-1}	1.02×10^{-4}		66.50	1.98×10^{-1}	2.95×10^{-5}
2	IP	29.86	30.12	8.77×10^{-3}	2.44×10^{-6}	55.52	56.19	1.21×10^{-2}	2.51×10^{-6}
	SQP		29.95	3.17×10^{-3}	2.48×10^{-6}		55.87	6.43×10^{-3}	2.64×10^{-6}
	AS		30.31	1.53×10^{-1}	2.26×10^{-6}		56.55	1.86×10^{-2}	2.27×10^{-6}
3	IP	29.83	30.26	1.43×10^{-2}	2.99×10^{-6}	55.53	55.17	6.42×10^{-3}	2.87×10^{-6}
	SQP		30.24	1.36×10^{-2}	3.02×10^{-6}		55.13	7.12×10^{-3}	2.90×10^{-6}
	AS		30.24	1.36×10^{-2}	3.02×10^{-6}		55.13	7.12×10^{-3}	2.90×10^{-6}
Total	IP	29.84	30.38	1.81×10^{-2}	1.61×10^{-4}	55.52	55.69	8.53×10^{-3}	1.77×10^{-5}
	SQP		30.32	1.59×10^{-2}	2.02×10^{-4}		55.57	6.87×10^{-3}	1.14×10^{-5}
	AS		31.34	5.02×10^{-2}	2.59×10^{-3}		59.39	7.45×10^{-2}	7.63×10^{-3}

Table 8. The harmonic responsibilities and error statistics (Case 2).

Segment No.	Algorithm	Harmonic Responsibility of Load 1				Harmonic Responsibility of Load 2			
		Theoretical Value (%)	Calculated Value (%)	The Mean of the Relative Error	The Variance of the Relative Error	Theoretical Value (%)	Calculated Value (%)	The Mean of the Relative Error	The Variance of the Relative Error
1	IP	29.822026	30.082117	1.13×10^{-2}	7.12×10^{-5}	55.184464	55.330464	6.04×10^{-3}	3.65×10^{-5}
	SQP		30.081910	1.12×10^{-2}	7.11×10^{-5}		55.330081	6.04×10^{-3}	3.65×10^{-5}
	AS		30.081926	1.12×10^{-2}	7.11×10^{-5}		55.330075	6.04×10^{-3}	3.65×10^{-5}
2	IP	29.792924	32.418823	8.72×10^{-2}	1.20×10^{-4}	55.226745	59.116948	7.40×10^{-3}	5.48×10^{-5}
	SQP		32.418871	8.72×10^{-2}	1.20×10^{-4}		59.117040	7.40×10^{-3}	5.48×10^{-5}
	AS		32.418872	8.72×10^{-2}	1.20×10^{-4}		59.117016	7.40×10^{-3}	5.48×10^{-5}
3	IP	29.828965	30.810225	3.32×10^{-2}	4.03×10^{-5}	55.172325	57.453403	6.40×10^{-3}	4.10×10^{-5}
	SQP		30.814266	3.34×10^{-2}	4.03×10^{-5}		57.460951	6.40×10^{-3}	4.09×10^{-5}
	AS		30.814327	3.34×10^{-2}	4.03×10^{-5}		57.461059	6.40×10^{-3}	4.09×10^{-5}
Total	IP	29.814638	31.103722	4.39×10^{-2}	1.09×10^{-3}	55.194512	57.300272	3.98×10^{-2}	7.03×10^{-4}
	SQP		31.105015	4.39×10^{-2}	1.09×10^{-3}		57.302691	3.98×10^{-2}	7.03×10^{-4}
	AS		31.105042	4.39×10^{-2}	1.09×10^{-3}		57.302717	3.98×10^{-2}	7.03×10^{-4}

For the segment 1 of Case 1, the convergence curves of the three algorithms are shown in Figure 8. It can be observed that the minimum objective function values obtained by IP and SQP algorithms are approximately 173, while the minimum objective function value obtained by AS algorithm is around 184. Compared to IP and SQP, the calculation error of AS algorithm is slightly larger due to the fact it is subject to the premature convergence.

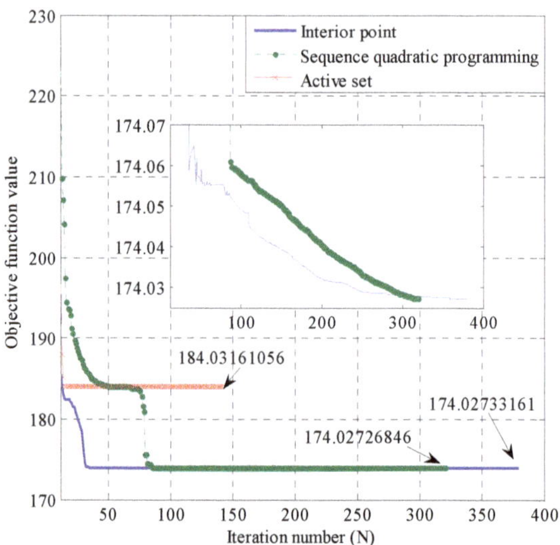

Figure 8. The convergence curves of the three algorithms.

For the segment 1 of Case 2, the harmonic responsibilities for each sample point of harmonic load 1 obtained by the three algorithms are shown in Figure 9, where Figure 9a shows the results of all the 4800 sampling points, and Figure 9b shows the results of the first 480 sampling points. In order to compare with the conventional regression analysis methods, the harmonic responsibilities obtained by the least square method [12] and the robust least square method [24] are also shown in Figure 9. As shown in Figure 9, the computational accuracy of the proposed method is superior to that of least square method and robust least square method.

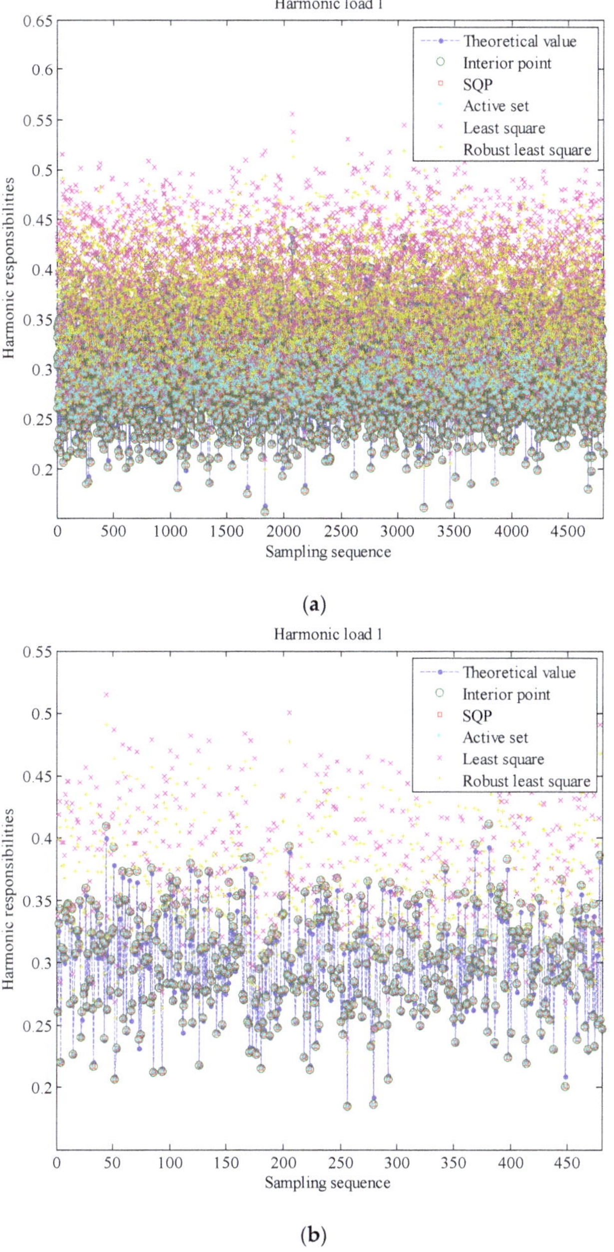

Figure 9. The harmonic responsibilities for each sample point of harmonic load 1: (**a**) The results of all the 4800 sampling points; (**b**) The results of the first 480 sampling points.

The results above illustrate that the proposed approach can get accurate and stable assessment results as the means and variances of the relative error are small. In addition, the calculated values obtained by the three algorithms in Case 2 are similar and close to the theoretical values, which indicate that the impacts of harmonic data fluctuation are insignificant to the three algorithms. In accordance with the test results, the IP and the SQP algorithm are recommended with the priority.

7. Conclusions

In existence of the utility harmonic impedance variations, the harmonic responsibility cannot be calculated directly using the linear regression method. Thus, this article proposes a technique for harmonic responsibility assessment by combining wavelet packet transform with piecewise bound constrained optimization approach on the condition of utility harmonic impedance changes. The first contribution lies in the determination of change times of the utility harmonic impedance by the wavelet packet transform, which aims to accurately segment the measured harmonic data according to different utility harmonic impedances. Secondly, the piecewise bound constrained optimization model is established to evaluate the harmonic responsibility of each data segment, which can provide accurate assessment results. Furthermore, the interior point, the sequential quadratic programming and the active set algorithms are utilized to solve this optimization model. Based on the results, the interior point and the sequential quadratic programming can deliver better performance compared with the active set. In the simulation process, the time variation characteristics of the harmonics have been considered. The proposed method has good robustness against the harmonic data fluctuation. Except for the measurement data of harmonic voltage and current, no additional data is required by the proposed method. The future works may focus on the adaptive modeling method for the piecewise bound constrained optimization model, which can conveniently calculate the harmonic responsibility of multiple harmonic loads.

Acknowledgments: The authors acknowledge the support by the National Natural Science Foundation for Distinguished Young Scholars of China under Grant 51525702, the National Natural Science Foundation of China under Grant 51407150, Innovative Think-tank Foundation for Young Scientists of China Association for Science and Technology under Grant DXB-ZKQN-2017-042, Open Project of Key Laboratory of electric Power Big Data of Guizhou Province and Guizhou Fengneng Science and Technology Development Co., Ltd.

Author Contributions: Tianlei Zang and Zhengyou He conceived and designed the experiments; Tianlei Zang and Yan Wang performed the experiments; Tianlei Zang and Ling Fu analyzed the data; Qingquan Qian contributed analysis tools; Tianlei Zang, Zhengyou He and Zhiyu Peng wrote the paper. All authors have read and approved the final manuscript.

Conflicts of Interest: The authors declare no conflict of interest.

Appendix A. The Introduction of IEEE 13-Bus Distribution System

The IEEE 13-bus system is a medium-sized industrial plant, which is extracted from a typical test system in the IEEE Color Book series. The system has been employed as a benchmark system to develop novel harmonic analysis approaches. The system is fed from a utility supply at 69 kV and the local distribution system operates at 13.8 kV. Because of the balanced nature of the test system, only the data of positive sequence is given. Capacitance of all cables and the short overhead lines are ignored. In addition, frequency dependence of model resistance and transformer magnetizing branch effects is ignored. The main parameters of the IEEE 13-bus distribution system are listed in Table A1.

Table A1. The main parameters of IEEE 13-bus distribution system.

Bus No.	Load		Bus No.		Line Impedance		Transformer Impedance	
	P (p.u.)	Q (p.u.)	From	To	R_L (p.u.)	X_L (p.u.)	R_T (p.u.)	X_T (p.u.)
1	0.0000	0.0000	3	4	0.00122	0.00243	0.00000	0.00000
2	0.0000	0.0000	1	2	0.00139	0.00296	0.00000	0.00000
3	0.2240	0.2000	3	6	0.00075	0.00063	0.00000	0.00000
4	0.0000	0.0000	3	9	0.00157	0.00131	0.00000	0.00000
5	0.0600	0.0530	3	12	0.00109	0.00091	0.00000	0.00000
6	0.0000	0.0000	2	3	0.00000	0.00000	0.00313	0.05324
7	0.5150	0.8290	4	5	0.00000	0.00000	0.06395	0.37796
8	0.1310	0.1130	6	7	0.00000	0.00000	0.05918	0.35510
9	0.0000	0.0000	6	8	0.00000	0.00000	0.04314	0.34514
10	0.0810	0.0800	9	10	0.00000	0.00000	0.05829	0.37887
11	0.0370	0.0330	12	11	0.00000	0.00000	0.05575	0.36240
12	0.0000	0.0000	12	13	0.00000	0.00000	0.01218	0.14616
13	0.2800	0.2500						

References

1. Yang, N.C.; Le, M.D. Three-phase harmonic power flow by direct ZBUS method for unbalanced radial distribution systems with passive power filters. *IET Gener. Transm. Distrib.* **2016**, *10*, 3211–3219. [CrossRef]
2. Zong, X.; Gray, P.A.; Lehn, P.W. New metric recommended for IEEE Standard 1547 to limit harmonics injected into distorted grids. *IEEE Trans. Power Deliv.* **2016**, *31*, 963–972. [CrossRef]
3. Ye, G.; Nijhuis, M.; Cuk, V.; Cobben, J.S. Stochastic residential harmonic source modeling for grid impact studies. *Energies* **2017**, *10*, 372. [CrossRef]
4. Sun, Z.M.; He, Z.Y.; Zang, T.L.; Liu, Y.L. Multi-interharmonic spectrum separation and measurement under asynchronous sampling condition. *IEEE Trans. Instrum. Meas.* **2016**, *65*, 1902–1912. [CrossRef]
5. Teng, J.H.; Liao, S.H.; Leou, R.C. Three-phase harmonic analysis method for unbalanced distribution systems. *Energies* **2014**, *7*, 365–384. [CrossRef]
6. Bagheri, P.B.; Xu, W.; Ding, T.Y. A distributed filtering scheme to mitigate harmonics in residential distribution systems. *IEEE Trans. Power Deliv.* **2016**, *31*, 648–656. [CrossRef]
7. Pfajfar, T.; Blazic, B.; Papic, I. Harmonic contributions evaluation with the harmonic current vector method. *IEEE Trans. Power Deliv.* **2008**, *23*, 425–433. [CrossRef]
8. Xu, W.; Liu, Y.L. A method for determining customer and utility harmonic contributions at the point of common coupling. *IEEE Trans. Power Deliv.* **2010**, *15*, 804–811.
9. Zebardast, A.; Mokhtari, H. Technique for online tracking of a utility harmonic impedance using by synchronising the measured samples. *IET Gener. Transm. Distrib.* **2016**, *10*, 1240–1247. [CrossRef]
10. Yang, H.; Pirotte, P.; Robert, A. Harmonic emission levels of industrial loads statistical assessment. In Proceedings of the International Conference on Large High Voltage Electric Systems, Paris, France, 21–26 August 1996; pp. 36–306.
11. Song, K.L.; Yuan, X.D.; Chen, B.; Zhao, S. A method for assessing customer harmonic emission level based on improved fluctuation method. In Proceedings of the 2012 China International Conference on Electricity Distribution (CICED), Shanghai, China, 5–6 September 2012; pp. 1–5.
12. Hui, J.; Freitas, W.; Vieira, J.C.M.; Yang, H.G.; Liu, Y.M. Utility harmonic impedance measurement based on data selection. *IEEE Trans. Power Deliv.* **2012**, *27*, 2193–2202. [CrossRef]
13. Matos, E.O.D.; Soares, T.M.; Bezerra, U.H.; Tostes, M.E.D.L.; Manito, A.R.A.; Costa, B.C. Using linear and non-parametric regression models to describe the contribution of non-linear loads on the voltage harmonic distortions in the electrical grid. *IET Gener. Transm. Distrib.* **2016**, *10*, 1825–1832. [CrossRef]
14. Wang, Y.; Mazin, H.E.; Xu, W.; Huang, B. Estimating harmonic impact of individual loads using multiple linear regression analysis. *Int. Trans. Electr. Energy Syst.* **2016**, *26*, 809–824. [CrossRef]
15. Karimzadeh, F.; Esmaeili, S.; Hosseinian, S.H. Method for determining utility and consumer harmonic contributions based on complex independent component analysis. *IET Gener. Transm. Distrib.* **2016**, *10*, 526–534. [CrossRef]

16. Zhao, X.; Yang, H.G. A new method to calculate the utility harmonic impedance based on FastICA. *IEEE Trans. Power Deliv.* **2016**, *31*, 381–388. [CrossRef]

17. Hernandez, J.-C.; Ortega, M.-J.; Medina, A. Statistical characterisation of harmonic current emission for large photovoltaic plants. *Int. Trans. Electr. Energy Syst.* **2014**, *24*, 1134–1150. [CrossRef]

18. Ruiz-Rodriguez, F.-J.; Hernandez, J.-C.; Jurado, F. Harmonic modelling of PV systems for probabilistic harmonic load flow studies. *Int. J. Circuit Theory Appl.* **2015**, *43*, 1541–1565. [CrossRef]

19. Zang, T.L.; He, Z.Y.; Fu, L.; Wang, Y.; Qian, Q.Q. Adaptive method for harmonic contribution assessment based on hierarchical K-means clustering and Bayesian partial least squares regression. *IET Gener. Transm. Distrib.* **2016**, *10*, 3220–3227. [CrossRef]

20. Shojaie, M.; Mokhtari, H. A method for determination of harmonics responsibilities at the point of common coupling using data correlation analysis. *IET Gener. Transm. Distrib.* **2014**, *8*, 142–150. [CrossRef]

21. Kandev, N.P.; Chenard, S. Method for determining customer contribution to harmonic variations in a large power network. In Proceedings of the 14th International Conference on Harmonics and Quality of Power (ICHQP 2010), Bergamo, Italy, 26–29 September 2010.

22. Mehrasa, M.; Pouresmaeil, E.; Akorede, M.F.; Jørgensen, B.N.; Catalão, J.P. Multilevel converter control approach of active power filter forharmonics elimination in electric grids. *Energy* **2015**, *84*, 722–731. [CrossRef]

23. Godina, R.; Rodrigues, E.; Pouresmaeil, E.; Matias, J.C.O.; Catalão, J.P.S. Model predictive control technique for energy optimization in residential sector. In Proceedings of the 16th IEEE International Conference on Environment and Electrical Engineering (EEEIC 2016), Florence, Italy, 7–10 June 2016.

24. Sun, Y.Y.; Li, J.Q.; Yin, Z.M. Quantifying harmonic impacts for concentrated multiple harmonic sources using actual data. *Proc. Chin. Soc. Electr. Eng.* **2014**, *34*, 2164–2171.

25. Do, H.T.; Zhang, X.; Nguyen, N.V.; Li, S.S.; Chu, T. Passive-islanding detection method using the wavelet packet transform in grid-connected photovoltaic systems. *IEEE Trans. Power Electr.* **2015**, *31*, 6955–6967. [CrossRef]

26. Hazewinkel, M. *Encyclopaedia of Mathematics: Cosine Theorem*; Springer: Berlin/Heidelberg, Germany, 2001.

27. Morini, B.; Simoncini, V.; Tani, M. Spectral estimates for unreduced symmetric KKT systems arising from Interior Point methods. *Numer. Linear Algebra Appl.* **2016**, *23*, 776–800. [CrossRef]

28. Izmailov, A.F.; Solodov, M.V.; Uskov, E.I. Globalizing stabilized sequential quadratic programming method by smooth primal-dual exact penalty function. *J. Optim. Theory Appl.* **2016**, *169*, 148–178. [CrossRef]

29. Buerger, J.; Cannon, M.; Kouvaritakis, B. Active set solver for min-max robust control with state and input constraints. *Int. J. Robust Nonlinear Control* **2016**, *26*, 3209–3231. [CrossRef]

30. Coleman, T.; Branch, M.; Grace, A. *Optimization Toolbox for Use with MATLAB, User's Guide*, version 3; MathWorks: Natick, MA, USA, 1999.

31. Urban, K. *Wavelets in Numerical Simulation: Wavelet Bases*; Springer: Berlin/Heidelberg, Germany, 2002; pp. 1–81.

32. Abu-Hashim, R.; Burch, R.; Chang, G.; Grady, M.; Gunther, E.; Halpin, M.; Harziadonin, C.; Liu, Y.; Marz, M.; Ortmeyer, T. Test systems for harmonics modeling and simulation. *IEEE Trans. Power Deliv.* **1999**, *14*, 579–587. [CrossRef]

33. Costa, V.M.D.; Martins, N.; Pereira, J.L.R. Developments in the Newton Raphson power flow formulation based on current injections. *IEEE Trans. Power Syst.* **1999**, *14*, 1320–1326. [CrossRef]

Article

Double-Carrier Phase-Disposition Pulse Width Modulation Method for Modular Multilevel Converters

Fayun Zhou [1], An Luo [1], Yan Li [1,2,*], Qianming Xu [1], Zhixing He [1] and Josep M. Guerrero [3]

[1] College of Electrical and Information Engineering, Hunan University, Changsha 410082, China; hnuzfy@126.com (F.Z.); an_luo@126.com (A.L.); hnuxqm@foxmail.com(Q.X.); hezhixingmail@163.com (Z.H.)
[2] School of Information Science and Engineering, Central South University, Changsha 410083, China
[3] Department of Energy Technology, Aalborg University, DK-9220 Aalborg East, Denmark; joz@et.aau.dk
*   Correspondence: liyanly@csu.edu.cn; Tel.: +86-731-8882-3964

Academic Editor: Gabriele Grandi
Received: 12 February 2017; Accepted: 20 April 2017; Published: 23 April 2017

Abstract: Modular multilevel converters (MMCs) have become one of the most attractive topologies for high-voltage and high-power applications. A double-carrier phase disposition pulse width modulation (DCPDPWM) method for MMCs is proposed in this paper. Only double triangular carriers with displacement angle are needed for DCPDPWM, one carrier for the upper arm and the other for the lower arm. Then, the theoretical analysis of DCPDPWM for MMCs is presented by using a double Fourier integral analysis method, and the Fourier series expression of phase voltage, line-to-line voltage and circulating current are deduced. Moreover, the impact of carrier displacement angle between the upper and lower arm on harmonic characteristics is revealed, and further the optimum displacement angles are specified for the circulating current harmonics cancellation scheme and output voltage harmonics minimization scheme. Finally, the proposed method and theoretical analysis are verified by simulation and experimental results.

Keywords: modular multilevel converter; double-carrier phase-disposition pulse width modulation; double Fourier integral analysis; harmonic characteristic; carrier displacement angle

---

1. Introduction

Nowadays, modular multilevel converters (MMCs) have become one of the most attractive multilevel converter topologies available for high-voltage and high-power applications such as voltage-sourced converter high-voltage direct current (VSC-HVDC) transmission [1–6], static synchronous compensators (STATCOMs) [7], unified power flow controllers (UPFCs) [8], active power filters (APFs) [9], medium voltage motor drives [10], integration of renewable energy sources into the electrical grid [11–13] and battery energy storage systems [14,15]. Compared to other multilevel converter topologies, the salient features of MMC include: high degree of modularity, high efficiency, superior harmonic performance, high reliability and absence of dc-link capacitors [5].

Many academic papers have focused on modeling [16–19], control [20–24], and modulation techniques [25–41] for MMCs. The multilevel converter pulse width modulation technique is one of the key technologies for MMCs as it affects the harmonic characteristics, voltage balancing and system efficiency. Various pulse width modulation techniques have been applied for MMCs, and each technique has advantages and drawbacks [25,26]. The selective harmonic elimination-pulse width modulation (SHE-PWM) method can provide good harmonic features with low switching frequency of sub-modules (SMs). However, the calculation of angles increases significantly as the number of output voltage levels increases [27–29]. The space vector modulation (SVM) method can provide more flexibility to optimize switching waveforms. However, when the number of voltage levels increases, the complexity of the algorithm for SVM grows exponentially [4]. In [30], a simplified SVM scheme was proposed for MMCs, which reduces the computation demands and can be used for any level MMC. The main advantage of the nearest level modulation (NLM) method is its simple implementation. However, the NLM method generates poor quality waveforms with small numbers of SMs [31,32]. The application scope of NLM is extended by introducing one SM operating in the PWM mode [33,34]. The phase-shifted carrier pulse-width modulation (PSCPWM) method achieves even power distribution among the SMs [35,36]. However, dedicated capacitor voltage balancing controllers for each SM are mandatory, which reduces the harmonic performance of the output voltage. Compared with PSCPWM, the phase-disposition pulse width modulation (PDPWM) method has superior harmonic characteristics by placing significant harmonic energy into the first carrier component in the phase voltage and relying upon the elimination of this component when the line-to-line voltages are created [25]. The main drawback of the PDPWM is the uneven loss distribution among the SMs, which can be solved by the voltage balancing method based on sorting [37–40]. In [41], an improved PDPWM method using a single carrier was proposed for MMCs, which reduces the control hardware requirement. However, the upper arm and lower arm use the same carrier, and the impact of the carrier displacement angle between the upper arm carrier and the lower arm carrier on the harmonic characteristics has not been considered. The circulating current and output voltage for MMCs are determined by the interactions between the upper arm voltage and the lower arm voltage [42]. Therefore, the carrier for the lower arm and the carrier for the upper arm need to be analyzed separately with an interleaved displacement angle. The displacement angle will influence the high-frequency interactions between the upper arm and lower arm, and further affect the harmonic characteristics of MMCs [43].

A double-carrier phase-disposition pulse width modulation (DCPDPWM) method for MMC is proposed in this paper. Only double triangular carriers with displacement angle are needed for DCPDPWM, one carrier for the upper arm and the other for the lower arm. The theoretical analysis of DCPDPWM for MMC is presented based on double Fourier integral analysis method, and the Fourier series expression of phase voltage, line-to-line voltage and circulating current are deduced. Moreover, the impact of carrier displacement angle between the upper and lower arms on harmonic characteristics is revealed, and the optimum displacement angles are specified for the circulating current harmonics cancellation scheme and output voltage harmonics minimization scheme.

The paper is organized as follows: Section 2 introduces the topology and mathematical model of MMC. Section 3 proposes the DCPDPWM method for MMC. Section 4 presents the theoretical analysis of DCPDPWM for MMC by using double Fourier integral analysis method, and the optimum displacement angles are specified for the circulating current harmonics cancellation scheme and output voltage harmonics minimization scheme. Sections 5 and 6 show the simulation and experimental results, respectively. The conclusions are summarized in Section 7.

2. Topology and Mathematical Model of MMC

The schematic diagram of a three phase MMC is shown in Figure 1. The MMC comprises upper and lower arms per phase-leg. Each arm consists of N series-connected, nominally identical SMs and a series buffer inductor. The buffer inductors for the upper and lower arms can be chosen as coupled or separate ones. The coupled inductor is adopted in this paper as it has lighter weight and smaller size than two separate inductors [34,35]. The power loss of SMs and the resistances of the inductors are ignored. Based on Kirchhoff's voltage law, the following equations can be obtained:

$$L_p \frac{di_{pj}}{dt} + M_u \frac{di_{nj}}{dt} = \frac{U_{dc}}{2} - u_{pj} - u_j \tag{1}$$

$$L_n \frac{di_{nj}}{dt} + M_u \frac{di_{pj}}{dt} = \frac{U_{dc}}{2} - u_{nj} + u_j \tag{2}$$

where U_{dc} is the dc-link voltage. u_j denotes the output voltage phase-j ($j = a, b, c$). u_{pj} and u_{nj} are the output voltage of the upper and lower arms, respectively. i_{pj} and i_{nj} refer to the current of the upper arm and the lower arm, respectively. L_p and L_n are the self-inductances of the coupling inductance for the upper and lower arms, respectively. M_u is the mutual inductance, assuming the two inductors are closely coupled and the leakage inductance can be ignored (i.e., $L_p = L_n = M_u = L$).

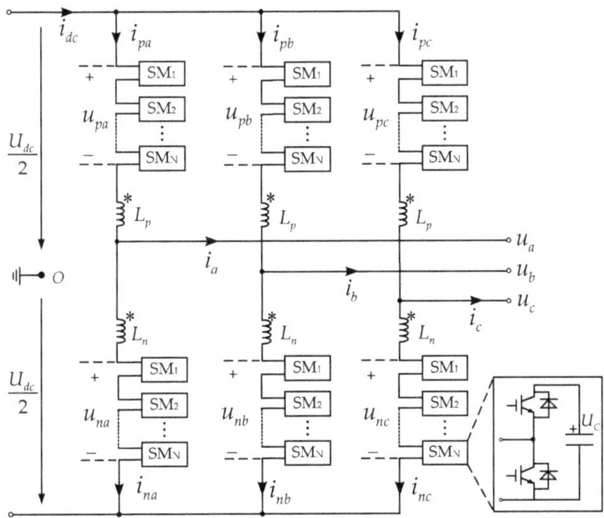

Figure 1. Schematic diagram of three phase modular multilevel converter (MMC). SM: sub-module.

According to Kirchhoff's current law, the upper and lower arm current of phase-j can be expressed as:

$$i_{pj} = i_{cj} + \frac{i_j}{2} \tag{3}$$

$$i_{nj} = i_{cj} - \frac{i_j}{2} \tag{4}$$

where i_j and i_{cj} are the output current and circulating current of phase-j, respectively.

Combining (1)–(4), the output voltage and circulating current of phase-j can be derived as:

$$u_j = \frac{1}{2}\left(u_{nj} - u_{pj}\right) \tag{5}$$

$$i_{cj} = \frac{1}{2}\left(i_{pj} + i_{nj}\right) \tag{6}$$

Combining (1), (2) with (6), the following equation can be obtained as:

$$4L\frac{di_{cj}}{dt} = U_{dc} - u_{pj} - u_{nj} \tag{7}$$

According to (7), the circulating current of phase-j can be calculated as:

$$i_{cj} = I_{cj} + \int_0^t \frac{U_{dc} - u_{pj} - u_{nj}}{4L}dt \tag{8}$$

where I_{cj} is the dc component of circulation current i_{cj}.

3. Implementation of Double-Carrier Phase Disposition Pulse Width Modulation for MMCs

The principle of DCPDPWM for MMCs is shown in Figure 2, where N is the number of SMs for each arm (e.g., $N = 10$). Only two carriers with displacement angle are needed for DCPDPWM, one carrier for the upper arm and the other for lower arm. Where θ is defined as the displacement angle between the upper arm carrier and lower arm carrier, and the range of θ can be obtained as $[0, 2\pi)$. Note that the displacement angle θ between upper arm carrier and lower arm carrier has a significant impact on the harmonic characteristics of MMC, which will be analyzed in Section 4.

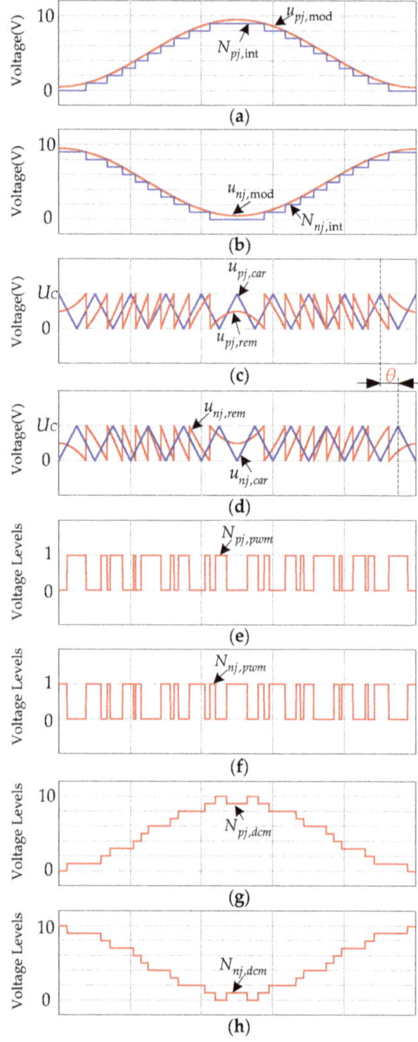

Figure 2. Principle of double-carrier phase-disposition pulse width modulation (DCPDPWM) for MMCs: (**a**) integer portion of the upper arm modulation signal; (**b**) integer portion of the lower arm modulation signal; (**c**) modulation of the remainder for the upper arm; (**d**) modulation of the remainder for the lower arm; (**e**) pulse width modulation (PWM) signal of the upper arm; (**f**) PWM signal of the lower arm; (**g**) number of on-state SMs for the upper arm; and (**h**) number of on-state SMs for the lower arm.

The reference voltages of the upper and lower arm for phase-*j* can be expressed as:

$$u_{pj,ref} = \frac{U_{dc}}{2}\left[1 + M\cos\left(\omega_o t + \pi + \phi_j\right)\right] \tag{9}$$

$$u_{nj,ref} = \frac{U_{dc}}{2}\left[1 + M\cos\left(\omega_o t + \phi_j\right)\right] \tag{10}$$

where M ($0 < M \leq 1$) denotes the modulation index. ω_o is the angular frequency of output voltage. ϕ_j is the phase angle of phase-*j* ($\phi_a = 0$, $\phi_b = -2\pi/3$, $\phi_c = 2\pi/3$).

The modulation signals of the upper and lower arms for phase-*j* can be obtained as:

$$u_{pj,mod} = \frac{u_{pj,ref}}{U_C} = \frac{N}{2}\left[1 + M\cos\left(\omega_o t + \pi + \phi_j\right)\right] \tag{11}$$

$$u_{nj,mod} = \frac{u_{nj,ref}}{U_C} = \frac{N}{2}\left[1 + M\cos\left(\omega_o t + \phi_j\right)\right] \tag{12}$$

where U_C is the capacitor rated voltage of SMs. Given that $U_C = U_{dc}/N$. The range of modulation signals are $[-N/2, N/2]$.

As shown in Figure 2a,b, the integer portion of upper arm and low arm modulation signals can be calculated as:

$$N_{pj,\text{int}} = floor\left(\frac{u_{pj,ref}}{U_C}\right) \tag{13}$$

$$N_{nj,\text{int}} = floor\left(\frac{u_{nj,ref}}{U_C}\right) \tag{14}$$

In which the function $floor(x)$ obtains the largest integer that is less than or equal to x. $N_{pj,int}$, $N_{nj,int}$ are the integer portion of modulation signals, respectively.

As shown in Figure 2c,d, the remainder of reference voltage for upper and lower arms can be derived as:

$$u_{pj,rem} = u_{pj,ref} - U_C \times N_{pj,\text{int}} \tag{15}$$

$$u_{nj,rem} = u_{nj,ref} - U_C \times N_{nj,\text{int}} \tag{16}$$

where $u_{pj,rem}$ and $u_{nj,rem}$ are the remainder of reference voltage for upper arm and lower arm, respectively.

The expressions of upper and the lower arms carriers can be obtained as:

$$u_{pj,car} = \begin{cases} \frac{U_C}{\pi}\left(\omega_c t - \theta - 2k\pi\right) , & 2k\pi \leq \omega_c t - \theta < 2k\pi + \pi \\ -\frac{U_C}{\pi}\left(\omega_c t - \theta - 2k\pi - 2\pi\right) , & 2k\pi + \pi \leq \omega_c t - \theta < 2k\pi + 2\pi \end{cases} \tag{17}$$

$$u_{nj,car} = \begin{cases} \frac{U_C}{\pi}\left(\omega_c t - 2k\pi\right) , & 2k\pi \leq \omega_c t < 2k\pi + \pi \\ -\frac{U_C}{\pi}\left(\omega_c t - 2k\pi - 2\pi\right) , & 2k\pi + \pi \leq \omega_c t < 2k\pi + 2\pi \end{cases} \tag{18}$$

where $u_{pj,car}$ and $u_{nj,car}$ are the carriers of upper and lower arms, respectively. k is the number of carrier period ($k \in [0, 1, ..., n]$). ω_c is the angular frequency of triangular carrier.

As shown in Figure 2e,f, the PWM signals of upper and lower arms can be obtained by comparing the remainders with the carriers of upper and lower arms, respectively. The PWM signals of upper and lower arms can be calculated as:

$$N_{pj,pwm} = \begin{cases} 1, & u_{pj,rem} > u_{pj,car} \\ 0, & u_{pj,rem} \leq u_{pj,car} \end{cases} \tag{19}$$

$$N_{nj,pwm} = \begin{cases} 1, & u_{nj,rem} > u_{nj,car} \\ 0, & u_{nj,rem} \leq u_{nj,car} \end{cases} \tag{20}$$

As shown in Figure 2g,h, the number of on-state SMs for each arm can be obtained by adding the integer portion and corresponding PWM portion. The number of on-state SMs for upper and lower arms can be obtained as:

$$N_{pj,dcm} = N_{pj,\text{int}} + N_{pj,pwm} \tag{21}$$

$$N_{nj,dcm} = N_{nj,\text{int}} + N_{nj,pwm} \tag{22}$$

where $N_{pj,dcm}$, $N_{nj,dcm}$ are the number of on-state SMs for upper and lower arms, respectively.

The block diagram of double-carrier phase-disposition pulse width modulation with a capacitor voltage balancing algorithm is shown in Figure 3. Firstly, the number of on-state SMs for each arm is obtained through the DCPDPWM. Then, the selection of the SMs is performed based on the reducing switching frequency (RSF) voltage balancing algorithm [44], which can achieve capacitor voltage balancing of SMs and reduce the average device switching frequency. $u_{pj}[i]$ ($i = 1, 2, \ldots, N$) and $u_{nj}[i]$ are the capacitor voltages of SMs for the upper and lower arms, respectively.

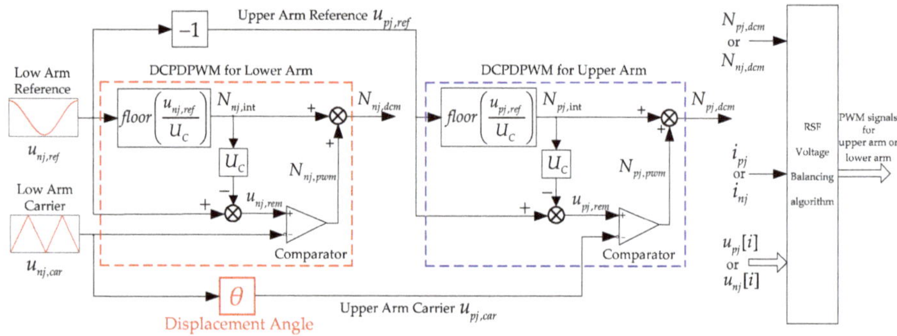

Figure 3. Block diagram of double-carrier phase-disposition pulse width modulation for MMCs.

As the circulating current control can influence several performance features of MMCs such as suppressing the low frequency circulating current, increasing the switching frequency, reducing the capacitor voltage ripple and so on, for simplicity, it is assumed that the circulating current control method is not applied in the block diagram as shown in Figure 3. Note that this assumption is reasonable as the circulating currents flowing through the three phase legs of the MMC caused by the voltage differences among the dc-link voltage and three phase legs, which will not affect the output voltages and currents [5,22]. Moreover, this paper mainly focuses on the high frequency (switching frequency) circulating current caused by DCPDPWM method for MMCs, whereas the circulating current control method focuses on low frequency circulating currents (mainly second-order harmonic currents), which can be suppressed by adding a circulating current control method [36,43]. Thus the circulating current control method affects significantly the low frequency harmonic circulating current and has a relatively small influence on switching frequency circulating current. The switching frequency harmonics caused by circulating current control method can be suppressed by selecting the arm inductance properly [24].

4. Theoretical Analysis of DCPDPWM Method for MMCs

The double Fourier integral analysis method is the most well-known analytical method for determining the harmonic components of a PWM method [45]. In this section, the theoretical analysis of DCPDPWM for MMCs is presented using the double Fourier integral analysis method. The Fourier series expression of phase voltage, line-to-line voltage and circulating current are deduced. Moreover, the influence of carrier displacement angle between upper and lower arms on the harmonic characteristics is revealed, and the optimum displacement angles are specified for the circulating current harmonics cancellation scheme and output voltage harmonics minimization scheme.

4.1. Influence of the Carrier Displacement Angle on Harmonic Characteristic of Output Voltage and Circulating Current for MMC

Note that the following harmonics analysis focuses on the switching harmonics produced by DCPDPWM, while the low-frequency harmonics (e.g., second-order harmonic for circulating current) caused by the energy oscillation between the upper arm and lower arm are not included.

Assuming N is even, and the range of modulation index M can be obtained as $[0, 1]$. For simplicity, it is assumed that the capacitance of each submodule is large enough and all the capacitor voltages of SMs are naturally balanced (i.e., $U_C = U_{dc}/N$), so that the capacitor voltage ripple can be ignored. Note that this assumption is reasonable as the capacitor voltage ripple is generally a relatively small portion comparing with the reference voltages of upper and lower arms [36,43].

The analytical technique for determining the spectral components of multilevel PWM method proposed in [46,47] is applied for DCPDPWM. The phase voltage can be derived as follows (see Appendix A):

$$
\begin{aligned}
u_j = \ & \frac{MU_{dc}}{2}\cos(\omega_o t + \phi_j) + \frac{8U_{dc}}{N\pi^2}\sum_{m=0}^{\infty}\frac{C_0}{2m+1}\sin\left[\frac{(2m+1)\theta}{2}\right]\cos\left[(2m+1)\omega_c t + \frac{(2m+1)\theta}{2} - \frac{\pi}{2}\right] \\
& + \frac{4U_{dc}}{N\pi^2}\sum_{m=0}^{\infty}\sum_{\substack{n=-\infty \\ (n\neq 0)}}^{\infty}\frac{C_1}{2m+1}\sin\left[\frac{(2m+1)\theta}{2}\right]\cos\left[(2m+1)\omega_c t + 2n(\omega_o t + \phi_j) + \frac{(2m+1)\theta}{2} - \frac{\pi}{2}\right] \\
& + \frac{2U_{dc}}{N\pi}\sum_{m=1}^{\infty}\sum_{n=-\infty}^{\infty}\frac{C_2}{2m}\cos(m\theta)\times\cos\left[2m\omega_c t + (2n-1)(\omega_o t + \phi_j) + m\theta\right]
\end{aligned}
\tag{23}
$$

where m denotes the carrier index variable and n refers to the baseband index variable.

The line-to-line voltage can be calculated as:

$$
\begin{aligned}
u_{ab} = \ & u_a - u_b = \frac{\sqrt{3}MU_{dc}}{2}\cos\left(\omega_o t + \frac{\pi}{6}\right) \\
& + \frac{8U_{dc}}{N\pi^2}\sum_{m=0}^{\infty}\sum_{\substack{n=-\infty \\ (n\neq 0)}}^{\infty}\frac{C_1}{2m+1}\sin\left(\frac{2n\pi}{3}\right)\sin\left[\frac{(2m+1)\theta}{2}\right]\cos\left[(2m+1)\omega_c t + 2n\left(\omega_o t - \frac{\pi}{3}\right) + \frac{(2m+1)\theta}{2}\right] \\
& + \frac{4U_{dc}}{N\pi}\sum_{m=1}^{\infty}\sum_{n=-\infty}^{\infty}\frac{C_2}{2m}\sin\left[\frac{(2n-1)\pi}{3}\right]\cos(m\theta)\times\cos\left[2m\omega_c t + (2n-1)\left(\omega_o t - \frac{\pi}{3}\right) + m\theta + \frac{\pi}{2}\right]
\end{aligned}
\tag{24}
$$

Substituting (23) and (24) into (7), the circulating current can be obtained as:

$$
\begin{aligned}
i_{cj} = \ & \frac{I_{dc}}{3} + \frac{4U_{dc}}{NL\pi^2\omega_c}\sum_{m=0}^{\infty}\frac{C_0}{(2m+1)^2}\cos\left[\frac{(2m+1)\theta}{2}\right]\times\cos\left[(2m+1)\omega_c t + \frac{(2m+1)\theta}{2} + \frac{\pi}{2}\right] \\
& + \frac{2U_{dc}}{NL\pi^2}\sum_{m=0}^{\infty}\sum_{\substack{n=-\infty \\ (n\neq 0)}}^{\infty}\frac{C_1}{(2m+1)((2m+1)\omega_c + 2n\omega_o)}\cos\left[\frac{(2m+1)\theta}{2}\right] \\
& \times\cos\left[(2m+1)\omega_c t + 2n(\omega_o t + \phi_j) + \frac{(2m+1)\theta}{2} + \frac{\pi}{2}\right] \\
& + \frac{U_{dc}}{NL\pi}\sum_{m=1}^{\infty}\sum_{n=-\infty}^{\infty}\frac{C_2}{2m(2m\omega_c + (2n-1)\omega_o)}\sin(m\theta)\times\cos\left[2m\omega_c t + (2n-1)(\omega_o t + \phi_j) + m\theta + \frac{\pi}{2}\right]
\end{aligned}
\tag{25}
$$

The magnitudes of carrier harmonic components and associated sideband harmonic components for the phase voltage, line-to-line voltage, circulating current can be expressed as follows:

$$
U_{j,m,n} = \begin{cases}
P_{m,n}\times\left|\sin\frac{(2m+1)\theta}{2}\right|, & if\ \omega_j = (2m+1)\omega_c,\, 2m+1\in\{1,3,\ldots\} \\
P_{m,n}\times\left|\sin\frac{(2m+1)\theta}{2}\right|, & if\ \omega_j = (2m+1)\omega_c + 2n\omega_o,\, 2m+1\in\{1,3,\ldots\},\, 2n\in\{-\infty,\ldots,-2,2,\ldots,\infty\} \\
P_{m,n}\times\left|\cos m\theta\right|, & if\ \omega_j = 2m\omega_c + (2n-1)\omega_o,\, 2m\in\{2,4,\ldots\},\, 2n-1\in\{-\infty,\ldots,-1,1,\ldots,\infty\}
\end{cases}
\tag{26}
$$

$$U_{ll,m,n} = \begin{cases} 0, & if \ \omega_{ll} = (2m+1)\omega_c, 2m+1 \in \{1,3,\ldots\} \\ 2P_{m,n} \times \left|\sin \frac{(2m+1)\theta}{2}\right|, & if \ \omega_{ll} = (2m+1)\omega_c + 2n\omega_o, 2m+1 \in \{1,3,\ldots\}, 2n \notin \{0,3,6,\ldots\} \\ 0, & if \ \omega_{ll} = (2m+1)\omega_c + 2n\omega_o, 2m+1 \in \{1,3,\ldots\}, 2n \in \{3,6,\ldots\} \\ 2P_{m,n} \times |\cos m\theta|, & if \ \omega_{ll} = 2m\omega_c + (2n-1)\omega_o, 2m \in \{2,4,\ldots\}, 2n-1 \notin \{0,3,6,\ldots\} \\ 0, & if \ \omega_{ll} = 2m\omega_c + (2n-1)\omega_o, 2m \in \{2,4,\ldots\}, 2n-1 \in \{0,3,6,\ldots\} \end{cases} \tag{27}$$

$$I_{cj,m,n} = \begin{cases} Q_{m,n} \times \left|\cos \frac{(2m+1)\theta}{2}\right|, & if \ \omega_{cj} = (2m+1)\omega_c, 2m+1 \in \{1,3,\ldots\} \\ Q_{m,n} \times \left|\cos \frac{(2m+1)\theta}{2}\right|, & if \ \omega_{cj} = (2m+1)\omega_c + 2n\omega_o, 2m+1 \in \{1,3,\ldots\}, 2n \in \{-\infty,\cdots,-2,2,\ldots,\infty\} \\ Q_{m,n} \times |\sin m\theta|, & if \ \omega_{cj} = 2m\omega_c + (2n-1)\omega_o, 2m \in \{2,4,\ldots\}, 2n-1 \in \{-\infty,\ldots,-1,1,\ldots,\infty\} \end{cases} \tag{28}$$

where:

$$P_{m,n} = \begin{cases} \frac{8U_{dc}}{(2m+1)N\pi^2}|C_0|, & if \ \omega_j = (2m+1)\omega_c, 2m+1 \in \{1,3,\ldots\} \\ \frac{4U_{dc}}{(2m+1)N\pi^2}|C_1|, & if \ \omega_j = (2m+1)\omega_c + 2n\omega_o, 2m+1 \in \{1,3,\ldots\}, 2n \in \{-\infty,\cdots,-2,2,\ldots,\infty\} \\ \frac{U_{dc}}{mN\pi}|C_2|, & if \ \omega_j = 2m\omega_c + (2n-1)\omega_o, 2m \in \{2,4,\ldots\}, 2n-1 \in \{-\infty,\ldots,-1,1,\ldots,\infty\} \end{cases} \tag{29}$$

$$Q_{m,n} = \begin{cases} \frac{P_{m,n}}{2L\omega_c(2m+1)}, & if \ \omega_{cj} = (2m+1)\omega_c, 2m+1 \in \{1,3,\ldots\} \\ \frac{P_{m,n}}{2L\omega_c[(2m+1)\omega_c+2n\omega_o]}, & if \ \omega_{cj} = (2m+1)\omega_c + 2n\omega_o, 2m+1 \in \{1,3,\ldots\}, 2n \in \{-\infty,\ldots,-2,2,\ldots,\infty\} \\ \frac{P_{m,n}}{2L\omega_c[2m\omega_c+(2n-1)\omega_o]}, & if \ \omega_{cj} = 2m\omega_c + (2n-1)\omega_o, 2m \in \{2,4,\ldots\}, 2n-1 \in \{-\infty,\ldots,-1,1,\ldots,\infty\} \end{cases} \tag{30}$$

where $U_{j,m,n}$, $U_{ll,m,n}$ and $I_{cj,m,n}$ are the magnitudes of carrier harmonic components and associated sideband harmonic components for the phase voltage, line-to-line voltage and circulating current, respectively. ω_j, ω_{ll}, and ω_{cj} are the angular frequency of phase voltage, line-to-line voltage and circulating current, respectively.

As shown in Equations (23)–(30), the phase voltage consists of a fundamental component, odd carrier harmonic components, even sideband harmonic components of odd carrier groups, and odd sideband harmonic components of even carrier groups. The carrier harmonic components and triple sideband harmonic components are cancelled in the line-to-line voltage. The circulating current consists of dc component, odd carrier harmonic components, even sideband harmonic components of odd carrier groups, and odd sideband harmonic components of even carrier groups.

It can be seen that the magnitudes of carrier harmonic components and associated sideband harmonic components for the phase voltage, line-to-line voltage and the circulating current are the function of displacement angle θ, respectively. Figure 4 shows the magnitudes of the harmonic components for the phase voltage and circulating current with different displacement angles. Only the first four harmonic groups ($m \leq 4$) are studied here due to the limitations of the paper. Where $P_{1,2n}$, $P_{2,2n-1}$, $P_{3,2n}$, $P_{4,2n-1}$ and $Q_{1,2n}$, $Q_{2,2n-1}$, $Q_{3,2n}$, $Q_{4,2n-1}$ are the maximums of the first four harmonic groups for phase voltage and circulating current, respectively.

It is found that the changing tendency of the magnitudes for harmonic components in the phase voltage and circulating current are opposite. When the magnitudes of harmonic components for phase voltage and line-to-line voltage are at their minima, the magnitudes of the harmonic components for the circulating current are maximum, and vice versa. Therefore, the displacement angle θ should be specified according to the specific industry application. When the number of SMs for each arm is large, such as in HVDC applications, superior harmonics characteristics of the output voltage can be achieved, so reducing the harmonics of the circulating current is the main problem. On the other hand, when the number of SMs for each arm is small, such as in STATCOM and motor drive applications, reducing the harmonics of the output voltage is preferred.

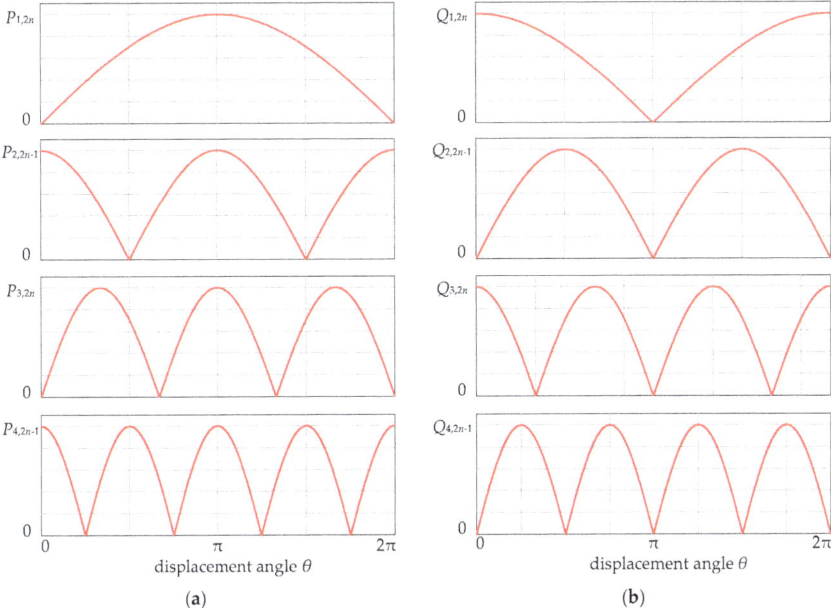

Figure 4. Magnitudes of the harmonic components for the phase voltage and circulating current with different displacement angles: (**a**) magnitudes of harmonic components for phase voltage; and (**b**) magnitudes of harmonic components circulating current.

4.2. Circulating Current Harmonics Cancellation Scheme for DCPDPWM Method

When displacement angle $\theta = \pi$, the circulating current harmonics cancellation scheme for DCPDPWM can be obtained. According to (23), the phase output voltage can be obtained as:

$$
\begin{aligned}
u_j = {} & \frac{MU_{dc}}{2}\cos(\omega_0 t + \phi_j) + \frac{8U_{dc}}{N\pi^2}\sum_{m=0}^{\infty}\frac{C_0}{2m+1}\cos[(2m+1)\omega_c t] \\
& + \frac{4U_{dc}}{N\pi^2}\sum_{m=0}^{\infty}\sum_{\substack{n=-\infty \\ (n\neq 0)}}^{\infty}\frac{C_1}{2m+1}\cos[(2m+1)\omega_c t + 2n(\omega_0 t + \phi_j)] \\
& + \frac{2U_{dc}}{N\pi}\sum_{m=1}^{\infty}\sum_{n=-\infty}^{\infty}\frac{C_2}{2m}\cos[2m\omega_c t + (2n-1)(\omega_0 t + \phi_j)]
\end{aligned}
\tag{31}
$$

According to (24), the line-to-line voltage can be expressed as:

$$
\begin{aligned}
u_{ab} = {} & \frac{\sqrt{3}MU_{dc}}{2}\cos\left(\omega_0 t + \frac{\pi}{6}\right) + \frac{8U_{dc}}{N\pi^2}\sum_{m=0}^{\infty}\sum_{\substack{n=-\infty \\ (n\neq 0)}}^{\infty}\frac{C_1}{2m+1}\sin\left(\frac{2n\pi}{3}\right)\times\cos\left[(2m+1)\omega_c t + 2n\left(\omega_0 t - \frac{\pi}{3}\right) + \frac{\pi}{2}\right] \\
& + \frac{4U_{dc}}{N\pi}\sum_{m=1}^{\infty}\sum_{n=-\infty}^{\infty}\frac{C_2}{2m}\sin\left[\frac{(2n-1)\pi}{3}\right]\cos\left[2m\omega_c t + (2n-1)\left(\omega_0 t - \frac{\pi}{3}\right) + \frac{\pi}{2}\right]
\end{aligned}
\tag{32}
$$

According to (25), the circulating current can be derived as:

$$
i_{cj} = \frac{I_{dc}}{3}
\tag{33}
$$

It can be seen that the magnitudes of the carrier harmonic components and associated sideband harmonic components for the circulating current are zero, which means that the carrier harmonic components and associated sideband harmonic components of the circulating current caused by DCPDPWM are completely cancelled out, leaving only the dc component. Therefore, the power loss and arm current stress are decreased. However, the magnitudes of the carrier harmonic components and sideband harmonic components for the phase voltage are at their maxima. The equivalent switching frequency (frequency of the lowest harmonic group) is f_{dcm}, where f_{dcm} denotes the carrier frequency of DCPDPWM.

4.3. Output Voltage Harmonics Minimization Scheme for the DCPDPWM Method

When carrier displacement angle $\theta = 0$, the output voltage harmonics minimization scheme for DCPDPWM can be obtained. According to (23), the phase output voltage can be derived as:

$$u_j = \frac{MU_{dc}}{2} \cos(\omega_0 t + \phi_j) + \frac{2U_{dc}}{N\pi} \sum_{m=1}^{\infty} \sum_{n=-\infty}^{\infty} \frac{C_2}{2m} \cos\left[2m\omega_c t + (2n-1)(\omega_0 t + \phi_j)\right] \tag{34}$$

According to (24), the line-to-line output voltage can be obtained as:

$$u_{ab} = \frac{\sqrt{3}MU_{dc}}{2} \cos\left(\omega_0 t + \tfrac{\pi}{6}\right) + \frac{4U_{dc}}{N\pi} \sum_{m=1}^{\infty} \sum_{n=-\infty}^{\infty} \frac{C_2}{2m} \sin\left[\frac{(2n-1)\pi}{3}\right] \times \cos\left[2m\omega_c t + (2n-1)\left(\omega_0 t - \tfrac{\pi}{3}\right) + \tfrac{\pi}{2}\right] \tag{35}$$

According to (25), the circulating current can be derived as:

$$\begin{aligned}
i_{cj} = {} & \frac{I_{dc}}{3} + \frac{4U_{dc}}{NL\pi^2\omega_c} \sum_{m=0}^{\infty} \sum_{k=1}^{\infty} \frac{C_0}{(2m+1)^2} \times \cos\left[(2m+1)\omega_c t + \tfrac{\pi}{2}\right] \\
& + \frac{2U_{dc}}{NL\pi^2} \sum_{m=0}^{\infty} \sum_{\substack{n=-\infty \\ (n \neq 0)}}^{\infty} \frac{C_1}{(2m+1)((2m+1)\omega_c + 2n\omega_0)} \times \cos\left[(2m+1)\omega_c t + 2n(\omega_0 t + \phi_j) + \tfrac{\pi}{2}\right]
\end{aligned} \tag{36}$$

It can be seen that the odd carrier harmonic components and the even sideband harmonic components of odd carrier groups for phase voltage are completely cancelled, leaving only the odd sideband harmonic components of even carrier groups. The equivalent switching frequency of phase voltage increases to $2 \times f_{dcm}$, which means that the better harmonic characteristics can be achieved for phase voltage and current. However, the magnitudes of odd carrier harmonic components and even sideband harmonic components of odd carrier groups for circulating current are maximized, which increases the current stress upon the power semiconductor devices and decreases the MMC efficiency.

When the carrier displacement angle $\theta = 0$, the carrier for the upper arm and carrier for the lower arm are the same, which means that only a single carrier is needed. Therefore, single carrier PDPWM is a special kind of DCPDPWM. Comparing Equations (34)–(36) with Equations (20), (34) and (21) in [43], it is found that when the equivalent switching frequency is the same (i.e., $f_{dcm} = N \times f_{psc}$, f_{psc} is the carrier frequency of PSCPWM), the phase voltage and line-to-line voltage of DCPDPWM have the same harmonic characteristics as PSCPWM, whereas the harmonics of the circulating current for DCPDPWM is different from PSCPWM. The harmonics of circulating current for DCPDPWM consist of odd carrier harmonics and odd sideband harmonics of even carrier groups, while the circulating current harmonics of the PSCPWM method comprise the sideband harmonic components of carrier groups.

5. Simulation Results

In order to verify the validity of the DCPDPWM method and the theoretical analysis in this paper, a MMC-based three phase inverter with ten SMs per arm was developed using PSIM software. The simulation parameters are listed in Table 1.

Comparison can be made between the proposed DCPDPWM method and PSCPWM method presented in [43] with circulating current harmonics cancellation scheme and output voltage harmonics minimization scheme. Note that the carrier frequency of DCPDPWM $f_{dcm} = N \times f_{psc}$, so that the average frequency of SMs for DCPDPWM method is basically equal to the PSCPWM method.

Table 1. Parameters of the simulation.

Parameter	Value
Number of SMs per arm	$N = 10$
Frequency of reference voltage	$f_o = 50$ Hz
Buffer inductors	$L_p = L_n = M_u = 0.5$ mH
Arm equivalent resistance	$0.1\ \Omega$
SMs capacitance	$C = 10$ mF
DC-link voltage	$U_{dc} = 10000$ V
Modulation index	$M = 0.95$
Carrier frequency of DCPDPWM	$f_{dcm} = 4000$ Hz
Carrier frequency of PSCPWM	$f_{psc} = 400$ Hz
Load inductance	$L_d = 2$ mH
Load resistance	$R_d = 80\ \Omega$

5.1. Comparison between DCPDPWM and PSCPWM Methods with Circulating Current Harmonics Cancellation Scheme

The comparison of simulation waveforms and harmonic spectra between the DCPDPWM and PSCPWM for MMC with circulating current harmonics cancellation scheme are shown in Figures 5 and 6, respectively. As shown in these figures, the voltage levels of phase voltage for both the PSCPWM and DCPDPWM methods are eleven. The equivalent switching frequency of phase voltage for PSCPWM method is the same with DCPDPWM method (i.e., $f_{equ} = f_{dcm} = N \times f_{psc} = 4000$ Hz).

It can be seen that the most significant harmonic for DCPDPWM is the first carrier harmonic component, which can be cancelled in the line-to-line voltage. The triplen sideband harmonics in the phase voltage for DCPDPWM are also cancelled in the line-to-line voltage. However, only the triplen sideband harmonic components of the phase voltage are eliminated in the line-to-line voltage for PSCPWM. The magnitudes of the sideband harmonic components of line-to-line voltage and phase current for DCPDPWM method are lower than for PSCPWM, which means that the DCPDPWM can achieve better harmonic performance than PSCPWM.

Moreover, it is found that the carrier harmonic components and associated sideband harmonic components of circulating current caused by DCPDPWM are completely cancelled, whereas the sideband harmonic components of circulating current caused by PSCPWM are also completely cancelled, leaving only dc components and low frequency harmonics (mainly second-order harmonics). Therefore, the circulating current harmonics for the PSCPWM method are similar to those of the DCPDPWM method with the circulating current harmonics cancellation scheme. Note that the low frequency harmonic components can be reduced by the circulating current control method. In order to ensure that the harmonic characteristics are only affected by the modulation method, the circulating current control method is not included in this paper. It can be concluded that the simulation results are completely consistent with the theoretical analysis with the circulating current cancellation scheme for the DCPDPWM method.

Note that only transitions caused by DCPDPWM and PSCPWM methods are considered in the following analysis when the circulating current control method is not applied for MMC. However, when the circulating current control method is applied for MMC, the transitions will be increased.

Comparison of the simulation results between the DCPDPWM and PSCPWM methods with circulating harmonics cancellation scheme are shown in Table 2. It can be seen that the total switching number per arm in 1 power grid period for PSCPWM and DCPDPWM are 80 and 79, respectively. The total harmonic distortion (*THD*) of line-to-line voltage and phase current for DCPDPWM are 6.89% and 3.91%, respectively. The *THD* of line-to-line voltage and phase current for PSCPWM are 9.77% and 7.01%, respectively. It is found that when the total switching frequency is basically the same, the DCPDPWM has better harmonic characteristics than PSCPWM with the circulating current harmonics cancellation scheme.

Figure 7 shows the *THD* of line-to-line voltage for DCPDPWM and PSCPWM for different modulation indexes with the circulating current harmonics cancellation scheme. It can be seen that DCPDPWM method has better line-to-line harmonic characteristics than the PSCPWM method in the whole modulation index region.

Table 2. Comparison of simulation results between DCPDPWM and PSCPWM methods with the circulating current harmonics cancellation scheme. *THD*: total harmonic distortion.

Modulation Methods	DCPDPWM	PSCPWM
THD of line-to-line output voltage (%)	6.89	9.77
THD of phase current (%)	3.91	7.01
Total switching number per arm (1 power grid period)	79	80

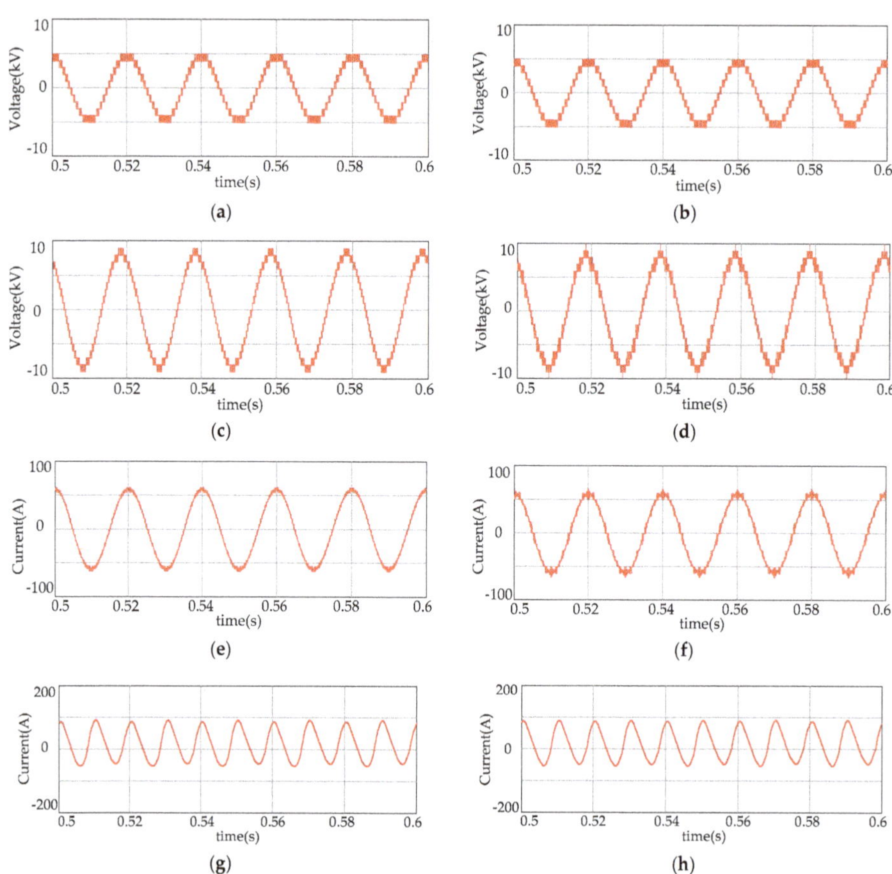

Figure 5. Comparison of simulation waveforms between DCPDPWM and phase-shifted carrier pulse-width modulation (PSCPWM) methods with the circulating current harmonics cancellation scheme: (**a**) phase voltage of DCPDPWM; (**b**) phase voltage of PSCPWM; (**c**) line-to-line voltage of DCPDPWM; (**d**) line-to-line voltage of PSCPWM; (**e**) phase current of DCPDPWM; (**f**) phase current of PSCPWM; (**g**) circulating current of PDPWM; and (**h**) circulating current of PSCPWM.

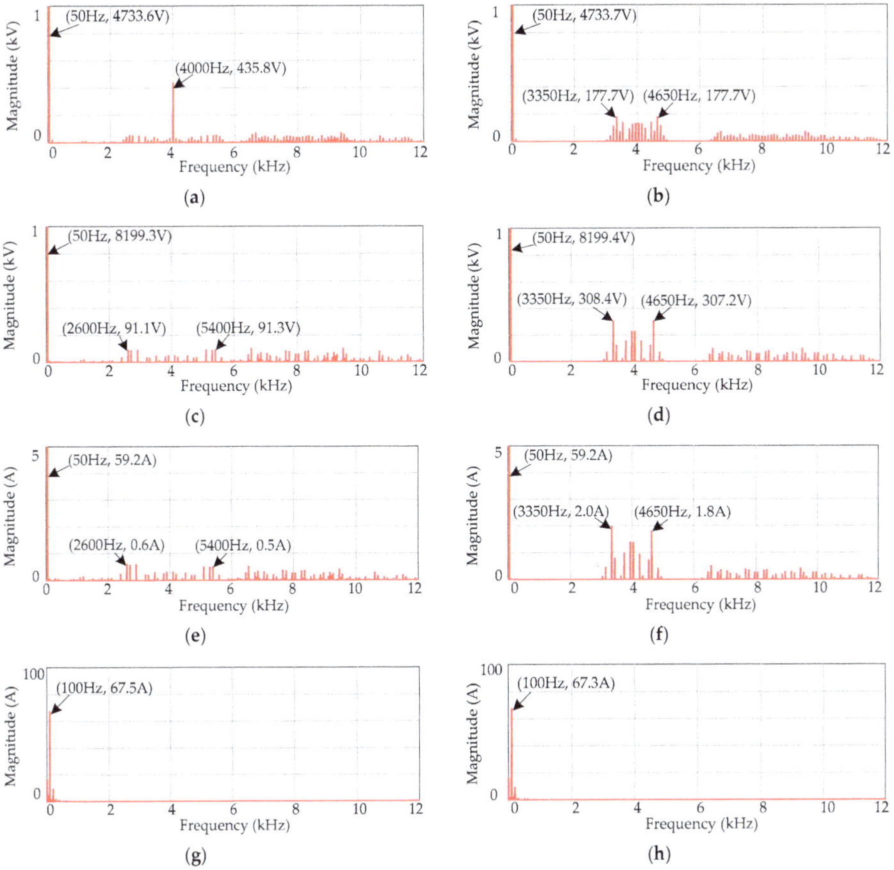

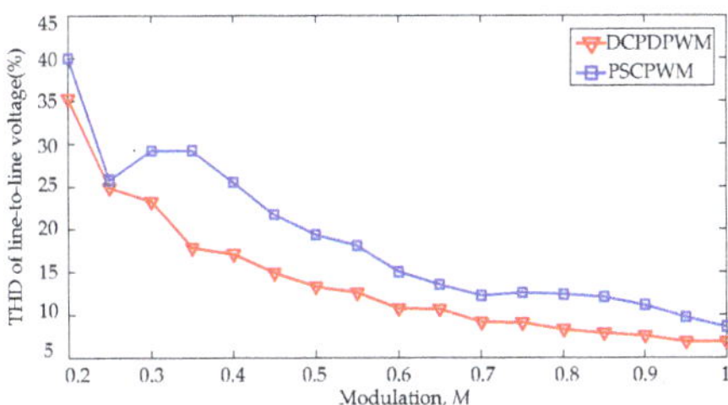

Figure 6. Comparison of harmonic spectra between DCPDPWM and PSCPWM methods with the circulating current harmonics cancellation scheme: (**a**) phase voltage of DCPDPWM; (**b**) phase voltage of PSCPWM; (**c**) line-to-line voltage of DCPDPWM; (**d**) line-to-line voltage of PSCPWM; (**e**) phase current of DCPDPWM; (**f**) phase current of PSCPWM; (**g**) circulating current of PDPWM; and (**h**) circulating current of PSCPWM.

Figure 7. *THD* of line-to-line voltage for DCPDPWM and PSCPWM methods in different modulation index with circulating current harmonics cancellation scheme.

5.2. Comparison between DCPDPWM and PSCPWM Method with Output Voltage Harmonics Minimization Scheme

The comparison of simulation waveforms and harmonic spectra between the DCPDPWM and PSCPWM methods for MMC with output voltage harmonics minimization scheme are presented in Figures 8 and 9, respectively.

It can be seen that the voltage levels of phase voltage for both the PSCPWM and DCPDPWM methods rise to twenty-one. The equivalent switching frequency increases to $f_{equ} = 2f_{dcm} = 2N \times f_{psc} = 8000$ Hz. It is found that the odd carrier harmonic components, and even sideband harmonic components of odd carrier groups in the phase voltage are completely eliminated in the line-to-line voltage for DCPDPWM, leaving only the even sideband harmonics of odd carrier groups. The sideband harmonic components of odd carrier groups in the phase voltage are cancelled for PSCPWM, leaving also the even sideband harmonics of odd carrier groups. It is found that the phase voltage, line-to-line voltage and phase current for PSCPWM and DCPDPWM methods have the same harmonic performance.

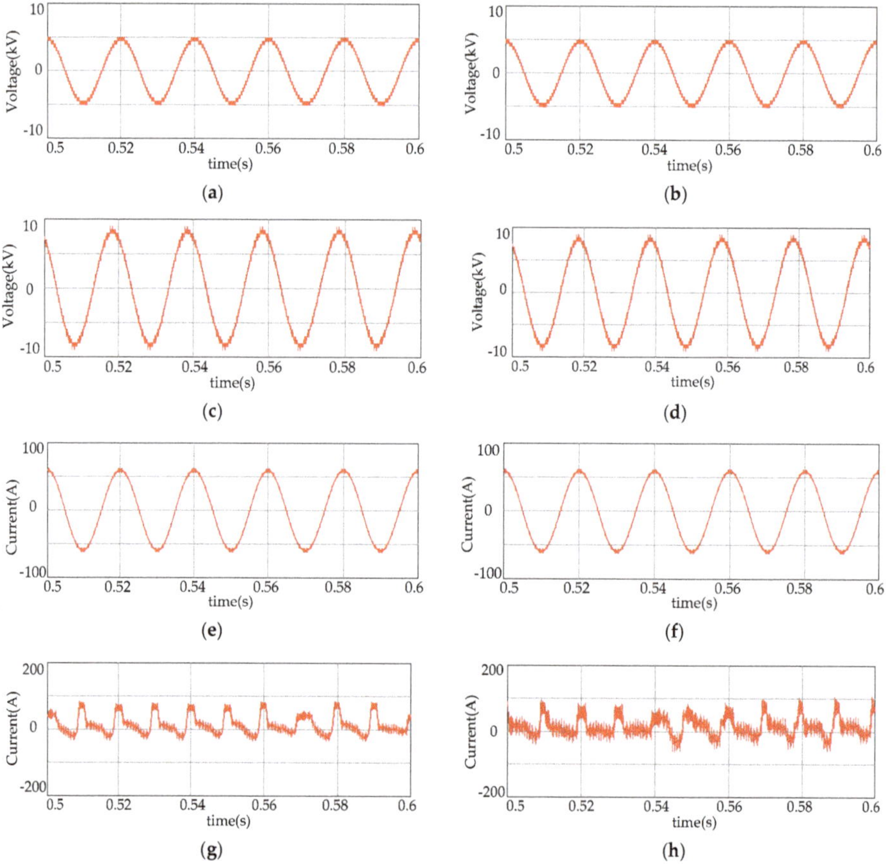

Figure 8. Comparison of simulation waveforms between the DCPDPWM and PSCPWM methods with output voltage harmonics minimization scheme: (**a**) phase voltage of DCPDPWM; (**b**) phase voltage of PSCPWM; (**c**) line-to-line voltage of DCPDPWM; (**d**) line-to-line voltage of PSCPWM; (**e**) phase current of DCPDPWM; (**f**) phase current of PSCPWM; (**g**) circulating current of PDPWM; and (**h**) circulating current of PSCPWM.

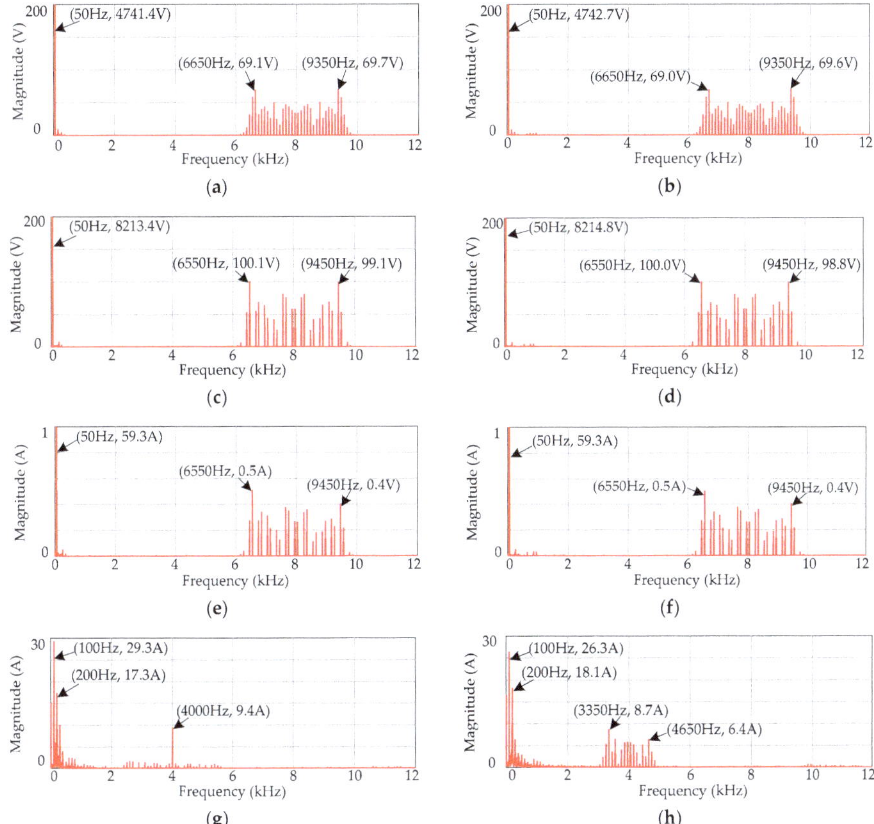

Figure 9. Comparison of harmonic spectra between DCPDPWM and PSCPWM methods with output voltage harmonics minimization scheme: (**a**) phase voltage of DCPDPWM; (**b**) phase voltage of PSCPWM; (**c**) line-to-line voltage of DCPDPWM; (**d**) line-to-line voltage of PSCPWM; (**e**) phase current of DCPDPWM; (**f**) phase current of PSCPWM; (**g**) circulating current of PDPWM; (**h**) circulating current of PSCPWM.

It can be seen that the harmonics of circulating currents between DCPDPWM and PSCPWM are different. The switching harmonics of circulating current for DCPDPWM includes odd carrier harmonic components and odd sideband harmonic components of even carrier groups. The main switching harmonics of DCPDWPM in the circulating current is the first carrier harmonic, while the switching harmonics of circulating current for PSCPWM consist of the sideband harmonics of carrier groups. It can be concluded that the simulation results completely agree with the theoretical analysis when applying the output voltage harmonics minimization scheme for DCPDPWM method.

Comparison of simulation results between the DCPDPWM and PSCPWM methods with output voltage harmonics minimization scheme are shown in Table 3. It can be seen that when the total switching number per arm in 1 power grid period are basically the same, the *THD* of line-to-line output voltage and phase current for both the DCPDPWM and PSCPWM are 4.78% and 2.44%, respectively.

Figure 10 shows the *THD* of line-to-line voltage for DCPDPWM and PSCPWM methods in different modulation index with output voltage harmonics minimization scheme. It can be seen that DCPDPWM method has the same line-to-line voltage harmonic characteristics with PSCPWM method in the whole modulation index region.

Table 3. Comparison of simulation results between DCPDPWM and PSCPWM method with the output voltage harmonics minimization scheme.

Modulation Methods	DCPDPWM	PSCPWM
THD of line-to-line output voltage (%)	4.78	4.78
THD of phase current (%)	2.44	2.44
Total number of switching per arm (1 power grid period)	79	80

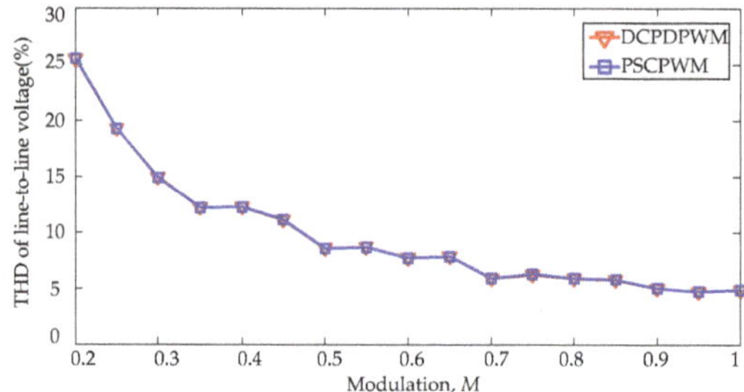

Figure 10. *THD* of line-to-line output voltage for DCPDPWM and PSCPWM methods in different modulation index with output voltage harmonics minimization scheme.

6. Experimental Verification

In order to further verify the proposed method and theoretical analysis, a three-phase MMC laboratory prototype was built. The parameters of the prototype are shown in Table 4. The dc-link voltage is 400 V, and the number of SMs per arm is $N = 4$. A coupled inductor is adopted as the arm inductor, and a resistance-inductor load is used.

Table 4. Parameters of Prototype.

Parameter	Value
DC-link voltage	$U_{dc} = 400$ V
Number of SMs per arm	$N = 4$
Frequency of reference voltage	$f_o = 50$ Hz
Arm inductor	$L_p = L_n = M_u = 1$ mH
SMs capacitance	$C = 2.2$ mF
Modulation index	$M = 0.9$
Carrier frequency of DCPDPWM	$f_{dcm} = 4000$ Hz
Carrier frequency of PSCPWM	$f_{psc} = 1000$ Hz
Load inductance	$L_d = 2$ mH
Load resistance	$R_d = 20\ \Omega$

Figures 11 and 12 show the experimental waveforms and harmonic spectra of DCPDPWM and PSCPWM with circulating current harmonics cancellation scheme, respectively. It is found that the switching harmonics of the circulating current are basically cancelled for DCPDPWM and PSCPWM methods. It can be seen that voltage level number of phase voltage is five, and the equivalent switching frequency of phase voltage for both the DCPDPWM and PSCPWM methods is 4000 Hz ($f_{equ} = f_{dcm} = N \times f_{psc}$). For the DCPDPWM method, the main harmonic component of phase voltage is the first carrier harmonic component, which is eliminated in the line-to-line voltage. The harmonic components magnitudes in the line-to-line voltage and phase current for DCPDPWM are lower than PSCPWM with circulating current harmonics cancellation scheme.

From Figure 12, it can be seen that the *THD* of line-to-line voltage and phase current for DCPDPWM are 18.70% and 4.55%, respectively. The *THD* of line-to-line voltage and phase current for PSCPWM are 30.30% and 9.89%, respectively. The experimental results show that DCPDPWM has better harmonic characteristics than PSCPWM with the circulating current harmonics cancellation scheme. It is found that the experimental results match well with the theoretical analysis and simulation results.

Figures 13 and 14 present the experimental waveforms and harmonic spectra of DCPDPWM and PSCPWM with output voltage harmonics minimization scheme, respectively. It can be seen that the voltage level number of the phase voltage increases to nine for both the DCPDPWM and PSCPWM methods, which means that the lower *THD* of phase voltage and line-to-line voltage can be achieved. It is found that the harmonic components of the first carrier groups are basically cancelled in the phase voltage, and the equivalent switching frequency of phase voltage for DCPDPWM and PSCPWM methods rises to 8000 Hz ($f_{equ} = 2f_{dcm} = 2 \times N \times f_{psc}$).

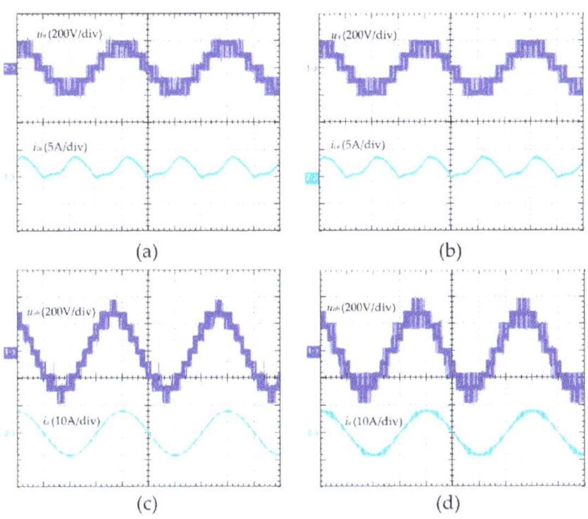

Figure 11. Experimental waveforms of DCPDPWM method and PSCPWM method with circulating current harmonics cancellation scheme: (**a**) phase voltage and circulating current of DCPDPWM; (**b**) phase voltage and circulating current of PSCPWM; (**c**) line-to-line voltage and phase current of DCPDPWM; and (**d**) line-to-line voltage and phase current of PSCPWM.

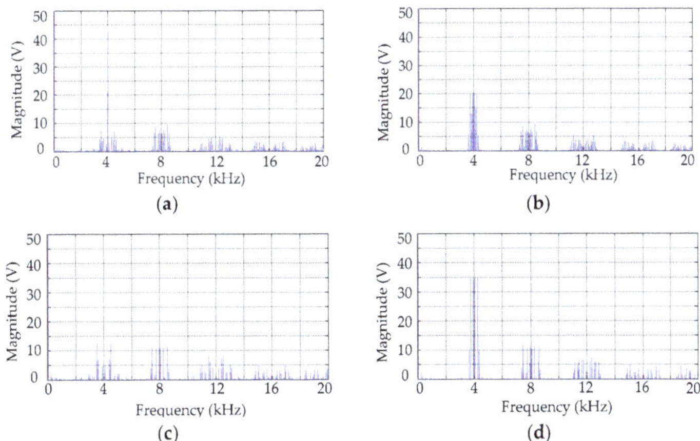

Figure 12. *Cont.*

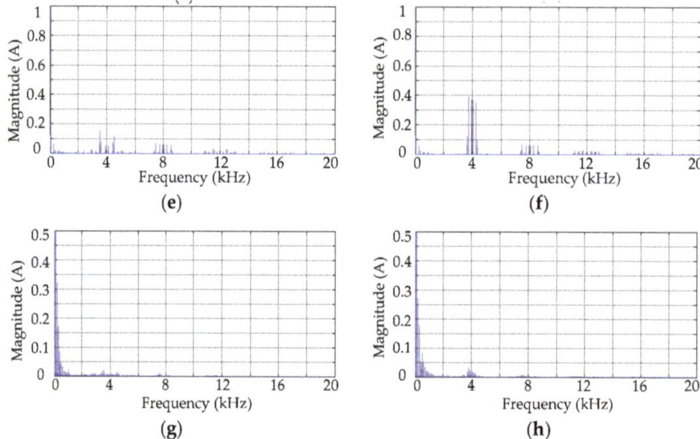

Figure 12. Harmonic spectra of experimental waveforms for the DCPDPWM and PSCPWM methods with circulating current harmonics cancellation scheme: (**a**) phase voltage of DCPDPWM; (**b**) phase voltage of PSCPWM; (**c**) line-to-line voltage of DCPDPWM; (**d**) line-to-line voltage of PSCPWM; (**e**) phase current of DCPDPWM; (**f**) phase current of PSCPWM; (**g**) circulating current of PDPWM; and (**h**) circulating current of PSCPWM.

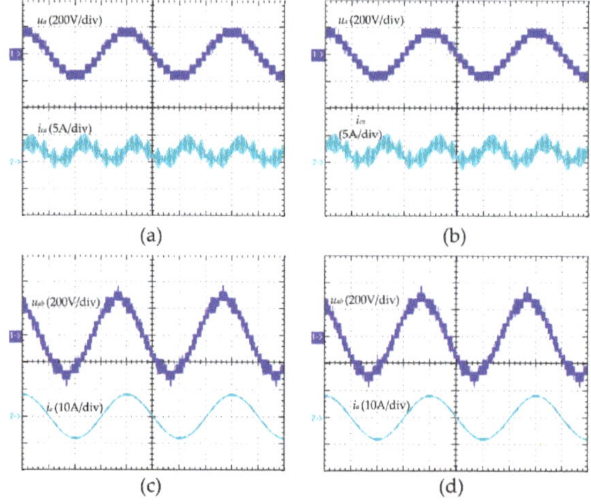

Figure 13. Experimental waveforms of DCPDPWM method and PSCPWM method with output voltage harmonics minimization scheme: (**a**) phase voltage and circulating current of DCPDPWM; (**b**) phase voltage and circulating current of PSCPWM; (**c**) line-to-line voltage and phase current of DCPDPWM; (**d**) line-to-line voltage and phase current of PSCPWM.

Meanwhile, it can be seen that there are many switching harmonics in the circulating current, which causes high switching ripples in the waveform of circulating current. The circulating current harmonics for DCPDPWM are different with PSCPWM. The main switching harmonic component in the circulating current of DCPDPWM is the first carrier harmonic component, while the switching harmonics in the circulating current for PSCPWM are the sideband harmonic of the carrier groups. From Figure 14, it can be seen that the *THD* of line-to-line voltage and phase current for DCPDPWM are 13.27% and 2.93%, respectively. The *THD* of line-to-line voltage and phase current for PSCPWM are 13.34% and 2.98%, respectively. The experimental results show that DCPDPWM has the same harmonic characteristics than PSCPWM with the output voltage harmonics minimization scheme. The experimental results agree with the theoretical analysis and simulation results.

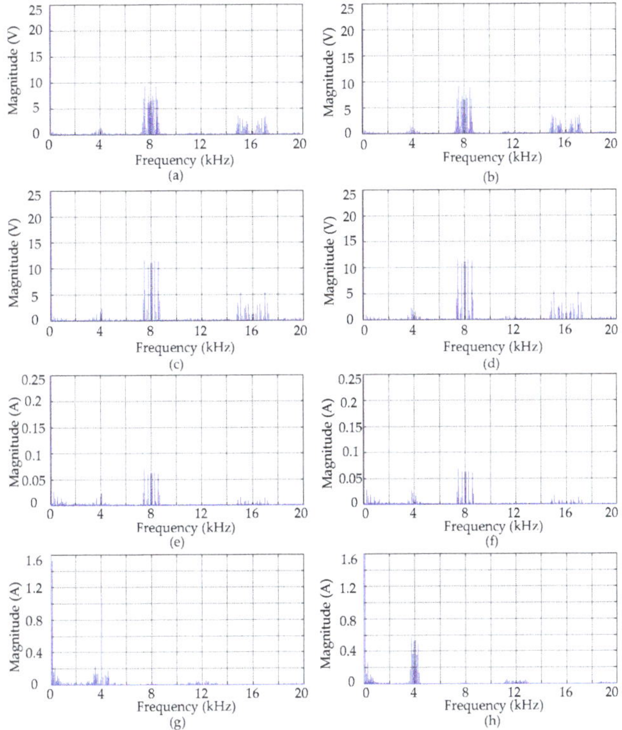

Figure 14. Harmonic spectra of experimental results for DCPDPWM and PSCPWM methods with output voltage harmonics minimization scheme: (**a**) phase voltage of DCPDPWM; (**b**) phase voltage of PSCPWM; (**c**) line-to-line voltage of DCPDPWM; (**d**) line-to-line voltage of PSCPWM; (**e**) phase current of DCPDPWM; (**f**) phase current of PSCPWM; (**g**) circulating current of PDPWM; and (**h**) circulating current of PSCPWM.

7. Conclusions

This paper has proposed a DCPDPWM method for MMCs. Only double triangular carriers with displacement angle are needed, one carrier for the lower arm, and the other carrier for the upper arm. The theoretical analysis of DCPDPWM for MMCs is presented by using a double Fourier integral analysis method. The Fourier series expression of phase voltage, line-to-line voltage and circulating current are deduced, and further the influence of carrier displacement angle between the upper and lower arms on the harmonic characteristics of the output voltage and circulating current is revealed. Furthermore, the optimum displacement angles are specified for the circulating current harmonics cancellation scheme and the output voltage harmonics minimization scheme. The proposed method and theoretical analysis are verified by simulations and experimental results.

It can be concluded that when applying the circulating current harmonics cancellation scheme, the carrier and associated sideband harmonics of the circulating current caused by DCPDPWM are completely cancelled, leaving only the dc component and low frequency components, which is similar to the circulating current harmonic characteristics for PSCPWM. The DCPDPWM method has better line-to-line voltage harmonic characteristics than the PSCPWM method with the same equivalent switching frequency. When applying the output voltage harmonics minimization scheme, the odd carrier harmonics and even sideband harmonics of odd carrier groups for phase voltage are eliminated, and the phase voltage and line-to-line voltage of DCPDPWM have the same harmonic characteristics as PSCPWM. However, the magnitudes of the odd carrier harmonics and odd sideband harmonics of even carrier groups for circulating current are at a maximum. The harmonic circulating current characteristics between the PSCPWM and DCPDPWM methods are different. The main harmonic caused by DCPDWPM in the circulating current is the first carrier harmonic, while the circulating current harmonics caused by PSCPWM consist of the carrier group sideband harmonics.

Acknowledgments: This work was supported by the National Natural Science Foundation of China (NSFC) under Grant No. 51477045.

Author Contributions: All the authors conceived and designed the study. Fayun Zhou performed the simulation and wrote the manuscript with guidance from An Luo and Yan Li. Fayun Zhou, Qianming Xu and Zhixing He performed the experiment. An Luo, Yan Li, Qianming Xu, Zhixing He and Josep M. Guerrero reviewed the manuscript and provided valuable suggestions.

Conflicts of Interest: The authors declare no conflict of interest.

Appendix A

The lower arm voltage can be obtained as:

$$
\begin{aligned}
u_{nj}(t) = \quad & \frac{U_{dc}}{2} + \frac{MU_{dc}}{2}\cos(\omega_o t + \phi_j) + \frac{8U_{dc}}{N\pi^2}\sum_{m=0}^{\infty}\frac{C_0}{2m+1}\cos[(2m+1)\omega_c t] \\
& + \frac{4U_{dc}}{N\pi^2}\sum_{m=0}^{\infty}\sum_{\substack{n=-\infty \\ (n\neq 0)}}^{\infty}\frac{C_1}{2m+1}\cos\left[(2m+1)\omega_c t + 2n(\omega_o t + \phi_j)\right] \\
& + \frac{2U_{dc}}{N\pi}\sum_{m=1}^{\infty}\sum_{n=-\infty}^{\infty}\frac{C_2}{2m}\cos\left[2m\omega_c t + (2n-1)(\omega_o t + \phi_j)\right]
\end{aligned}
\tag{A1}
$$

where m denotes the carrier index variable and n refers to the baseband index variable.

The coefficients C_0, C_1, C_2 are as follow:

$$
\begin{aligned}
C_0 = \quad & \sum_{k=0}^{\infty}\cos(k\pi)J_{2k+1}\left[\frac{(2m+1)N\pi M}{2}\right] \\
& \times\left\{\frac{1}{(2k+1)}\left[\sin\left((2k+1)\frac{\pi}{2}\right) + 2\sum_{h=1}^{\frac{N}{2}-1}\sin\left((2k+1)\cos\left(\frac{2h}{NM}\right)^{-1}\right)\cos(h\pi)\right]\right\}
\end{aligned}
\tag{A2}
$$

$$
\begin{aligned}
C_1 = \quad & \sum_{k=0}^{\infty}\cos(k\pi)J_{2k+1}\left(\frac{(2m+1)N\pi M}{2}\right) \\
& \times\left\{\begin{array}{l}
\frac{1}{2n-2k-1}\left[\sin\left((2n-2k-1)\frac{\pi}{2}\right) + 2\sum_{h=1}^{\frac{N}{2}-1}\sin\left((2n-2k-1)\cos^{-1}\left(\frac{2h}{NM}\right)\right)\cos(h\pi)\right] \\
+ \frac{1}{2n+2k+1}\left[\sin\left[(2n+2k+1)\frac{\pi}{2}\right] + 2\sum_{h=1}^{\frac{N}{2}-1}\sin\left[(2n+2k+1)\cos^{-1}\left(\frac{2h}{NM}\right)\right]\cos(h\pi)\right]
\end{array}\right\}
\end{aligned}
\tag{A3}
$$

$$
C_2 = J_{2n-1}(mN\pi M)\cos((n-1)\pi)
\tag{A4}
$$

where $J_n(\lambda)$ represents the Bessel coefficient of order n and argument λ.

The upper arm voltage can be derived as:

$$
\begin{aligned}
u_{pj} = \ & \frac{U_{dc}}{2} + \frac{MU_{dc}}{2}\cos\left(\omega_0 t + \pi + \phi_j\right) + \frac{8U_{dc}}{N\pi^2}\sum_{m=0}^{\infty}\frac{C_0}{2m+1}\cos\left[(2m+1)(\omega_c t + \theta)\right] \\
& + \frac{4U_{dc}}{N\pi^2}\sum_{m=0}^{\infty}\ \sum_{\substack{n=-\infty \\ (n\neq 0)}}^{\infty}\frac{C_1}{2m+1}\cos\left[(2m+1)(\omega_c t + \theta) + 2n\left(\omega_0 t + \pi + \phi_j\right)\right] \\
& + \frac{2U_{dc}}{N\pi}\sum_{m=1}^{\infty}\sum_{n=-\infty}^{\infty}\frac{C_2}{2m}\cos\left[2m(\omega_c t + \theta) + (2n-1)\left(\omega_0 t + \pi + \phi_j\right)\right]
\end{aligned}
\tag{A5}
$$

Substituting (A1) and (A5) into (8), the phase voltage can be derived as:

$$
\begin{aligned}
u_j = \ & \frac{MU_{dc}}{2}\cos\left(\omega_0 t + \phi_j\right) + \frac{8U_{dc}}{N\pi^2}\sum_{m=0}^{\infty}\frac{C_0}{2m+1}\sin\left[\frac{(2m+1)\theta}{2}\right]\cos\left[(2m+1)\omega_c t + \frac{(2m+1)\theta}{2} - \frac{\pi}{2}\right] \\
& + \frac{4U_{dc}}{N\pi^2}\sum_{m=0}^{\infty}\ \sum_{\substack{n=-\infty \\ (n\neq 0)}}^{\infty}\frac{C_1}{2m+1}\sin\left[\frac{(2m+1)\theta}{2}\right]\cos\left[(2m+1)\omega_c t + 2n\left(\omega_0 t + \phi_j\right) + \frac{(2m+1)\theta}{2} - \frac{\pi}{2}\right] \\
& + \frac{2U_{dc}}{N\pi}\sum_{m=1}^{\infty}\sum_{n=-\infty}^{\infty}\frac{C_2}{2m}\cos(m\theta)\times\cos\left[2m\omega_c t + (2n-1)\left(\omega_0 t + \phi_j\right) + m\theta\right]
\end{aligned}
\tag{A6}
$$

References

1. Lesnicar, A.; Marquardt, R. An innovative modular multilevel converter topology suitable for a wide power range. In Proceedings of the 2003 IEEE Bologna Power Tech Conference, Bologna, Italy, 23–26 June 2003.
2. Marquardt, R. Modular multilevel converter: An universal concept for HVDC-networks and extended DC-bus-applications. In Proceedings of the 2010 International Power Electronics Conference, Sapporo, Japan, 21–24 June 2010.
3. Akagi, H. Classification, terminology, and application of the modular multilevel cascade converter (MMCC). *IEEE Trans. Power Electron.* **2011**, *26*, 3119–3130. [CrossRef]
4. Perez, M.A.; Bernet, S.; Rodriguez, J.; Kouro, S.; Lizana, R. Circuit topologies, modeling, control schemes, and applications of modular multilevel converters. *IEEE Trans. Power Electron.* **2015**, *30*, 4–17. [CrossRef]
5. Debnath, S.; Qin, J.; Bahrani, B.; Saeedifard, M.; Barbosa, P. Operation, control, and applications of the modular multilevel converter: A review. *IEEE Trans. Power Electron.* **2015**, *30*, 37–53. [CrossRef]
6. Nami, A.; Liang, J.; Dijkhuizen, F.; Demetriades, G.D. Modular multilevel converters for HVDC applications: Review on converter cells and functionalities. *IEEE Trans. Power Electron.* **2015**, *30*, 18–36. [CrossRef]
7. Mohammadi, H.P.; Bina, M.T. A transformerless medium-voltage STATCOM topology based on extended modular multilevel converters. *IEEE Trans. Power Electron.* **2011**, *26*, 1534–1545.
8. Zhuo, G.; Jiang, D.; Lian, X. Modular multilevel converter for unified power flow controller application. In Proceedings of the Third International Conference on Digital Manufacturing and Automation (ICDMA), Guilin, China, 31 July–2 August 2012.
9. Shu, Z.; Liu, M.; Zhao, L.; Song, S.; Zhou, Q.; He, X. Predictive harmonic control and its optimal digital implementation for MMC-based active power filter. *IEEE Trans. Ind. Electron.* **2016**, *63*, 5244–5254. [CrossRef]
10. Hagiwara, M.; Nishimura, K.; Akagi, H. A medium-voltage motor drive with a modular multilevel PWM inverter. *IEEE Trans. Power Electron.* **2010**, *25*, 1786–1799. [CrossRef]
11. Liu, H.; Ma, K.; Poh, C.L.; Blaabjerg, F. Online fault identification based on an adaptive Observer for modular multilevel converters applied to wind power generation systems. *Energies* **2015**, *8*, 7140–7160. [CrossRef]
12. Liu, H.; Ma, K.; Qin, Z.; Loh, P.C.; Blaabjerg, F. Lifetime estimation of MMC for offshore wind power HVDC application. *IEEE J. Emerg. Sel. Topics Power Electron.* **2016**, *4*, 504–511. [CrossRef]
13. Mei, J.; Xiao, B.; Shen, K.; Tolbert, L.M.; Zheng, J.Y. Modular multilevel inverter with new modulation method and its application to photovoltaic grid-connected generator. *IEEE Trans. Power Electron.* **2013**, *28*, 5063–5073. [CrossRef]
14. Trintis, I.; Munk-Nielsen, S.; Teodorescu, R. A new modular multilevel converter with integrated energy storage. In Proceedings of the IECON 2011–37th Annual Conference on IEEE Industrial Electronics Society, Melbourne, Australia, 7–10 November 2011.

15. Konstantinou, G.; Pou, J.; Pagano, D.; Ceballos, S. A hybrid modular multilevel converter with partial embedded energy storage. *Energies* **2016**, *9*, 1012. [CrossRef]
16. Jovcic, D.; Jamshidifar, A.A. Phasor model of modular multilevel converter with circulating current suppression control. *IEEE Trans. Power Deliv.* **2015**, *30*, 1889–1897. [CrossRef]
17. Jamshidifar, A.; Jovcic, D. Small-signal dynamic DQ model of modular multilevel converter for system studies. *IEEE Trans. Power Deliv.* **2016**, *31*, 191–199. [CrossRef]
18. Lu, X.; Lin, W.; Wen, J.; Yao, W.; An, T.; Li, Y. Dynamic phasor modelling and operating characteristic analysis of Half-bridge MMC. In Proceedings of 2016 IEEE 8th International Power Electronics and Motion Control Conference (IPEMC-ECCE Asia), Hefei, China, 22–26 May 2016.
19. Mehrasa, M.; Pouresmaeil, E.; Zabihi, S.; Catalao, J.P.S. Dynamic model, control and stability analysis of MMC-HVDC transmission systems. *IEEE Trans. Power Deliv.* **2016**. [CrossRef]
20. Wang, C.; Ooi, B.T. Incorporating deadbeat and low-frequency harmonic elimination in modular multilevel converters. *IET Gener. Transm. Distrib.* **2015**, *9*, 369–378. [CrossRef]
21. Harnefors, L.; Antonopoulos, A.; Norrga, S.; Angquist, L.; Nee, H.P. Dynamic analysis of modular multilevel converters. *IEEE Trans. Ind. Electron.* **2013**, *60*, 2526–2537. [CrossRef]
22. Darus, R.; Pou, J.; Konstantinou, G.; Ceballos, S.; Agelidis, V.G. Circulating current control and evaluation of carrier dispositions in modular multilevel converters. In Proceedings of the 2013 IEEE Energy Conversion Congress and Exhibition (ECCE Asia), Melbourne, Australia, 3–6 June 2013.
23. Pou, J.; Ceballos, S.; Konstantinou, G.; Agelidis, V.G.; Picas, R.; Zaragoza, J. Circulating current injection methods based on instantaneous information for the modular multilevel converter. *IEEE Trans. Ind. Electron.* **2015**, *62*, 777–788. [CrossRef]
24. Li, Y.; Jones, E.A.; Wang, F. Circulating current suppressing control's impact on arm inductance selection for modular multilevel converter. *IEEE J. Emerg. Sel. Topics Power Electron.* **2017**, *5*, 182–188. [CrossRef]
25. Leon, J.I.; Kouro, S.; Franquelo, L.G.; Rodriguez, J.; Wu, B. The essential role and the continuous evolution of modulation techniques for voltage-source inverters in the past, present, and future power electronics. *IEEE Trans. Ind. Electron.* **2016**, *63*, 2688–2701. [CrossRef]
26. Konstantinou, G.; Pou, J.; Ceballos, S.; Darus, R.; Agelidis, V.G. Switching frequency analysis of staircase modulated modular multilevel converters and equivalent PWM techniques. *IEEE Trans. Power Deliv.* **2016**, *31*, 28–36. [CrossRef]
27. Konstantinou, G.; Ciobotaru, M.; Agelidis, V.G. Selective harmonic elimination pulse-width modulation of modular multilevel converters. *IET Power Electron.* **2013**, *6*, 96–107. [CrossRef]
28. Dahidah, M.S.A.; Konstantinou, G.; Agelidis, V.G. A review of multilevel selective harmonic elimination PWM: Formulations, solving algorithms, implementation and applications. *IEEE Trans. Power Electron.* **2015**, *30*, 4091–4106. [CrossRef]
29. Ilves, K.; Antonopoulos, A.; Staffan, N.; Nee, H.P. A new modulation method for the modular multilevel converter allowing fundamental switching frequency. *IEEE Trans. Power Electron.* **2012**, *27*, 3482–3494. [CrossRef]
30. Deng, Y.; Wang, Y.; Teo, K.H.; Harley, R.G. A simplified space vector modulation scheme for multilevel converters. *IEEE Trans. Power Electron.* **2016**, *31*, 1873–1886. [CrossRef]
31. Tu, Q.; Xu, Z. Impact of sampling frequency on harmonic distortion for modular multilevel converter. *IEEE Trans. Power Deliv.* **2011**, *26*, 298–306. [CrossRef]
32. Hu, P.; Jiang, D. A level-increased nearest level modulation method for modular multilevel converters. *IEEE Trans. Power Electron.* **2015**, *30*, 1836–1842. [CrossRef]
33. Rohner, S.; Bernet, S.; Hiller, M.; Sommer, R. Modulation, losses, and semiconductor requirements of modular multilevel converters. *IEEE Trans. Ind. Electron.* **2010**, *57*, 2633–2642. [CrossRef]
34. Li, Z.; Wang, P.; Zhu, H.; Chu, Z.; Li, Y. An improved pulse width modulation method for chopper-cell-based modular multilevel converters. *IEEE Trans. Power Electron.* **2012**, *27*, 3472–3481. [CrossRef]
35. Hagiwara, M.; Akagi, H. Control and experiment of pulse width-modulated modular multilevel converters. *IEEE Trans. Power Electron.* **2009**, *24*, 1737–1746. [CrossRef]
36. Lu, S.; Yuan, L.; Li, K.; Zhao, Z. An improved phase-shifted carrier modulation scheme for a hybrid Modular Multilevel Converter. *IEEE Trans. Power Electron.* **2017**, *32*, 81–97. [CrossRef]
37. Saeedifard, M.; Iravani, R. Dynamic performance of a modular multilevel back-to-back HVDC system. *IEEE Trans. Power Deliv.* **2010**, *25*, 2903–2912. [CrossRef]

38. Darus, R.; Konstantinou, G.; Pou, J.; Ceballos, S.; Agelidis, V.G. Comparison of phase-shifted and level-shifted PWM in the modular multilevel converter. In Proceedings of the International Power Electronics Conference (IPEC–ECCE Asia), Hiroshima, Japan, 18–21 May 2014.

39. Mei, J.; Shen, K.; Xiao, B.; Tolbert, L.M.; Zheng, J. A new selective loop Bias mapping phase disposition PWM with dynamic voltage balance capability for modular multilevel converter. *IEEE Trans. Ind. Electron.* **2014**, *61*, 798–807. [CrossRef]

40. Darus, R.; Pou, J.; Konstantinou, G.; Ceballos, S.; Picas, R.; Agelidis, V.G. A modified voltage balancing algorithm for the modular multilevel converter: Evaluation for staircase and phase-disposition PWM. *IEEE Trans. Power Electron.* **2015**, *30*, 4119–4127. [CrossRef]

41. Fan, S.; Zhang, K.; Xiong, J.; Xue, Y. An improved control system for modular multilevel converters with new modulation strategy and voltage balancing control. *IEEE Trans. Power Electron.* **2015**, *30*, 358–371. [CrossRef]

42. Ilves, K.; Antonopoulos, A.; Norrga, S.; Nee, H.P. Steady-state analysis of interaction between harmonic components of arm and line quantities of modular multilevel converters. *IEEE Trans. Power Electron.* **2012**, *27*, 57–68. [CrossRef]

43. Li, B.; Yang, R.; Xu, D.; Wang, G.; Wang, W.; Xu, D. Analysis of the phase-shifted carrier modulation for modular multilevel converters. *IEEE Trans. Power Electron.* **2015**, *30*, 297–310. [CrossRef]

44. Tu, Q.; Xu, Z.; Xu, L. Reduced switching-frequency modulation and circulating current suppression for Modular Multilevel Converters. *IEEE Trans. Power Deliv.* **2011**, *26*, 2009–2017.

45. Holmes, D.G.; Lipo, T.A. *Pulse Width Modulation for Power Converters: Principles and Practice*; IEEE Press: Piscataway, NJ, USA, 2003.

46. McGrath, B.P.; Holmes, D.G. An analytical technique for the determination of spectral components of multilevel carrier-based PWM methods. *IEEE Trans. Power Deliv.* **2002**, *49*, 847–857. [CrossRef]

47. McGrath, B.P.; Holmes, D.G. Multicarrier PWM Strategies for Multilevel Inverters. *IEEE Trans. Ind. Electron.* **2002**, *49*, 858–867. [CrossRef]

Article

Impedance Decoupling in DC Distributed Systems to Maintain Stability and Dynamic Performance

Ahmed Aldhaheri * and Amir Etemadi

Department of Electrical and Computer Engineering, The George Washington University, Washington, DC 20052, USA; etemadi@email.gwu.edu
* Correspondence: aaldhaheri@gwmail.gwu.edu; Tel.: +1-202-372-6444

Academic Editor: Birgitte Bak-Jensen
Received: 10 February 2017; Accepted: 28 March 2017; Published: 2 April 2017

Abstract: DC distributed systems are highly reliable and efficient means of delivering DC power or adopting renewable energy resources. However, DC distributed systems are prone to instability and dynamic performance degradation due to the negative incremental input impedance of DC-DC converts. In this paper, we propose a generic method to eliminate the impact of the negative input impedance on DC systems by shaping the source output impedance such that its bode-plot is restricted in the area that is contained below the product of the source's duty ratio and its characteristic impedance. The performance deterioration originates whenever the output impedance of the source exceeds, in magnitude, the input impedance of the load converter due to deficiency in stability margins. Hence, confining the impedance in the proposed region helps decouple the interaction between the converters and preserve their own dynamic performances. The proposed method was proven by analytical analysis, time-based simulation, and practical experiments. All of their outcomes were in agreement, proving the effectiveness of the proposed method in preserving the dynamic performance of distributed systems.

Keywords: active damping; DC distributed systems; dynamic performance; impedance decoupling; impedance overlap; minor loop gain; non-causal system; stability margins

1. Introduction

DC distributed power systems are a remarkable application of power electronics and include a wide range of applications, such as in DC microgrids, motor drive systems, hybrid vehicles, aircrafts, ships, submarines, and satellites [1,2]. Most such applications are size-limited, so they require highly reliable power sources with high efficiency and power density based on DC distributed systems, to fulfill their power requirement [3]. A typical distributed DC system consists of a DC-DC load converter connected in series with a DC-DC source converter or an input filter, as illustrated in Figure 1 [4], which is also referred to as a cascaded DC system [5]. Each converter in cascaded systems represents a module that can be designed individually and then integrated with the rest of the components of the system [6]. Thus, cascaded systems are modular, scalable, and capable of meeting various load requirements, which adds to their attractiveness.

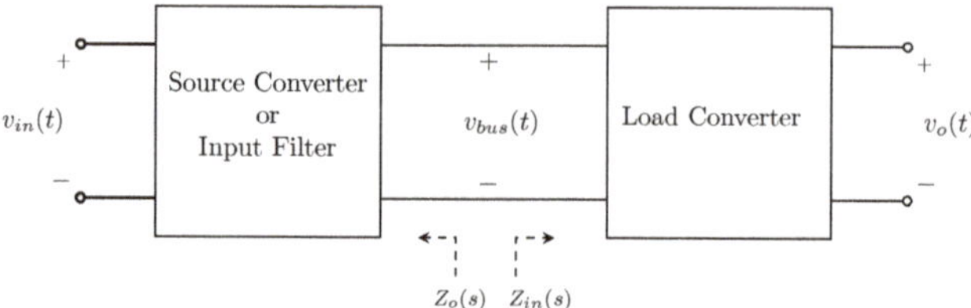

Figure 1. A typical cascaded system configuration illustrating the source output impedance (Z_o) and the load input impedance (Z_{in}).

Despite these appealing characteristics of cascaded DC systems, they are susceptible to dynamic performance degradation and instability problems. These problems arise because the load converter tightly regulates its output current or voltage, regardless of the voltage or current variations at the DC bus [7–9]. As a consequence, the load converter supplies its load with constant power, so the load converter acts as a power sink attached to the source [10,11]. The small-signal ac analysis of DC-DC converters shows that they have negative input impedance within the bandwidth of their controllers. Therefore, the source converter or the input filter sees the load converter as a negative impedance [12], whereas they are designed to supply non-negative resistive loads [13].

The performance of cascaded systems was first investigated by R.D. Middlebrook in the 1970s. He studied a cascaded system that consisted of a load converter with an input filter. His study concluded that adding the input filter reduces the relative stability margins of the load converter. If the output impedance magnitude ($|Z_o(s)|$) of the filter exceeded the input impedance of the load ($|Z_{in}(s)|$), the system would become unstable [14]. On the other hand, $Z_o(s)$ represents the closed-loop output impedance, if the source was a DC-DC converter. The ratio of the impedances is referred to as the minor loop gain ($T_m(s) = Z_o(s)/Z_{in}(s)$), and plays an important role in stability analysis of the overall system.

Several approaches and criteria have been proposed in the literature to ensure stability and proper dynamic performance of cascaded systems. R.D. Middlebrook proposed a criterion that confines the polar plot of $T_m(s)$ within a circle with a radius of $1/GM$ in order to ensure stability without degradation in the dynamic performance, where GM is the gain margin of $T_m(s)$ [14]. The criterion is artificially stringent [8]; it requires high capacitance at the DC bus to be fulfilled. Therefore, other criteria such as gain margin phase margin (GMPM) [15], the opposing argument [16], Energy Source Analysis Consortium (ESAC) [17], three-steps [18], and passivity-based [19] were introduced to relax the conservativeness of the Middlebrook criterion, or to expand its approach to cover a load system that consists of multiple load modules. All criteria are sufficient to ensure stability and/or dynamic performance [20]. Applying the Nyquist criterion to $T_m(s)$ is the only necessary and sufficient condition to ensure both stability and dynamic performance, because all criteria presume that every converter in the system is standalone stable except the three-steps criterion, which defines its minor loop gain differently. Since every converter is standalone stable, the polar plot of $T_m(s)$ must not encircle $(-1, j0)$—in the complex plane— in order to ensure stability [21].

The Middlebrook, opposing argument, GMPM, and ESAC are design-oriented criteria [8]. They assume that the output impedance of the source converter is known; hence, the load system must be designed accordingly [22]. This assumption limits the modularity feature of cascaded systems because the dynamic performance would not be guaranteed for any other load converters. Thus, passive damping methods were introduced to damp cascaded systems using bulky passive components connected to the DC bus [7]. Nevertheless, they incur extra power loss, and increase both the system size and weight. As a result, the efficiency, reliability, and power-density of the cascaded systems are compromised.

In order to tackle the drawbacks of passive damping methods, active damping methods have become a popular substitute. The system stability and/or dynamic performance are preserved by modifying the control loop of the source or the load converters. These methods do not typically incur extra power losses; however, some of them reduce the power-density of cascaded systems, such as the solution proposed in [23]. Several noticeable limitations or disadvantages are observed in the proposed solutions in the literature, which can be classified into ensuring stability with poor dynamics (i.e., long settling time) [24], limiting the controller application to a single converter topology (e.g., buck converters) [25,26], and inability to handle multiple converter load systems [27–30].

In this paper, we propose an active damping method that preserves both the stability and dynamic performance of cascaded systems. The approach is compatible with any minimum-phase DC-DC converter configuration and with any linear feedback control scheme. It is based on shaping the output impedance of the source converter such that its bode-plot is consistently less than the region below the product of the source characteristic impedance and its duty ratio. In addition, a quantifiable approach is proposed to shrink the impedance in that area in order to preserve the dynamic performance without artificial conservativeness.

This paper discusses the dynamic performance of cascaded systems in Section 2. Then, the proposed controller is demonstrated in Section 3. Section 4 illustrates the proposed reshaping method of the source output impedance. Next, a prototype cascaded system is analytically discussed in Section 5, and the analytical results are validated by simulations and experiments in Section 6. Finally, Section 7 sums up the main conclusions and outcomes.

2. Analysis of Cascaded System Dynamics

2.1. Dynamic Performance

The impact of the negative input impedance on cascaded systems can be studied using the canonical model [12] of a single-load-single-source, as shown in Figure 1. For such a system, the input-to-output voltage relationship can be described as

$$\frac{\tilde{v}_o}{\tilde{v}_{in}} = \frac{G_{vg,S}G_{vg,L}}{(1+T_S)(1+T_L)(1+T_m)} \tag{1}$$

where $\tilde{v}_{in}$ is the input voltage of the source converter, $\tilde{v}_o$ is the output voltage of the load converter, and $G_{vg,S}$ and $G_{vg,L}$ are the input-to-output voltage transfer functions of the source and the load converters, respectively. T_S and T_L denote the voltage loop gains of the source and load converters, respectively. Interconnecting the converters adds more poles to the system due to the added term $(1+T_m)$ in Equation (1), which alters the overall dynamic response of the system [31]. According to Equation (1), if $|T_m| \ll 1$, each converter of the cascaded system will operate as it was individually designed. Hence, the Middlebrook criterion mathematically describes the change in the load converter control-to-output transfer function ($G_{vd,L}$) as:

$$G_{vd,L}^S = G_{vd,L} \frac{1 + \overbrace{\dfrac{Z_o}{Z_{in}}}^{T_m}}{1 + \dfrac{Z_o}{Z_{in,o}}} \tag{2}$$

where $G_{vd,L}^S$ is the source-affected control-to-output transfer function of the load, and $Z_{in,o}$ is the load input open-loop impedance. In order to minimize the loading impact and eliminate the source-load dynamic coupling,

$$\begin{cases} |Z_o| \ll |Z_{in}| \\ |Z_o| \ll |Z_{in,o}| \end{cases} \tag{3}$$

should hold for the entire range of frequency according to the Middlebrook criterion. Consequently, the extra poles in (1) will be eliminated and $G^S_{vd,L} \approx G_{vd,L}$. Thus, each converter will operate as initially designed [32]. The terms $|Z_o/Z_{in}|$ and $|Z_o/Z_{in,o}|$ are approximately equal to each other within the load controller bandwidth (f_{L_BW}) [33,34]. Diminishing $|T_m|$ would preserve the dynamic performance of the load converter.

2.2. Impedance Interaction and Instability

T_m is crucial to ensuring both stability and dynamic performance. In order to ensure the stability of the system, $(1 + T_m)$ in (1) must have no zeros in the right-half-plane (RHP). These undesirable zeros would occur [20] if and only if:

$$\begin{cases} \left|\dfrac{Z_o}{Z_{in}}\right| \geq 1 \\ \varphi(Z_o) - \varphi(Z_{in}) \geq 180° \end{cases} \tag{4}$$

where $\varphi(Z_o)$ and $\varphi(Z_{in})$ are the phase angles of source output and load input impedances, respectively. Figure 2 shows typical plots of Z_o and Z_{in} for $f_{L\_BW} = 1.60$ kHz. Any impedance overlap that occurs within the controller bandwidth would satisfy (4) because $\varphi(Z_{in}) = -180°$ and $|Z_o/Z_{in}| > 1$, which implies that the overall cascaded system will have unstable poles.

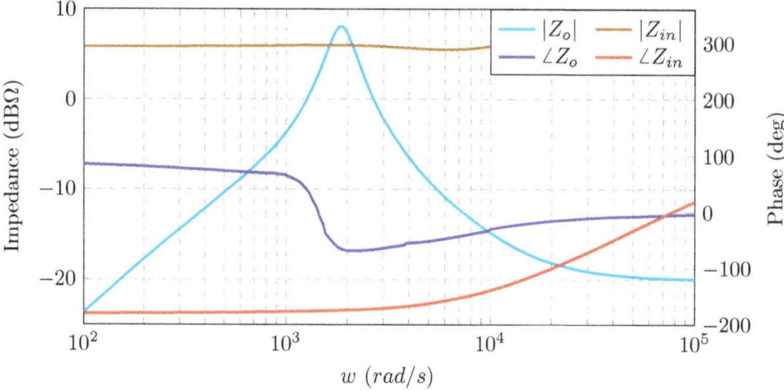

Figure 2. Typical plots of the source output impedance $|Z_o(s)|$ and the load input impedance $|Z_{in}(s)|$, depicting an impedance interaction.

2.3. The Source Performance

The Middlebrook criterion was initially developed to study the impact of the input filter on the performance of DC-DC converters. Replacing the filter with a source DC-DC converter has an impact on the load converter, as described in (2). On the other hand, the loop gain of the source converter is modified by the load converter, due to the loading impact given by [32].

$$T^L_S = \frac{T_S}{1 + T_m(1 + T_S)} \tag{5}$$

where T^L_S is the loaded loop gain of the source converter. Similarly, diminishing $|T_m|$ would preserve the dynamic performance of the source because $T^L_S \approx T_S$. In (5), T^L_S will have RHP poles if and only if T_m satisfies (4).

3. The Proposed Controller

In order to preserve the dynamic performance of voltage mode controlled converters in cascaded systems, the control loop should not change the relationship between the reference voltage ($\tilde{v}_{ref}$) and the output voltage of the converter ($\tilde{v}_o$), so their relationship has to be preserved as

$$\frac{\tilde{v}_o}{\tilde{v}_{ref}} = \frac{G_c G_{PWM} G_{vd,s}}{1 + G_c G_{PWM} G_{vd,s}} = \frac{T_S}{1 + T_S} \tag{6}$$

where G_c is the controller transfer function, and G_{PWM} is the pulse-width modulator. In addition, the controller must be capable of actively shaping the output impedance in order to eliminate the impedance interaction between the source and the load converters. Thus, we used the controller topology that is introduced in [35,36], which modified a conventional control scheme, as in Figure 3a, to the controller that is shown in Figure 3b. Then, we modified it to comply with our proposal.

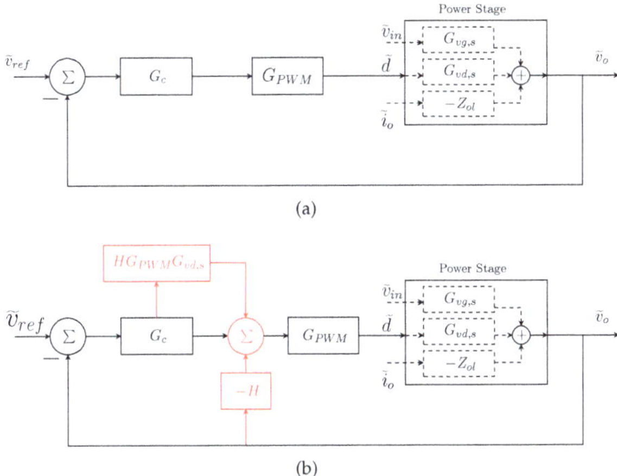

Figure 3. The block diagram of (**a**) the conventional controller; and (**b**) the modified controller emphasizing the modifications in red. PWM: pulse-width modulator.

The output voltage of a standalone converter implementing this controller (displayed in Figure 3b) is expressed as:

$$\tilde{v}_o = \tilde{v}_{ref} \frac{T_S}{1 + T_S} - \tilde{i}_o \frac{Z_o}{1 + HG_{PWM}G_{vd,s}} + \tilde{v}_{in} \frac{G_{vg,s}}{(1 + HG_{PWM}G_{vd,s})(1 + T_S)} \tag{7}$$

where H is the transfer function that is used to shape Z_o, and Z_o is derived using the open-loop output impedance (Z_{ol}) as $Z_o = Z_{ol}/(1 + T_s)$. The variations in the input voltage ($\tilde{v}_{in}$) can be ignored, because the source voltage is assumed to be ideal. It is noteworthy that we aim to preserve (7), because once the source converter is loaded, the dynamics of the system will be changed by substituting (5) into (7), resulting in the following:

$$\tilde{v}_o = \tilde{v}_{ref} \frac{T_S^L}{1 + T_S^L} - \tilde{i}_o \frac{Z_o}{1 + HG_{PWM}G_{vd,s}} \tag{8}$$

The dynamics of the system in (6) would have been modified if $|Z_o| \ll |Z_{in}|$ is violated according to (8). In order to eliminate the interaction, we propose to shape the output impedance according to:

$$Z_{o,new} = \frac{Z_o}{1 + \lambda} \tag{9}$$

where $Z_{o,new}$ is the reshaped output impedance of the source, and λ is a constant (its selection method is illustrated in the next section). Reducing $|Z_o|$ as proposed in (9) is challenging because $(1 + HG_{PWM}G_{vd})$ in (8) is a transfer function whose magnitude should be converted into a constant that equals $1 + \lambda$. Choosing $H = \lambda G_{PWM}^{-1}G_{vd,s}^{-1}$ is not a practical choice because the reciprocation results in a transfer function that has more zeros than poles; $G_{vd,s}$ is a strictly proper transfer function for any DC-DC converter. A transfer function of more zeros than poles is non-realistic. In control theory, such transfer functions describe non-causal systems because their current output depends on their future outcomes. To overcome this impediment, we propose to realize $G_{vd,s}^{-1}$ within a certain frequency band using low-pass filters. The transfer function of a low-pass filter (G_F) is described [37] as:

$$G_F = \frac{1}{(1 + \frac{s}{w_c})^\alpha} \tag{10}$$

where w_c is the cut-off frequency of the filter and α is a constant that dictates the order of the low-pass filter to be used. The order of the filter is chosen to ensure $\lambda G_F G_{PWM}^{-1}G_{vd,s}^{-1}$ as a strictly proper transfer function. For instance, a buck converter has two storage elements (a capacitor and an inductor), so its $G_{vd,s}$ will be of second order; α must be at least equal to 2. Other converters, like the minimum-phase fourth-order buck DC-DC converter [38] have four storage elements; therefore, α must be at least 4. Setting:

$$H = \lambda G_F G_{PWM}^{-1}G_{vd,s}^{-1} \tag{11}$$

and then substituting (11) in (8) changes the dominator of the impedance $(1 + HG_{PWM}G_{vd,s})$ to:

$$1 + \lambda G_F \; \cancel{G}_{PWM}^{-1} \; \cancel{G}_{PWM} \; \cancel{G}_{vd,s}^{-1} \; \cancel{G}_{vd,s} \tag{12}$$

which can be interpreted as:

$$1 + \lambda \quad \omega \leq \omega_c \tag{13}$$

Thus, the output voltage of the proposed controller can be re-expressed as:

$$\tilde{v}_o \approx \tilde{v}_{ref} \frac{T_S^L}{1 + T_S^L} - \tilde{i}_o \frac{Z_o}{1 + \lambda} \quad \omega \leq \omega_c \tag{14}$$

Figure 4a shows the reshaped output impedance $(Z_{o,new})$, using a low-pass filter with $\omega_c = 10^5$ rad/s, subjected to different values of λ. With these proposed modifications to the controller, the magnitude of the source output impedance can be precisely controlled within the filter bandwidth; however, beyond ω_c, the magnitude of $Z_{o,new}$ consistently converges to coincide with the magnitude of Z_o.

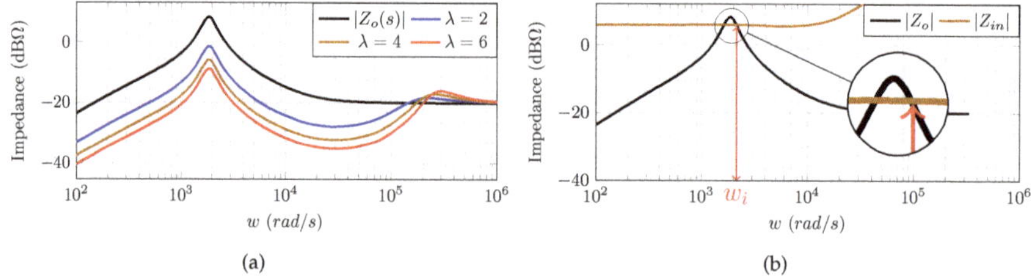

Figure 4. (**a**) The shaped output impedance magnitudes using a low-pass filter with $\omega_c = 10^5$ rad/s compared to the original output impedance $|Z_o|$; and (**b**) demonstrates the selection of ω_i.

Thus, in order to effectively reshape Z_o such that the source and the load converters are decoupled, ω_c should be greater than ω_i, as shown in Figure 4b. Otherwise the shaping will fail to improve the dynamic performance. Ultimately, the implemented controller with the proposed modifications is shown in Figure 5.

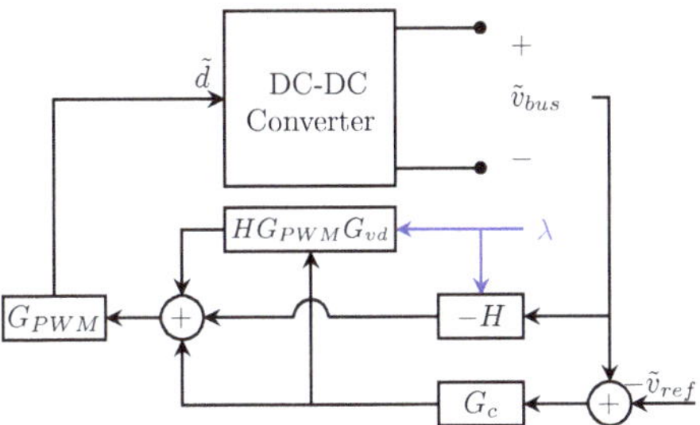

Figure 5. The proposed controller, emphasizing the modification to the controller in blue.

4. Impedance Reshaping for Decoupling the Source-Load Interaction

The impedance overlap in cascaded systems tends to occur in the vicinity of the peak impedance of the source converter [10]. The dynamic response of the system is compromised due to the impedance interaction around that region. The peak impedance occurs either at the cut-off frequency of the source controller (f_{s_BW}) due to low phase margins, or at the source resonant frequency ($f_o = 1/\sqrt{LC}$) if the the bandwidth of the controller was less than the resonance frequency of the converter. In the latter case, the peaking is more severe because DC-DC converters are designed to be highly efficient, which lightly damps the converter. The peaking of the output impedance can be expressed [39] as:

$$|Z_{o,max}| = \frac{|Z_{ol}(f_{s\_BW})|}{\sqrt{(2 - 2\cos(\phi_m))}} \quad f_{s\_BW} > f_o \tag{15}$$

$$|Z_{o,max}| = \frac{|Z_{ol}(f_o)|}{1 - 10^{(-\Psi/20)}} \quad f_{s\_BW} < f_o \tag{16}$$

where ϕ_m and Ψ are the phase and the gain margins of the source controller, respectively. A low gain or phase margin causes severe peaking, as depicted in Figure 6. For instance, a phase margin of $30°$ adds 5.8 dB to the output impedance, while a gain margin of 5 dB adds 7.20 dB.

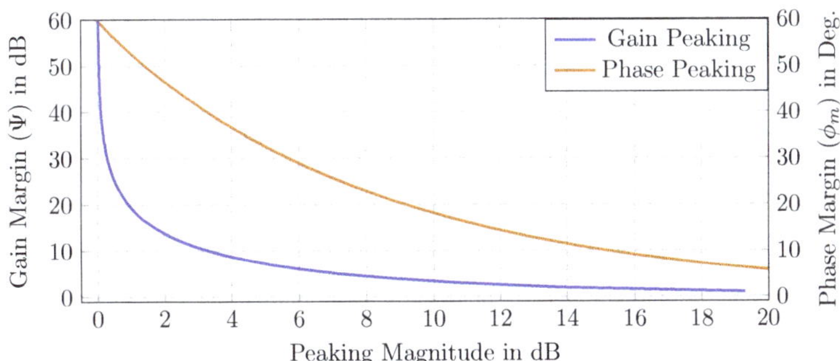

Figure 6. The peaking caused by the low relative stability margins.

Reducing the magnitude of Z_o below Z_{in} to ensure stability is mathematically quantifiable because Z_{in}, within the load controller bandwidth, can be described as [40]:

$$Z_{in} = -\frac{V_{bus}^2}{P_o} \quad \forall f < f_{L\_BW} \tag{17}$$

where V_{bus} is the DC bus voltage, and P_o is the power consumed by the load. Hence, the system will be stable as long as $|Z_o| < |Z_{in}|$. However, the system dynamic performance would still be compromised if $|Z_o| \lll |Z_{in}|$, so we are proposing a method to quantify $(\lll)$. The earlier mentioned criteria are used to design the load input impedance. Yet, we propose to shape the output impedance of the source converter to be confined in the region that is below the characteristic impedance of the source converter $(Z_x = D\sqrt{L/C})$, as demonstrated in Figure 7. Shrinking the impedance in the proposed region ensures the dynamic performance of the system, regardless of the attached load. The peaking is more pronounced as the relative stability margins decrease; as a result, the peak impedance will exceed the stability limit. The dynamic performance can be ensured if $|Z_{o,max}| < |Z_x|$, so $|Z_o|$ can be reduced to the proposed region of Figure 7 by dividing Z_o by:

$$\rho = \frac{|Z_{o,max}|}{|Z_x|} \tag{18}$$

where ρ is the reduction factor in the magnitude of Z_o, and $Z_{o,max}$ is the peak impedance of Z_o. Hence, equating the dominators of (Z_o/ρ)—which falls in the proposed region—with (9) yields:

$$\lambda = \rho - 1 \tag{19}$$

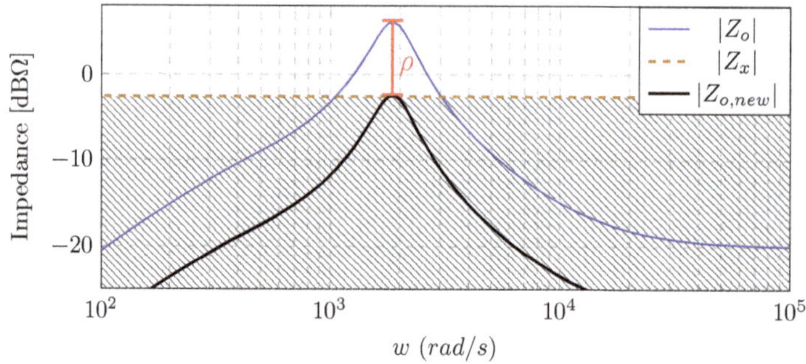

Figure 7. Plots of $|Z_o|$ and $|Z_x|$ showing the reduction factor ρ with the shaped impedance $|Z_{o,new}|$, where $|Z_{o,new}|$ is proposed to be in the hatched area.

Mathematical Validation of Performance Preservation

The shaped output impedance of the source can be expressed as:

$$Z_{o,new} = \frac{Z_o}{\rho} = \frac{Z_o\, D}{|Z_{o,max}|}\sqrt{\frac{L}{C}} \tag{20}$$

As a result, the new minor loop gain (T_m^{new}) is given by:

$$T_m^{new} = \frac{Z_{o,new}}{Z_{in}} = \frac{Z_o\, D}{Z_{in}|Z_{o,max}|}\sqrt{\frac{L}{C}} = \frac{T_m\, D}{|Z_{o,max}|}\sqrt{\frac{L}{C}} \tag{21}$$

so substituting (21) in (5) modifies T_S^L to T_S^{LR} as follows:

$$T_S^{LR} = \frac{T_S}{1 + \dfrac{T_m(1 + T_S)\, D}{|Z_{o,max}|}\sqrt{\frac{L}{C}}} \tag{22}$$

where T_S^{LR} is the reshaped loop gain by implementing the proposed shaping in (18).

In order to prove $T_S^{LR} \approx T_S$, the denominator of (22) is evaluated where $|Z_{o,max}|$ occurs, where:

$$|T_m| = |Z_{o,max}/Z_{in}|$$ (23)

$$\delta = |1 + T_S| = \begin{cases} \sqrt{(2 - 2\cos(\phi_m))} & f_{s\_BW} > f_o \\ 1 - 10^{(-\Psi/20)} & f_{s\_BW} < f_o \end{cases}$$ (24)

In addition, the characteristic equation can be rewritten using (18) as:

$$\sqrt{\frac{L}{C}} = \frac{|Z_{o,max}|}{\rho D}$$ (25)

Plugging $|Z_{in}|$, (23), (24) and (25) into (22) yields:

$$T_S^{LR} = \frac{T_S}{1 + \dfrac{\delta}{\rho V_{bus^2}} Z_{o,max} P_o}$$ (26)

By inspecting the second term of the denominator, the term $\delta/(\rho V_{bus}^2) \ll 1$, and hence, $1 + \delta Z_{o,max} P_o/(\rho V_{bus}^2)) \approx 1$. As a result, $T_S^{LR} \approx T_S$, which validates the proposed shaping method.

5. Theoretical Analysis

The effectiveness of the proposed solution to retain the dynamic performance of cascaded systems has been examined using a prototype that consists of two buck converters connected in series, as shown in Figure 8. The source converter tightly regulates the bus voltage at 7 V, while the load converter was designed to supply a resistive load of 16 W, with an output voltage of 4 V. G_{PWM} was chosen to be $1\ V^{-1}$. Each converter was devised to be standalone stable, as the polar plots of their voltage loop gains, T_S and T_L, imply in Figure 9. However, cascading the converters was not possible, because the overall system would suffer from instability as impedance overlap occurred, as shown in Figure 10a.

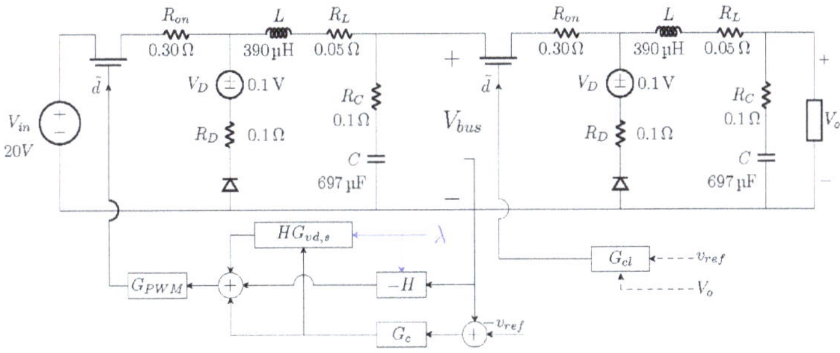

Figure 8. The prototype cascaded system used for analysis, simulation, and experiment.

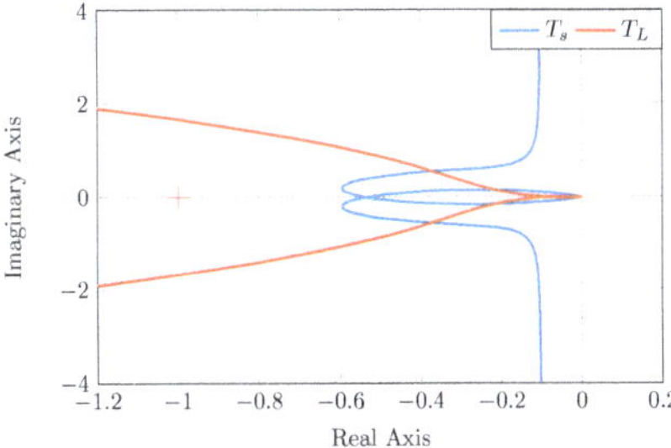

Figure 9. The loop gains of the source and the load, showing their standalone stability.

Impedance Reshaping and Performance Improvement

In order to preserve both the stability and the dynamic performance of cascaded systems, we propose to shape Z_o according to (20). The characteristic impedance of the source converter is 748.02 mΩ, and the peak impedance of the source is 11.5 dBΩ, which corresponds to 3.76 Ω. Thus, the resultant divisor factor ρ, using (18), is 14.5, which yields $\lambda = 13.5$. Using Figure 10a, ω_i was found to be 2.2×10^3 rad/s, so w_c was set to 10^5 rad/s. As buck converters have two storage elements, α was chosen to be 2 in order to guarantee that H is invertible. Plugging ω_c and α in (10) gives:

$$G_F = \frac{1}{\left(1 + \dfrac{s}{10^5}\right)^2} \tag{27}$$

so H, using (11), will be:

$$H = \frac{6.597 \times 10^6 s^2 + 7.612 \times 10^9 s + 2.427 \times 10^{13}}{s^3 + 2.143 \times 10^5 s^2 + 1.287 \times 10^{10} s + 1.435 \times 10^{14}} \tag{28}$$

As a result, the shaped output impedance of the source converter ($Z_{o,new}$) is substantially reduced compared to Z_o, which emphasizes the ability of the proposed method to decouple the impedances, as shown in Figure 10b.

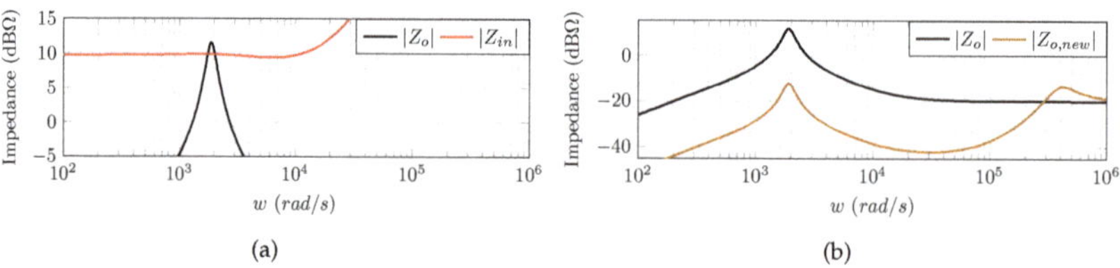

Figure 10. The impedance analysis of the prototype: (**a**) showing the interaction between $|Z_o|$ and $|Z_{in}|$; and (**b**) comparing $|Z_o|$ with the shaped impedance $|Z_{o,new}|$.

In order to show the improvement in the dynamic performance, Figure 11 compares the unity feedback closed loop response of T_S to its counterpart T_S^{LR}. The response of T_S^{LR} closely tracks its reference response, which highlights the effectiveness of the proposed shaping to stabilize and improve the dynamic performance of cascaded systems.

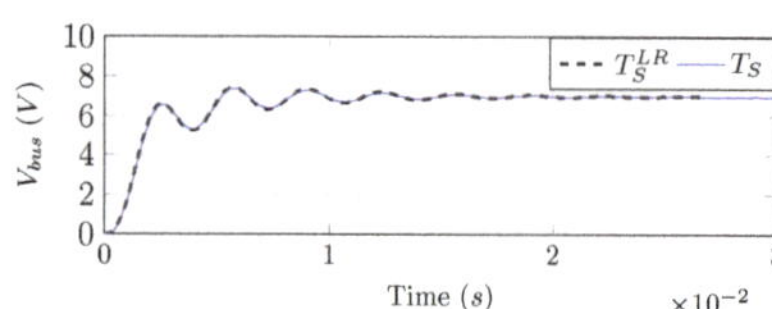

Figure 11. The unity feedback step response of T_S^{LR} compared to T_S.

6. Simulation and Experiment Case Studies

6.1. Time-Domain Simulations

The system illustrated in Figure 8 was simulated using PLECS Standalone Package (Plexim, Zurich, Switzerland) in order to evaluate the effectiveness of the proposed controller. Two test cases were carried out. The first test was conducted while the system was at the verge of instability; the load converter was supplying 8 W. Once the system reached steady state, the other 8 W load was added at $t = 0.15$ s. The second test was conducted by starting the system while the load converter was supplying the 16 W in order to compare its response to the theoretical response of T_S.

The system equipped with the conventional controller (as in Figure 3a) was tested in order to demonstrate the impact of the negative input impedance on the system dynamics. For the first test, the system was stable at the start-up with compromised dynamic response because the settling time was 50 ms compared to 20 ms in Figure 11. Connecting the other 8 W destabilized the bus voltage permanently, as shown in Figure 12a. Then, the second test was run; Figure 12b shows that the system was also unstable. These tests agree with the analytical analysis, which expected the instability of the system using the conventional controller.

The system was then equipped with the proposed controller as in Figure 5 or Figure 8. The first test was conducted, and the system reached the steady state in 20 ms. In addition, increasing the output power by 8 W had negligible impact on the system as Figure 13a depicts. Conducting the second test, as shown in Figure 13b, shows that the system response was similar to the of T_S in Figure 11, which proves the effectiveness of the proposed reshaping to stabilize and improve the dynamic performance of cascaded systems.

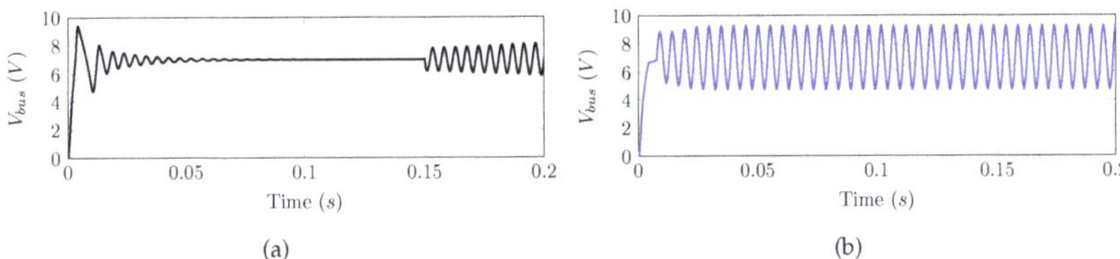

(a) (b)

Figure 12. The instability of the system with the conventional controller: (**a**) sequential loading; and (**b**) full loading.

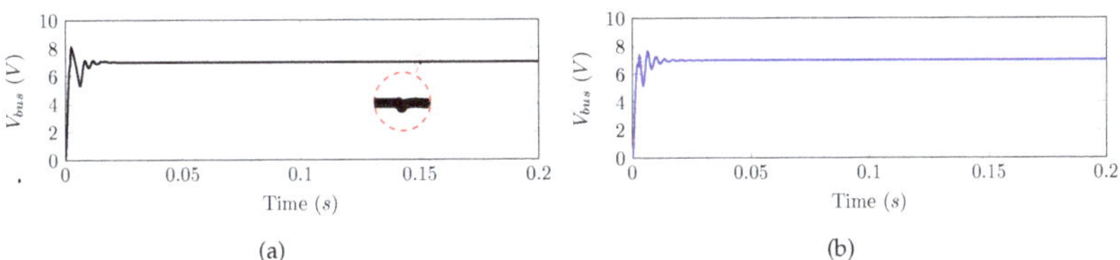

(a) (b)

Figure 13. The improved dynamic response using the proposed method: (**a**) sequential loading; and (**b**) full loading.

6.2. Experimental Cases

The prototype cascaded system was physically built as shown in Figure 14. The digital controller platform NIcRio 9024 (National Instruments, Austin, TX, USA) was used to control the output voltages of the converters, with a sampling rate of 30 Ksample/s. Similar to the simulation, the system with the conventional controller was tested using the two test procedures, where v_{bus} denotes the DC bus voltage. Increasing the load sequentially—as in the first test—destabilized the bus voltage as soon as the load power reached 16 W, as shown in Figure 15a. A similar result was yielded by starting up the system supplying the entire load, as shown in Figure 15b. Hence, the determined impact of the negative input impedance is obvious.

The proposed controller was implemented to demonstrate its effectiveness in preserving the dynamic performance of the cascaded system. The first test was run, and Figure 15c shows the improvement in the response at the start up, while increasing the load had no impact on the stability of the system. Moreover, the second test was conducted, and the response of the system at the start-up while supplying the entire load was identical to the simulated and theoretical responses of T_S as Figure 15d depicts. Hence, the experimental results have proven the efficacy of the proposed shaping in preserving the dynamic response of cascaded systems.

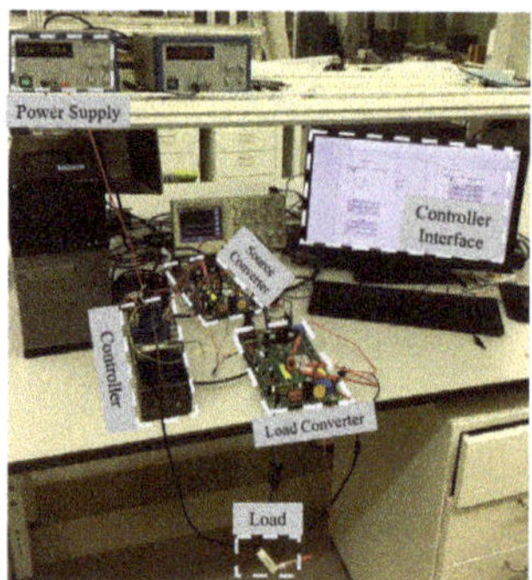

Figure 14. Experimental prototype set-up.

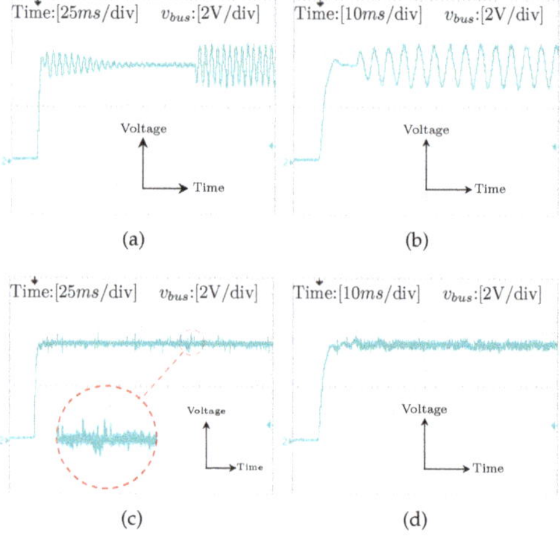

Figure 15. Experimental results: (**a**) The instability using the conventional controller by sequential loading; (**b**) The instability at full load with the conventional controller; (**c**) The improved performance using the proposed controller with sequential loading; and (**d**) The response of the system at full loading using the proposed controller.

7. Conclusions

In this paper, we proposed an active damping method to stabilize and retain the dynamic performance of cascaded systems by reshaping the impedance of the source converter. The magnitude of the shaped impedance was absolutely restricted to be in the region limited by the source characteristic impedance in order to prevent dynamic performance degradation. The proposed method is compatible with any linearized feedback control scheme and any DC-DC converter topology. The mathematical approach was developed in order to evaluate every parameter required to shrink the impedance of the source in the proposed region.

A prototype DC cascaded system was built and analyzed, where the performance of the system with and without the proposed method were compared. Analytical analyses, time-domain simulations, and practical experiments have proven the ability of the proposed method to decouple the interaction of the source and the load converters, and hence preserve their dynamic performances. In contrast, the conventional control method suffered from permanent oscillations at the DC bus.

Author Contributions: The authors have participated equally in this work. Analyses, simulations, and experiments were conducted and analyzed by both of the authors.

Conflicts of Interest: The authors declare no conflict of interest.

References

1. Dragicevic, T.; Lu, X.; Vasquez, J.C.; Guerrero, J.M. DC microgrids—Part I: A review of control strategies and stabilization techniques. *IEEE Trans. Power Electron.* **2016**, *31*, 4876–4891.
2. Xu, C.D.; Cheng, K.W.E. A survey of distributed power system—AC versus DC distributed power system. In Proceedings of the 2011 4th International Conference on Power Electronics Systems and Applications, Hong Kong, China, 8–10 June 2011.
3. Luo, S. A review of distributed power systems part I: DC distributed power system. *IEEE Aerosp. Electron. Syst. Mag.* **2005**, *20*, 5–16.
4. Du, W.; Zhang, J.; Zhang, Y.; Qian, Z. Stability criterion for cascaded system with constant power load. *IEEE Trans. Power Electron.* **2013**, *28*, 1843–1851.
5. Ahmadi, R. Dynamic modeling, stability analysis, and controller design for DC distribution systems. Ph.D. Thesis, Missouri University of Science and Technology, Rolla, MO, USA, 2013.
6. Riccobono, A.; Santi, E. Comprehensive review of stability criteria for DC power distribution systems. *IEEE Trans. Ind. Appl.* **2014**, *50*, 3525–3535.
7. Cespedes, M.; Xing, L.; Sun, J. Constant-power load system stabilization by passive damping. *IEEE Trans. Power Electron.* **2011**, *26*, 1832–1836.
8. Sudhoff, S.D.; Corzine, K.A.; Glover, S.F.; Hegner, H.J.; Robey, H.N. DC link stabilized field oriented control of electric propulsion systems. *IEEE Trans. Energy Convers.* **1998**, *13*, 27–33.
9. Emadi, A.; Ehsani, M.; Miller, J.M. *Vehicular Electric Power Systems: Land, Sea, Air, and Space Vehicles*; Marcel Dekker: New York, NY, USA, 2004.
10. Basso, C.P. *Designing Control Loops for Linear and Switching Power Supplies: A Tutorial Guide*; Artech House: Boston, MA, USA, 2012.
11. Ioinovici, A. *Power Electronics and Energy Conversion Systems*; Wiley Academic: Hoboken, New York, NJ, USA, 2013.
12. Erickson, R.W.; Dragan, M. *Fundamentals of Power Electronics*; Kluwer Academic: Secaucus, NJ, USA, 2001.
13. Ang, S.S. *Power-Switching Converters*; Dekker, M., Ed.; Marcel Dekker Inc.: New York, NY, USA, 1995.
14. Middlebrook, R.D. Input filter considerations in design and application of switching regulators. In Proceedings of the IEEE Industry Applications Society Annual Meeting, Chicago, IL, USA, 11–14 October 1976.
15. Wildrick, C.M.; Lee, F.C.; Cho, B.H.; Choi, B. A method of defining the load impedance specification for a stable distributed power system. In Proceedings of the 24th Annual IEEE Power Electronics Specialists Conference (PESC '93 Record), Seattle, WA, USA, 20–24 June 1993; pp. 826–832.
16. Feng, X.; Ye, Z.; Xing, K.; Lee, F.C.; Borojevic, D. Impedance specification and impedance improvement for DC distributed power system. In Proceedings of the 30th Annual IEEE Power Electronics Specialists Conference (PESC), Charleston, SC, USA, 1 July 1999; Volume 2, pp. 889–894.
17. Sudhoff, S.D.; Glover, S.F.; Lamm, P.T.; Schmucker, D.H.; Delisle, D.E. Admittance space stability analysis of power electronic systems. *IEEE Trans. Aerosp. Electron. Syst.* **2000**, *36*, 965–973.
18. Wang, X.; Yao, R.; Rao, F. Three-step impedance criterion for small-signal stability analysis in two-stage DC distributed power systems. *IEEE Power Electron. Lett.* **2003**, *1*, 83–87.
19. Riccobono, A.; Santi, E. A novel Passivity-Based Stability Criterion (PBSC) for switching converter DC distribution systems. In Proceedings of the 2012 Twenty-Seventh Annual IEEE Applied Power Electronics Conference and Exposition (APEC), Orlando, FL, USA, 5–9 February 2012; pp. 2560–2567.
20. Hankaniemi, M. Dynamical Profile of Switched-Mode Converter—Fact or Fiction? Ph.D. Thesis, University of Tampere, Tampere, Finland, 2007.
21. Suntio, T. *Dynamic Profile of Switched-Mode Converter: Modeling, Analysis and Control*; Wiley-VCH: Weinheim, Germany, 2009.
22. Sudhoff, S.D.; Crider, J.M. Advancements in generalized immittance based stability analysis of DC power electronics based distribution systems. In Proceedings of the 2011 IEEE Electric Ship Technologies Symposium, Alexandria, VA, USA, 10–13 April 2011.

23. Zhang, X.; Ruan, X.; Kim, H.; Tse, C.K. Adaptive active capacitor converter for improving stability of cascaded DC power supply system. *IEEE Trans. Power Electron.* **2013**, *28*, 1807–1816.

24. Wu, M.; Lu, D.D.C. A novel stabilization method of LC input filter with constant power loads without load performance compromise in DC microgrids. *IEEE Trans. Ind. Electron.* **2015**, *62*, 4552–4562.

25. Cai, W.; Yi, F.; Cosoroaba, E.; Fahimi, B. Stability optimization method based on virtual resistor and nonunity voltage feedback loop for cascaded DC-DC converters. *IEEE Trans. Ind. Appl.* **2015**, *51*, 4575–4583.

26. Zhang, L.; Ren, X.; Ruan, X. A bandpass filter incorporated into the inductor current feedback path for improving dynamic performance of the front end DC-DC converter in two-stage inverter. *IEEE Trans. Ind. Electron.* **2014**, *61*, 2316–2325.

27. Ahmadi, R.; Ferdowsi, M. Improving the performance of a line regulating converter in a converter-dominated DC microgrid system. *IEEE Trans. Smart Grid* **2014**, *5*, 2553–2563.

28. Zhang, X.; Zhong, Q.C.; Ming, W.L. Stabilization of a cascaded DC converter system via adding a virtual adaptive parallel impedance to the input of the load converter. *IEEE Trans. Power Electron.* **2016**, *31*, 1826–1832.

29. Zhang, X.; Ruan, X.; Zhong, Q.C. Improving the stability of cascaded DC/DC converter systems via shaping the input impedance of the load converter with a parallel or series virtual impedance. *IEEE Trans. Ind. Electron.* **2015**, *62*, 7499–7512.

30. Ahmadi, R.; Ferdowsi, M. Controller design method for a cascaded converter system comprised of two DC-DC converters considering the effects of mutual interactions. In Proceedings of the 2012 Twenty-Seventh Annual IEEE Applied Power Electronics Conference and Exposition (APEC), Orlando, FL, USA, 5–9 February 2012; pp. 1838–1844.

31. Pidaparthy, S.K.; Choi, B. Stability analysis of PWM converters connected to general load subsystems. In Proceedings of the 2015 9th International Conference on Power Electronics and ECCE Asia (ICPE-ECCE Asia), Seoul, Korea, 1–5 June 2015; pp. 1033–1040.

32. Li, P.; Lehman, B. Performance prediction of DC-DC converters with impedances as loads. *IEEE Trans. Power Electron.* **2004**, *19*, 201–209.

33. Rivetta, C.H.; Emadi, A.; Williamson, G.A.; Jayabalan, R.; Fahimi, B. Analysis and control of a buck DC-DC converter operating with constant power load in sea and undersea vehicles. *IEEE Trans. Ind. Appl.* **2006**, *42*, 559–572.

34. Zamierczuk, M.K.; Cravens, R.C.; Reatti, A. Closed-loop input impedance of PWM buck-derived DC-DC converters. In Proceedings of the IEEE International Symposium on Circuits and Systems (ISCAS '94), London, UK, 30 May–2 June 1994; Volume 6, pp. 61–64.

35. Cao, L.; Loo, K.H.; Lai, Y.M. Systematic derivation of a family of output-impedance shaping methods for power converters—A case study using fuel cell-battery-powered single-phase inverter system. *IEEE Trans. Power Electron.* **2015**, *30*, 5854–5869.

36. Cao, L.; Loo, K.H.; Lai, Y.M. Output-impedance shaping of bidirectional DAB DC-DC converter using double-proportional-integral feedback for near-ripple-free DC bus voltage regulation in renewable energy systems. *IEEE Trans. Power Electron.* **2016**, *31*, 2187–2199.

37. Manolakis, D.G.; Ingle, V.K. *Applied Digital Signal Processing: Theory and Practice*; Cambridge University Press: Cambridge, UK, 2011.

38. Veerachary, M. Analysis of minimum-phase fourth-order buck DC-DC converter. *IEEE Trans. Ind. Electron.* **2016**, *63*, 144–154.

39. Wildrick, C.M. Stability of Distributed Power Supply Systems. Ph.D. Thesis, Virginia Polytechnic Institute and State University, Blacksburg, VA, USA, 1993.

40. Ahmadi, R.; Paschedag, D.; Ferdowsi, M. Analyzing stability issues in a cascaded converter system comprised of two voltage-mode controlled DC-DC converters. In Proceedings of the 2011 Twenty-Sixth Annual IEEE Applied Power Electronics Conference and Exposition (APEC), Fort Worth, TX, USA, 6–11 March 2011; pp. 1769–1775.

Article

Consideration of Reactor Installation to Mitigate Voltage Rise Caused by the Connection of a Renewable Energy Generator

Yeonho Ok [1,2], Jaewon Lee [1] and Jaeho Choi [2,*]

1   Research Institute, Power 21 Corporation, Daejeon 305-509, Korea; oyh@power21.co.kr (Y.O.);
     ljw@power21.co.kr (J.L.)
2   School of Electrical Engineering, Chungbuk National University, Chungbuk 28644, Korea
*   Correspondence: choi@cbnu.ac.kr; Tel.: +82-43-261-2425; Fax: +82-43-276-7217

Academic Editor: Birgitte Bak-Jensen
Received: 13 January 2017; Accepted: 2 March 2017; Published: 10 March 2017

Abstract: This paper describes the detailed analysis of a reactor application for a power plant to mitigate the voltage rise of a distribution line (DL) caused by the connection of distributed resources (DRs). The maximum capacity of renewable energy generators (REGs) that meets the acceptable voltage rise of a DL and the necessary capacity of the reactor to mitigate that voltage rise according to the different types of REGs are analyzed. The re-coordination of a protection relay and the loss of generation revenue as well as the installation location of a reactor are described. Finally, the ON/OFF conditions of the reactor, such as the magnitudes of the grid voltage and generator voltage, and the duration time of the voltage rise are analyzed. As the voltage rise is mitigated and self-limited in small power plants, it is confirmed that the capacity of the DRs connected to the DL can be increased through a field demonstration.

Keywords: renewable energy; reactor; voltage rise; distribution line; distribution resource; grid connection; generator; power factor; coordination of protection relay

1. Introduction

The locations of power plants sourced by renewable energy, such as wind or solar power generation, are restricted by the applicable natural conditions. Therefore, it is necessary to consider how and where to connect them to the power grid. While the large-capacity power plants deliver power through a dedicated power line, the small-scale generators, such as distributed resources (DRs), are linked directly to the close distribution line (DL) to decrease the transmission loss and the construction cost. For example, in South Korea, a 10 MW or larger power plant has to be linked through a dedicated line [1]. On the other hand, renewable energy generators (REGs) which are generally smaller than 10 MW are connected to the DL directly. This paper describes what should be considered when connecting the small-scale REGs to the DL directly.

To connect REGs to the DL, a technological understanding of anti-islanding and power quality solutions against the harmonics and line voltage fluctuations, etc., are necessary [2]. When any faults occur in the grid, REGs should stop supplying power to the grid line using the appropriate anti-islanding method. Nevertheless, with the spread of microgrid technologies, the operation of REGs in stand-alone mode in all types of line fault conditions has attracted recent attention [3]. The voltage fluctuations are caused mostly by load variations in the grid, and the application of a static synchronous compensator (STATCOM) has been studied to improve voltage regulation [4–6]. Harmonics problems have become more serious with the increased use of power electronics equipment in power utility applications. The harmonic distortion of currents and voltages occurs at the Direct Current/Alternating

Current (DC/AC) inverter that feeds power from the REGs and the AC/DC converter that supplies power to the DC load. To reduce these harmonics problems, passive filters are installed conventionally as a simple method, but the active power filter (APF) has also been applied as an improved form of technology [7]. Although the performance of STATCOM and APF is good, the system is expensive and the controller is complicated.

The increase in REG output influences the voltage rise of the DL, and the voltage rise has a direct influence on the load linked to the DL. This voltage rise caused by the interconnection of DRs has been standardized by the IEEE Std. 1547 and there have been many studies aiming to solve grid voltage rise problems [8–11]. The rate of grid voltage rise related to DRs is limited to 2% in South Korea [1]. The allowable capacity of DRs for connecting to DL can be calculated, but it applies less capacity than the available renewable energy to meet the regulation of the voltage rise. Therefore, to maximize the utilization of renewable energy, it is important to suppress the voltage rise in accordance with the capacity increase of DRs using some technologies.

The grid voltage rise in the DL is dependent on the capacity of DRs, the characteristics of loads, and the line parameters of DL. The voltage rise of DL could be adjusted by controlling the load [12]. This was based on the policy that customers adjust loads themselves according to the voltage of the DL, with DRs to regulate the voltage, and the benefit of electricity charges was given to customers who accepted this policy. However, it was difficult to apply this system in a normal DL because of the separate electricity rate system. Some studies have been performed to mitigate the voltage rise with the tap changer of transformers of DL with DRs [13–16], but these are not directly applicable to systems that attempt to solve the voltage rise problems inside the power plant. Evaluations of the allowable REG capacities to connect to the DL have been carried out [17–21], but these studies were usually applied to take less capacity than the available energy. Therefore, they could not satisfy the original intention to use the available energy as much as possible.

To mitigate the voltage rise of the DL without downsizing the generating capacity, additional facilities, such as the Static Var Compensator (SVC) [22–24], STATCOM [25–28], and reactors, have been considered for installation in a power plant. The SVC and STATCOM are generally applied in large-capacity power plants over 60 MVA, but their system costs are expensive compared to the construction costs of small-capacity REGs [29,30]. A reactor has been installed in a small-capacity generator system, but a detailed study of the installation capacity, location, and coordination of protection relays has not been performed [31,32]. Therefore, its applications have been very limited in spite of its merits as a simple and cheap method.

This paper describes the detailed analyses of a reactor application of a power plant to mitigate the voltage rise due to the connection to DL of REGs. Firstly, the maximum capacity of REGs that meet the acceptable voltage rise of the DL without a reactor and the necessary capacity of the reactor to mitigate the voltage rise of the DL according to each type of REG is analyzed. Secondly, the installation locations of the reactor are examined by the comparative analysis of the advantages and disadvantages at each location considering the necessity of re-coordination of a protection relay and the loss of generation revenue according to the installation locations. Finally, the ON/OFF conditions of the reactor, such as the magnitude of the grid voltage and the generator output and the duration time of the set values, are investigated.

The effects of installing of the reactor are analyzed through field application tests by comparing the vector diagrams, the daily connection voltage data, and the monthly average grid voltage data before and after the installing the reactor.

2. Considerations for Reactor Installation

The voltage rise of DL with REGs depends on the load condition of DL. Therefore, the no load condition of DL is assumed in this study, which provides the maximum voltage rise with REGs [23].

2.1. Capacity

The capacity of REGs to be installed in DL is dependent on the capacity of the existing DRs in DL, the allowable load capacity of DL, the available voltage rise rate of DL, and the type of generators to be newly added. Figure 1 shows the single line and vector diagrams based on the connection point voltage when a REG is added to DL [5], and the voltage rise rate and DL voltage are described as Equations (1)–(4).

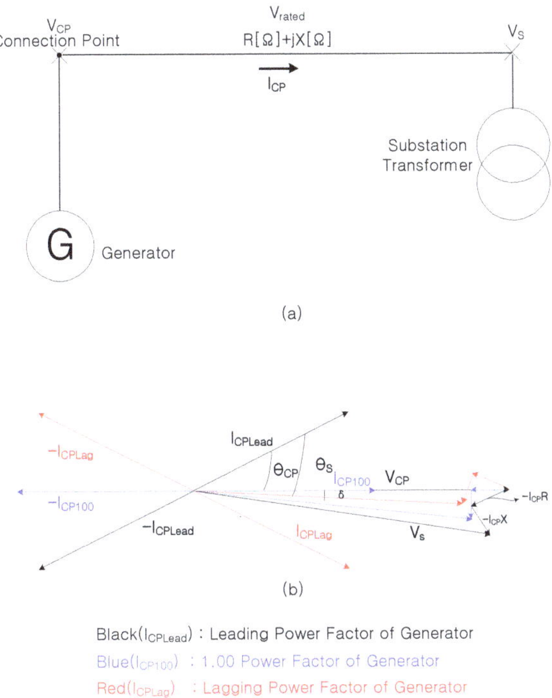

Figure 1. Grid-connected renewable energy generators (REGs): (**a**) Single line diagram; (**b**) vector diagram. V_{Rated}: Rated voltage of DL (V); $\vec{V}_{CP}$: Connection point voltage (V); $\vec{I}_{CP}$: Connection point current (A); DL: Distribution line; $\vec{V}_S$: Station voltage (V).

$$\%V_{VRR} = \frac{V_{CP} - V_S}{V_{Rated}} \times 100 \tag{1}$$

$$\vec{V}_S = \vec{V}_{CP} - \sqrt{3}\,\vec{I}_{CP}(R + jX) \tag{2}$$

$$\vec{I}_{CP} = I_{CP}(\cos\theta_{CP} \pm j\sin\theta_{CP}) \tag{3}$$

$$V_S = \sqrt{\left\{V_{CP} - \sqrt{3}I_{CP}(R\cos\theta_{CP} \mp X\sin\theta_{CP})\right\}^2 + \left\{\sqrt{3}I_{CP}(X\cos\theta_{CP} \pm R\sin\theta_{CP})\right\}^2}$$
$$= \sqrt{V_{CP}^2 + 3I_{CP}^2(R^2 + X^2) - 2\sqrt{3}V_{CP}I_{CP}(R\cos\theta_{CP} \mp X\sin\theta_{CP})} \tag{4}$$

where $\%V_{VRR}$ is the voltage rise rate (%), V_{Rated} is the rated voltage of DL (V), $\vec{I}_{CP}$ is the connection point current (A), $\vec{V}_{CP}$ is the connection point voltage (V), $\cos\theta_{CP}$ is the power factor of the connection point, R and X are the line parameters (Ω), and $\vec{V}_S$ is the station voltage (V).

According to Figure 1 and Equation (4), the voltage rise of DL is dependent on the power factor under the given value of the line constants of DL and the voltage and current of REG. The lower the generator lagging power factor, the higher the voltage rise, and the lower the generator leading power factor, the lower the voltage rise. In addition, the power factor of REGs that does not exceed the allowable voltage rise rate in DL can be calculated:

$$kVA_R = kVA_{REG}(\sin\theta_1 - \sin\theta_2) \tag{5}$$

On the other hand, the power factor of REGs is usually fixed excluding the synchronous generator, so it may be possible to calculate the capacity of the reactor to compensate for the power factor, as expressed in Equation (5), where kVA_R is the capacity of the reactor, kVA_{REG} is the capacity of REG, θ_1 is the tuned power factor angle, and θ_2 is the REG power factor angle.

Figure 2 presents the single line and vector diagrams of DL with the reactor, where the red parts show the changes after installing a reactor. The blue θ_{CPN} of Figure 2 corresponds to θ_1 in Equation (5). The limited values of I_{CPN}, V_{sN}, and θ_{CPN}, shown in blue in Figure 2 that do not exceed the allowable voltage rise rate in DL are then derived. If the reactor with the value calculated from Equation (5) is installed, θ_{CPN} is the same, but I_{CPN} and V_{sN} are different from I_{CPR} and V_{sR}, such as in the red parts in Figure 2. Therefore, the voltage rise in DL may be higher than the allowable voltage rise. A larger reactor should be installed to cover this additional voltage rise.

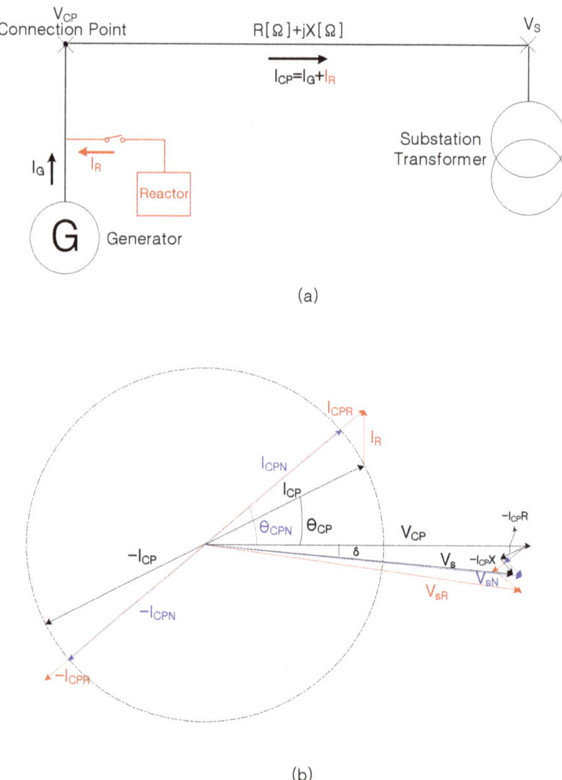

Figure 2. Grid-connected REG with reactor; (**a**) single line diagram; (**b**) vector diagram.

2.2. Selection of Reactor Installing Location

In this paper, the installation locations of the reactor are examined by the comparative analysis of the advantages and disadvantages at each location considering the necessity of re-coordination of a protection relay and the loss of generation revenue according to the installation locations. For these analyses, the reactor locations are selected firstly depending on where they are located, before or after CT_1 or CT_2. Therefore, there are only four locations where reactors can be installed in the plant as shown in Figure 3, and the most suitable place can be chosen among them after considering the coordination of the protective relays and economic issues. Table 1 lists the currents of CT_1 and CT_2 as the ON/OFF operation of reactor for each different location. The currents through CTs are not the same, so the protection relay must be coordinated differently according to the reactor ON/OFF condition. Figure 4 gives an example of the re-coordination for overcurrent relay (OCR) of CT_2, when the reactor is installed at the location of ①, ②, or ③ [33].

Location ① is between the generator and the generator CT. The reactor is installed to suppress the voltage rise, but the line voltage is decreased too much if the reactor is turned on when the line voltage is not high. Therefore, the reactor should be turned on or off considering the status of line voltage. For a small synchronous generator, CT_1 is usually attached close to the generator; it is difficult to install the reactor between the generator and CT_1 practically.

Location ② is between the generator CT and the low-voltage side of the transformer. When the reactor is turned on with the generator start-up, protection relay 1 will detect only the generator current but protection relay 2 will perform the summing current of the generator and reactor. Therefore, 51 (overcurrent) and 87 (percent differential) of protection relay 2 should be re-coordinated, considering the two elements carefully.

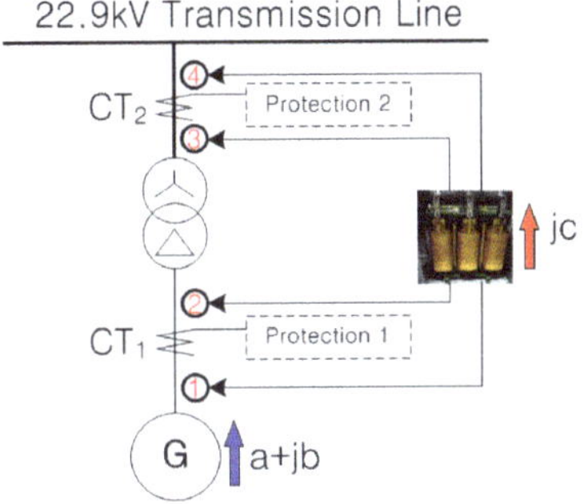

Figure 3. Candidate for location of the reactor.

Table 1. CT currents according to locations of the reactor.

Location of Reactor	CT$_1$		CT$_2$	
	ON	**OFF**	**ON**	**OFF**
①	a + j(b + c)	a + jb	a + j(b + c)	a + jb
②	a + jb	a + jb	a + j(b + c)	a + jb
③	a + jb	a + jb	a + j(b + c)	a + jb
④	a + jb	a + jb	a + jb	a + jb

a is the active current of generator; b is the reactive current of the generator; and c is the reactive current of the reactor.

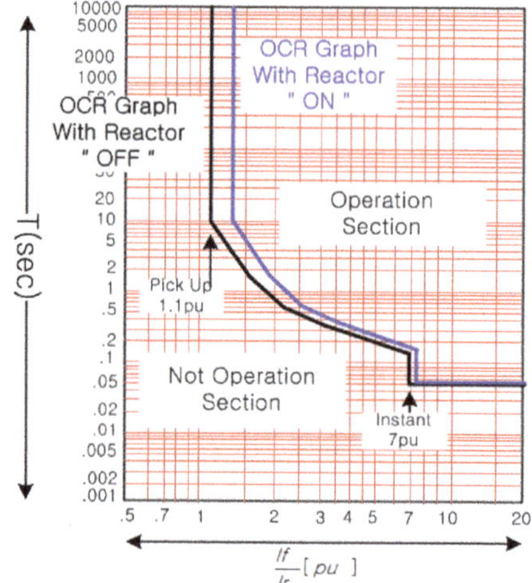

Figure 4. Example of re-coordination of overcurrent relay (OCR) due to installing the reactor.

Location ③ is between the high-voltage side of the transformer and the transformer CT. The re-coordination process should be applied, the same as for location ②. The more serious issue with respect to installing a reactor at this location is that the price of the reactor is about twice as much as that of reactors for locations ① or ②.

Location ④ is at DL side of the transformer CT. If a reactor is installed at this location, then there is no difference at protection relays 1 and 2 as the reactor is in ON/OFF operation. Therefore, it is sufficient to consider the reactor ON/OFF conditions only.

3. Design and Application of Reactors

3.1. Design of Reactor

Figure 5 shows the single line diagram of the application site. Two 1.5-MVA small hydro synchronous generators are connected in the grid, of which the output powers are limited to two-thirds of the rated output power due to the voltage rise. To increase the generator output powers with the accessible voltage rise, two 215-kVA reactor banks are installed at each generator. As mentioned in Section 2, it is recommended to install the reactor at the DL side of the transformer CT (location ④). However, they are installed between the generator CT and the low-voltage side of the transformer (location ②), because there is no space for the reactor to be installed in location ④.

Table 2 lists the voltage rise rate corresponding to the power factor of the generator. As shown in Table 2, the leading power factor to connect the 3-MVA generator to DL should be 0.9385 or less to meet the 2% permissible voltage rise rate in Korea. On the other hand, the leading power factor of the applied generators is 0.99 and the 3.79% voltage rise rate is more than the permissible range. Therefore, it is necessary to install reactors to make the leading power factor less than 0.9385. The required capacity of the reactor can be calculated from Equation (5) and Figure 2, as shown in Table 3. Table 3 shows the voltage rise rates depending on the capacity of the reactor.

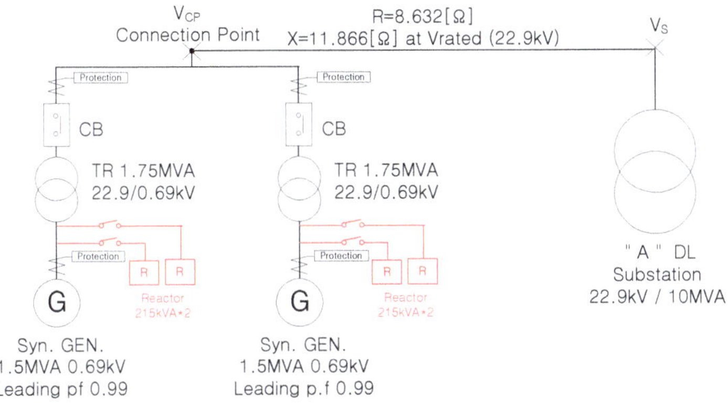

Figure 5. Single line diagram of application site.

Table 2. Voltage rise rate according to power factor. $\%V_{VRR}$: voltage rise rate (%).

$\cos\theta_{CP}$	$\sin\theta_{CP}$	V_S (V)	$\%V_{VRR}$	Remarks
1.000	0.000	21,824.6	4.93%	
0.990	0.141	22,065.2	3.78%	
0.980	0.199	22,170.2	3.29%	
0.940	0.341	22,443.5	2.03%	Gen. capacity: 1.5 MVA $\times$ 2 ea
0.9385	0.345	22,451.7	2.00%	Gen. voltage : 0.69 kV
0.930	0.368	22,496.8	1.79%	Connection point voltage: 22.9 kV $< 0°$
0.910	0.415	22,594.1	1.35%	Station voltage: V_S
0.900	0.436	22,639.0	1.15%	Rated voltage of DL: V_{Rated}
0.890	0.456	22,682.0	0.96%	
0.840	0.543	22,874.2	0.11%	
0.8325	0.554	22,900.5	0.00%	

Table 3. Voltage rise rate according to reactor capacity.

Characteristic	A	B	C	D
Capacity of REG (kVA)	3000	3145	3177	3235
Power Factor	0.9900	0.9442	0.9348	0.9180
Active Power (kW)	2970	2970	2970	2970
Reactive Power (kVar)	423	1036	1128	1283
Voltage (V)	22,900	22,900	22,900	22,900
Current (A)	75.6	79.3	80.1	81.6
Voltage-rise Rate (%)	3.78	2.22	2.00	1.61
Capacity of Reactor (kVA)	-	613	705	860

In Table 3, the reference for all values is the point at which the REG is connected to the DL. A is the case when the synchronous generators are connected without any reactors. The voltage rise rate is 3.78%, so it is necessary to install a reactor. The maximum capacity of the grid-connected synchronous generator to guarantee the permissible voltage rise rate without installing a reactor can be calculated using Equation (4) and 1.55 MVA. B is the case when the 613-kVA reactor calculated using Equation (5) is installed. The voltage rise rate is 2.22%, so the capacity of the reactor should be increased to meet the 2% voltage rise rate. C is this case and a 705-kVA reactor is installed. D is the case for the real implementation in this paper, in which an 860-kVA reactor is installed considering the margin because the power factor of REG is not fixed.

The inductive generator does not have an excitation system and cannot control the power factor. The generator has a fixed leading power factor of approximately 0.9, so the reactor is not necessary, as shown in Table 2. On the other hand, the solar generating system has a fixed unit power factor, so a 1144 kVA reactor calculated from Section 2.1 should be installed for the DL. The maximum capacity of the grid-connected solar generator to guarantee the permissible voltage rise rate without installing a reactor is calculated using Equation (5) and 1.21 MW.

3.2. Operation of Reactor

If the reactor is switched ON–OFF at once, the reactor current flows either fully or not. The grid voltage is then changed considerably and the peripheral equipment may be adversely affected due to the switching surge. Therefore, it would be better to install several small capacity reactors to decrease the abrupt voltage variation. Many divisions allow the system to be controlled linearly but this is not cost-effective. Therefore, it is recommended practically to operate the reactor in two steps. In addition, the duration (D_T) must be set to avoid repeating ON and OFF of the reactor in a very short time.

Based on the following information on the generator ON or OFF condition (G = 0 or 1), the number of ON reactors (R = 0 or 1 or 2), the active power of the generator (P_G), and the voltage of DL (V_{CP}) are detected to operate the reactor. The reactor is ON only if both conditions of the active power of the generator and the voltage of the DL are met. On the other hand, if one of the conditions is not met after being an ON reactor, the reactor is set to be switched OFF to prevent the voltage of DL becoming worse due to the operation of the reactor. Figures 6 and 7 present the recommended set values for the ON and OFF of the reactor, respectively. The duration should be determined by the control speed of the output. Therefore, it is recommended that the duration of solar or wind with a quick control speed be five minutes, and hydro with the late control speed duration be one minute.

Figure 6 shows the voltage set criteria for the ON and OFF operation of the reactor, which are the permissible maximum (V_{PMax}) and minimum (V_{PMin}) voltages of the DL. The voltage condition (V_{STRON}), in which two reactors are switched ON is a value obtained by subtracting the permissible voltage rise rate (%V_{PVRR}) from V_{PMax}. The voltage condition (V_{SORON}), in which one reactor is switched ON, is the value obtained by subtracting %V_{PVRR} from V_{STRON}. The voltage condition for the reactor OFF will be the opposite of the ON condition.

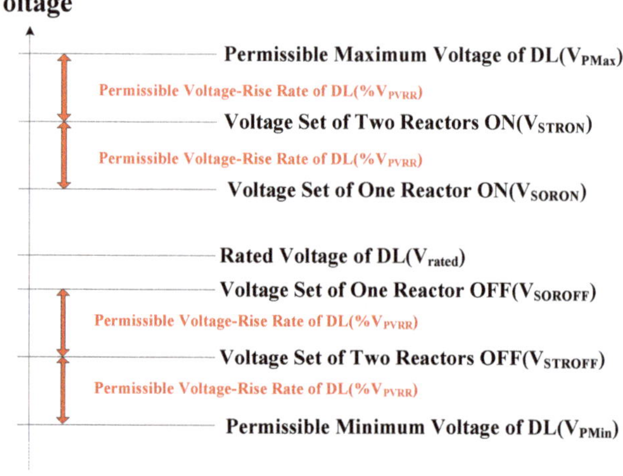

Figure 6. Voltage set of DL for ON and OFF of reactors.

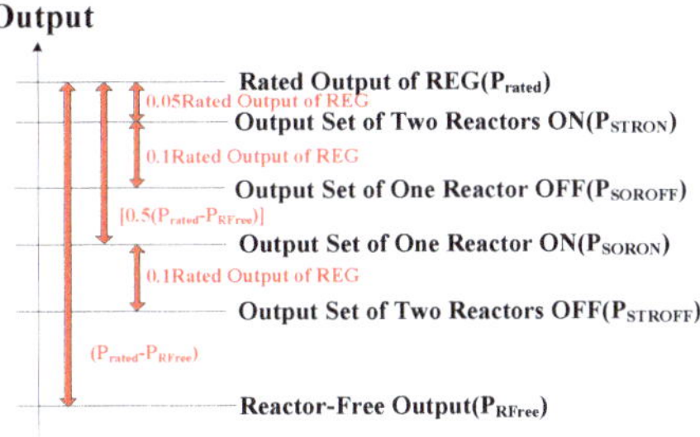

Figure 7. Output set of REG for ON and OFF of reactors.

Figure 7 presents the output set criteria of REG for the ON and OFF operation of the reactor. The output condition (P_{STRON}) in which two reactors are switched ON is 95% of the rated output (P_{rated}) of the REG. Half of the value obtained by subtracting the reactor-free output (P_{RFree}) of REG in Section 3.1 from P_{rated} is subtracted from P_{rated}. This is the output condition (P_{SORON}), in which one reactor is switched ON. The output condition for OFF of the reactor is below 10%P_{rated} of the ON condition.

Figures 8 and 9 present the flow chart for the ON and OFF operation of the reactor. If the voltage of DL exceeds V_{SORON} and the generator output exceeds P_{SORON} in the case of reactor OFF state, and if the duration time exceeds the setting value, one reactor is turned ON, as shown in the left part of Figure 8. If the voltage of DL exceeds V_{STRON} and the generator output exceeds P_{STRON} in case of one reactor ON state, and if the duration time exceeds the setting value, one reactor is additionally turned ON as shown in the right part of Figure 8. On the other hand, if the voltage of DL is lower than V_{SOROFF} or the generator output is lower than P_{SOROFF} in the case of two reactors in the ON state, and if the duration time exceeds the setting value, one reactor is turned OFF, as shown in the left part of Figure 9. If the voltage of DL is lower than V_{STROFF} or the generator output is lower than P_{STROFF} in case of one reactor ON state, and if the duration time exceeds the setting value, the reactors are all turned OFF, as shown in the right part of Figure 9.

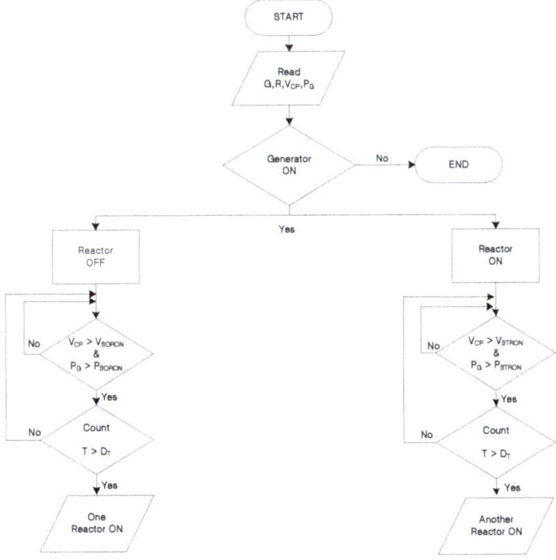

Figure 8. Flow chart for ON operation of reactors.

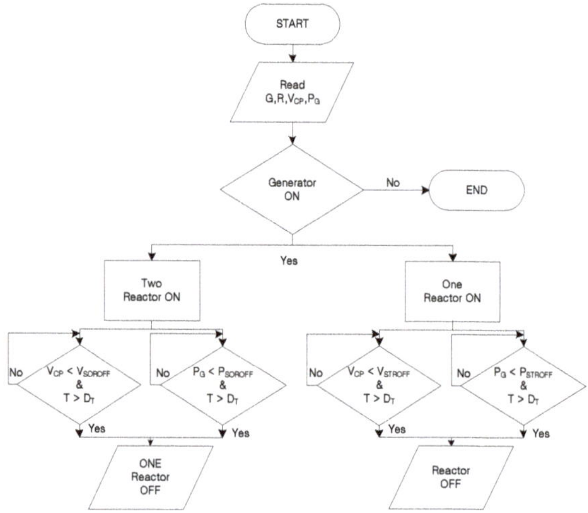

Figure 9. Flow chart for OFF operation of reactors.

4. Demonstration and Effectiveness

To demonstrate the research results, the grid-connected synchronous generators shown in Figure 5 are selected. The field application result is different from the result calculated from the theoretical equations. This is because the latter calculation has been done under the assumption of a no-load condition where the voltage rise is maximized by the connection of a generator to the grid, but the no-load condition cannot be provided in a field test. The load condition of the DL before and after installing the reactor cannot be the same because of the frequent variations of the load, DL voltage, and output in grid-connected DRs. Therefore, the effects of field application are analyzed in the following three ways. The first test is conducted in the connection point of REG, but the second and third tests are added in the terminal of generators. The data from the second and third tests are per second from the Supervisory Control and Data Acquisition (SCADA) system.

Firstly, the vector diagram of the no reactor operation and two reactors operation are compared, as shown in Figures 10a–d and 11a–d. Tests are conducted at similar active outputs for a one-day interval to meet the same test conditions as much as possible. The vector diagram shows the phase voltages (V_{CP}) and currents (I_{CP}) at the connection point of the REG to the DL. Compared to Figure 10, Figure 11 shows that the magnitude and phase angle of the current vector are increased due to the reactor operation. The substation voltage (V_S) is calculated because the substation is very far from the power plant. The test results are applied in Equations (1)–(4) and Table 4 lists the effectiveness of the reactor operation. The substation voltages are 22,459 V, 22,449 V, 22,435 V, 22,448 V from Equation (4) and the voltage rises due to the connection of REG in the two reactors OFF state are 539 V, 549 V, 542 V, 548 V, but the reactor is not switched ON because the DL voltage does not satisfy the condition of the reactor ON. The substation voltages are 23,125 V, 23,337 V, 23,324 V, 23,331 V using Equation (4) and the voltage rises due to the connection of REG in two reactors ON state are 1 V, 3 V, 3 V, 6 V. Accordingly, the mitigations of the voltage rise are 538 V, 546 V, 539 V, 536 V with the installation of the reactor under similar test conditions.

Table 4. Effectiveness of reactors.

Electric Values	Without Reactor				With Two Reactors			
	First	Second	Third	Fourth	First	Second	Third	Fourth
V_{CP} (V)	22,998	22,998	22,977	22,996	23,126	23,340	23,327	23,337
I_{CP} (A)	70.13	70.48	70.36	70.31	79.93	79.12	79.14	79.05
$\cos\theta_{CP}$ in leading	0.960	0.961	0.960	0.961	0.834	0.834	0.834	0.835
V_S by calculation (V)	22,459	22,449	22,435	22,448	23,125	23,337	23,324	23,331
V_{CP}-V_S by calculation (V)	539	549	542	548	1	3	3	6
%V_{VRR} by calculation (%)	2.36	2.40	2.37	2.39	0.00	0.01	0.01	0.03

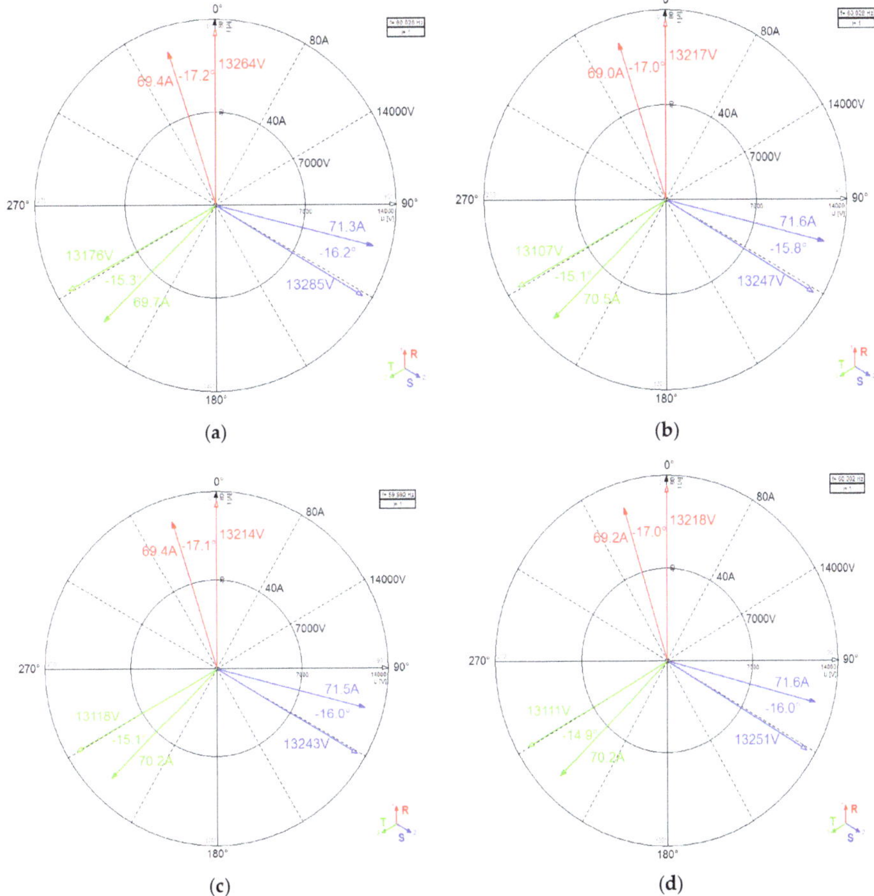

Figure 10. Non-reactor operations (**a**) first (**b**) second (**c**) third (**d**) fourth.

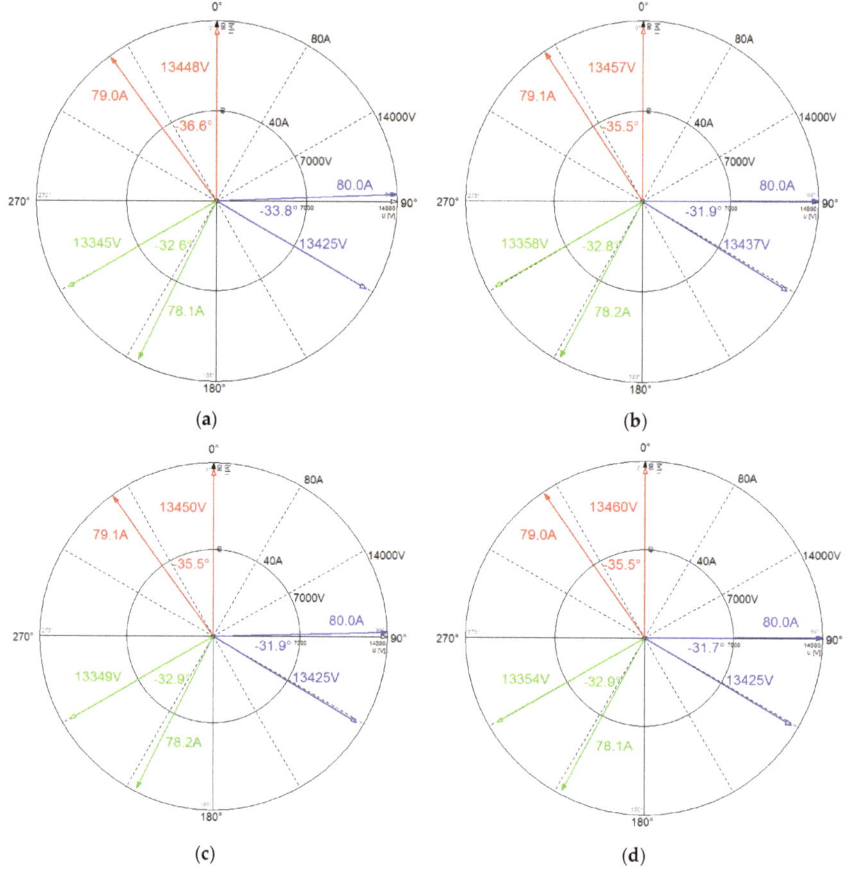

Figure 11. Two-reactor operations (**a**) first (**b**) second (**c**) third (**d**) fourth.

Secondly, the daily connection voltage data before and after installing the reactor are analyzed. To compare the test results before and after installing the reactor, tests are conducted on the daily-maximum generation day in the same season to provide the same condition. The excessive voltage rise occurs when the generator output is connected to the grid without a reactor so that the output-rise can be 1000 kW, which is two-thirds of the rated output. Table 5 and Figure 12a–c present the comparison data of the output active power and the daily grid voltage before and after installing the reactor. The voltage of the connection point increases with increasing the output of REGs. On the other hand, despite the increase in output, the voltage decreased due to the reactor operation. Figure 12a–c show that the generator outputs are increased by 455 kW and 494 kW on average, respectively, but the grid voltage is decreased by 56 V.

Thirdly, the monthly average grid voltage data before and after installing the reactor are analyzed. Similar to the second test, the same month is chosen to provide the same test condition. Figure 13a–c shows the monthly average grid voltage before and after installation of the reactor. The average grid voltage without the reactor is 23,279 V on average, whereas it is 23,047 V with the reactor, i.e., the reactor results in the voltage decrease of 233 V.

Table 5. Comparison of output active power and daily grid voltage before and after installing reactors.

No.	Active Power and Grid Voltage	First	Second	Third
①	No. 1 Active power without reactor (kW)	1001	986.4	971.7
②	No. 2 Active power without reactor (kW)	989	979.4	947.5
③	No. 1 Active power with reactor (kW)	1438	1446	1439
④	No. 2 Active power with reactor (kW)	1474	1471	1452
⑤	Grid voltage without reactor (V)	23,218	23,380	23,380
⑥	Grid voltage with reactor (V)	22,895	23,072	22,944

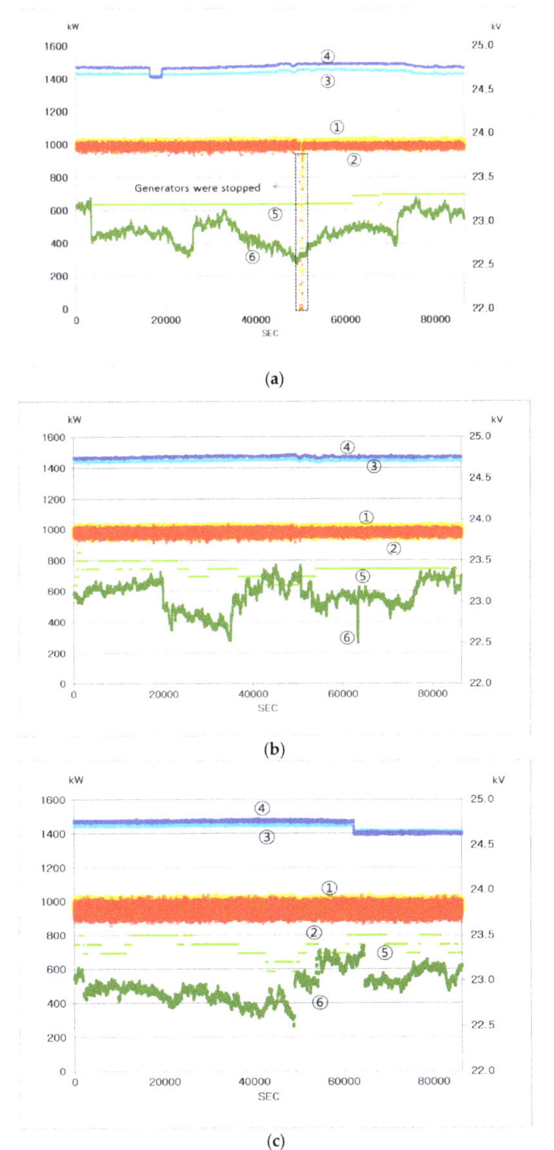

Figure 12. Output active power and daily grid voltage before and after installing reactors; locations ②, ③, ④, ⑤, ⑥ are same as in Table 5 (**a**) first (**b**) second (**c**) third.

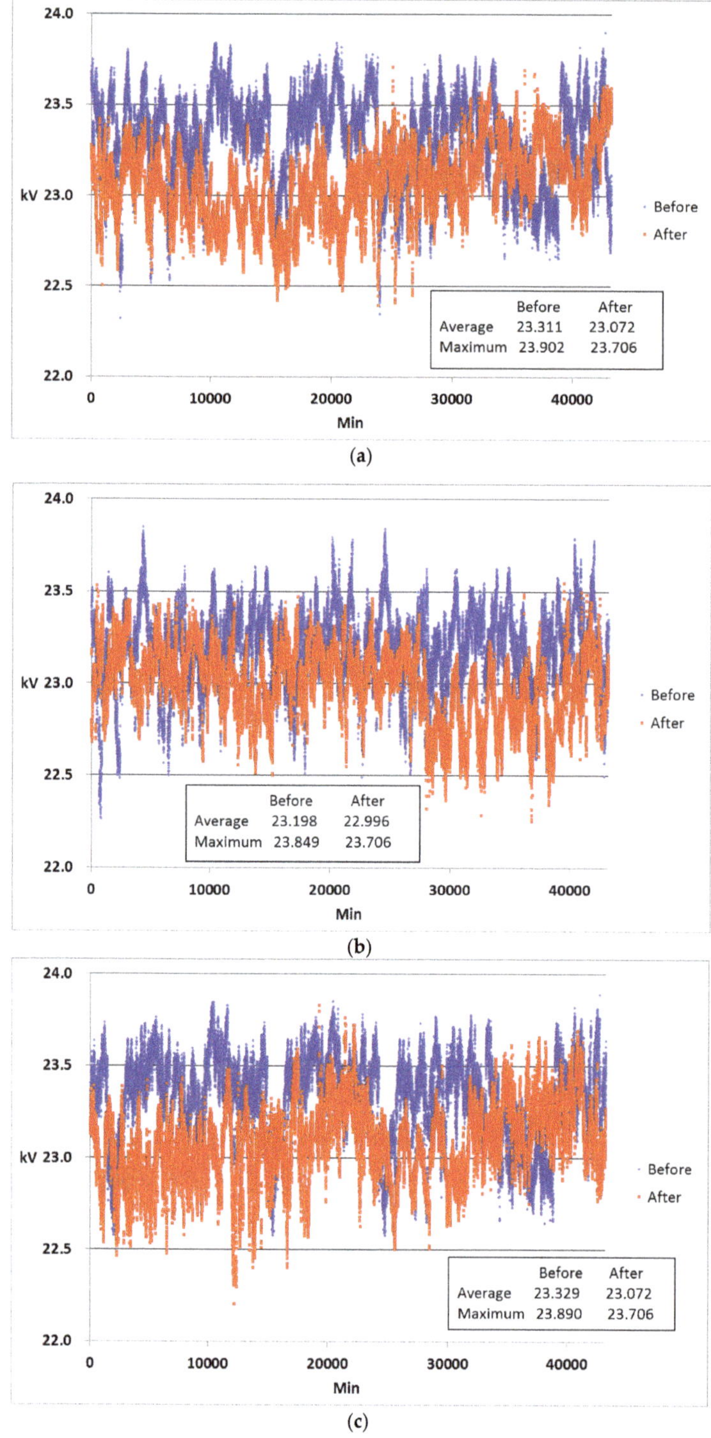

Figure 13. Comparison of monthly average grid voltage before and after installing the reactor. (**a**) first (**b**) second (**c**) third.

5. Conclusions

To mitigate the voltage rise caused by the connection of synchronous generators of DRs to DL, the adoption of a reactor of a reactor was studied in this paper. From the vector analysis, it was verified that the capacity of the reactor depends on the power factor of the generator. Therefore, the maximum grid-connected capacity of the DRs without a reactor can be calculated if the power factor and line constant are pre-determined, and the necessary capacity of a reactor to mitigate the voltage rise of the DL can be derived. The advantages and disadvantages at each location considering the necessity of re-coordination of a protection relay and the loss of generation revenue according to the installation locations were analyzed comparatively. Among four candidates for a reactor location, it was verified that the reactor at the grid-connected point did not affect the coordination of the protection relay and no additional protection was necessary. It was found that the ON–OFF switching of the large reactor caused an abrupt rise and fall of the DL voltage. Therefore, dividing the reactor into a couple of small reactors to be switched ON–OFF in two steps was recommended. The adoption of more steps might allow the easy and smooth control of the DL voltage, but it incurs a higher cost. In addition, the reactor should be switched ON in the 'AND' condition, satisfying both the grid voltage and the generator output, and be switched OFF in the 'OR' condition in the case of not satisfying one of the conditions. Furthermore, the system must be set to avoid the reactor being switched ON–OFF repeatedly in a short time frame.

From the field application data for power plants with voltage rising problems, this study concludes that the generator output could be increased from 1000 kW to 1500 kW, and the daily average voltage and the monthly average voltage at the connection point could be reduced 300 V and 200 V, respectively.

Acknowledgments: This study was supported by the KETEP (Korea Institute of Energy Technology Evaluation and Planning) through the (*Development for Microgrid Common Platform Technology, 20141010501870*) project.

Author Contributions: Yeonho Ok and Jaewon Lee conceived and designed the tests; Yeonho Ok and Jaeho Choi analyzed the data and wrote the paper.

Conflicts of Interest: The authors declare no conflict of interest.

References

1. Korea Electric Power Corporation. *The Standard for the Connection of the Distributed Energy Resource in the Distribution Power System*; KEPCO: Naju, Korea, 2015.
2. Barker, P.P.; De Mello, R.W. Determining the impact of distributed generation on power systems: Part 1—Radial distribution systems. In Proceedings of the IEEE Power Engineering Society Summer Meeting, Seattle, WA, USA, 16–20 July 2000; Volume 3, pp. 1645–1656.
3. Lasseter, R.H. Smart distribution: Coupled microgrids. *Proc. IEEE* **2011**, *99*, 1074–1082. [CrossRef]
4. Joos, G.; Ooi, B.T.; McGillis, D.; Galiana, F.D.; Marceau, R. The potential of distributed generation to provide ancillary services. In Proceedings of the IEEE Power Engineering Society Summer Meeting, Seattle, WA, USA, 16–20 July 2000; Volume 3, pp. 1762–1767.
5. Reed, G.F.; Takeda, M.; Iyoda, I. Improved power quality solutions using advanced solid-state switching and static compensation technologies. In Proceedings of the IEEE Power Engineering Society Winter Meeting, Edmonton, AB, Canada, 31 January–4 February 1999; Volume 2, pp. 1132–1137.
6. Sun, J.; Czarkowski, D.; Zabar, Z. Voltage flicker mitigation using PWM-based distribution STATCOM. In Proceedings of the IEEE Power Engineering Society Summer Meeting, Chicago, IL, USA, 21–25 July 2002; Volume 1, pp. 616–621.
7. Pogaku, N.; Green, T.C. Harmonic mitigation throughout a distribution system: A distributed-generator-based solution. *IEE Proc. Gener. Transm. Distrib.* **2006**, *153*, 350–358. [CrossRef]
8. Basso, T.S.; De Blasio, R. IEEE 1547 series of standards: Interconnection issues. *IEEE Trans. Power Electron.* **2004**, *19*, 1159–1162. [CrossRef]
9. Carrasco, J.M.; Franquelo, L.G.; Bialasiewicz, J.T.; Galvan, E.; Portillo Guisado, R.C.; Prats, A.M.; Leon, J.I.; Moreno-Alfonso, N. Power-electronic systems for the grid integration of renewable energy sources: A survey. *IEEE Trans. Ind. Electron.* **2006**, *53*, 1002–1016. [CrossRef]

10. Klapp, D.; Vollkommer, H.T. Application of an intelligent static switch to the point of common coupling to satisfy IEEE 1547 compliance. In Proceedings of the IEEE Power Engineering Society General Meeting, Tampa, FL, USA, 24–28 June 2007; pp. 1–4.

11. Kroposki, B.; Basso, T.S.; DeBlasio, R. Microgrid standards and technologies. In Proceedings of the IEEE Power and Energy Society General Meeting, Pittsburgh, PA, USA, 20–24 July 2008; pp. 1–4.

12. Scott, N.C.; Atkinson, D.J.; Morrell, J.E. Use of load control to regulate voltage on distribution networks with embedded generation. *IEEE Trans. Power Syst.* **2002**, *17*, 510–515. [CrossRef]

13. Liu, X.; Aichhorn, A.; Liu, L.; Li, H. Coordinated control of distributed energy storage system with tap changer transformers for voltage rise mitigation under high photovoltaic penetration. *IEEE Trans. Smart Grid* **2012**, *3*, 897–906. [CrossRef]

14. Choi, J.; Kim, J. Advanced voltage regulation method of power distribution systems interconnected with dispersed storage and generation systems. *IEEE Trans. Power Deliv.* **2001**, *16*, 329–334. [CrossRef]

15. Walling, R.A.; Saint, R.; Dugan, R.C.; Burke, J.; Kojovic, L.A. Summary of distributed resources impact on power delivery systems. *IEEE Trans. Power Deliv.* **2008**, *23*, 1636–1644. [CrossRef]

16. Liu, Y.; Bebic, J.; Kroposki, B.; De Bedout, J.; Ren, W. Distribution system voltage performance analysis for high-penetration PV. In Proceedings of the Energy 2030 Conference, Atlanta, GA, USA, 17–18 November 2008; pp. 1–8.

17. Keane, A.; O'Malley, M. Optimal allocation of embedded generation on distribution networks. *IEEE Trans. Power Syst.* **2005**, *20*, 1640–1646. [CrossRef]

18. Liew, S.N.; Strbac, G. Maximizing penetration of wind generation in existing distribution networks. *IEE Proc. Gener. Transm. Distrib.* **2002**, *149*, 256–262. [CrossRef]

19. Singh, R.K.; Goswami, S.K. Optimum allocation of distributed generations based on nodal pricing for profit, loss reduction, and voltage improvement including voltage rise issue. *Int. J. Electr. Power Energy Syst.* **2010**, *32*, 637–644. [CrossRef]

20. Harrison, G.P.; Wallace, A.R. Optimal power flow evaluation of distribution network capacity for the connection of distributed generation. *IEE Proc. Gener. Transm. Distrib.* **2005**, *152*, 115–122. [CrossRef]

21. Ochoa, L.F.; Dent, C.J.; Harrison, G.P. Distribution network capacity assessment: Variable DG and active networks. *IEEE Trans. Power Syst.* **2010**, *25*, 87–95. [CrossRef]

22. Divan, D.M.; Brumsickle, W.E.; Schneider, R.S.; Kranz, B.; Gascoigne, R.W.; Bradshaw, D.T.; Ingram, M.; Grant, I.S. A distributed static series compensator system for realizing active power flow control on existing power lines. *IEEE Trans. Power Deliv.* **2007**, *22*, 642–649. [CrossRef]

23. Lee, S.Y.; Wu, C.J.; Chang, W.N. A compact control algorithm for reactive power compensation and load balancing with static Var compensator. *Electr. Power Syst. Res.* **2001**, *58*, 63–70. [CrossRef]

24. Thukaram, D.; Lomi, A. Selection of static VAR compensator location and size for system voltage stability improvement. *Electr. Power Syst. Res.* **2000**, *54*, 139–150. [CrossRef]

25. Singh, B.; Saha, R.; Chandra, A.; Al-Haddad, K. Static synchronous compensators (STATCOM): A review. *IET Power Electron.* **2009**, *2*, 297–324. [CrossRef]

26. Giroux, P.; Sybille, G.; Le-Huy, H. Modeling and simulation of a distribution STATCOM using simulink's power system blockset. In Proceedings of the the 27th Annual Conference of the IEEE, Denver, CO, USA, 29 November–2 December 2001; Volume 2, pp. 990–994.

27. Chen, W.L.; Hsu, Y.Y. Direct output voltage control of a static synchronous compensator using current sensorless dq vector-based power balancing scheme. In Proceedings of the Transmission and Distribution Conference and Exposition, Dallas, TX, USA, 7–12 September 2003; Volume 2, pp. 545–549.

28. Freitas, W.; Morelato, A.; Xu, W.; Sato, F. Impacts of AC generators and DSTATCOM devices on the dynamic performance of distribution systems. *IEEE Trans. Power Deliv.* **2005**, *20*, 1493–1501. [CrossRef]

29. Xu, L.; Yao, L.; Sasse, C. Comparison of using SVC and STATCOM for wind farm integration. In Proceedings of the IEEE 2006 International Conference on Power System Technology, Chongqing, China, 22–26 October 2006; pp. 1–7.

30. El-Shimy, M.; Badr, M.L.; Rassem, O.M. Impact of large scale wind power on power system stability. In Proceedings of the Power System Conference, (IEEE MEPCON 2008), Aswan, Egypt, 12–15 March 2008; pp. 630–636.

31. Repo, S.; Laaksonen, H.; Jarventausta, P.; Huhtala, O.; Mickelsson, M. A case study of a voltage rise problem due to a large amount of distributed generation on a weak distribution network. In Proceedings of the IEEE Power Tech Conference, Bologna, Italy, 23–26 June 2003; Volume 4, pp. 1–6.

32. Ok, Y.; Choi, J.; Shin, M.; Choi, Y. Analysis of voltage rise phenomenon in distribution lines associated with grid-connected renewable energy system. In Proceedings of the IEEE International Conference on Power Electronics and ECCE Asia (ICPE-ECCE), Seoul, Korea, 1–5 June 2015; pp. 1790–1795.

33. Ok, Y.; Lee, J.; Choi, J. Analysis and Solution for Operations of Overcurrent Relay in Wind Power System. *Energies* **2016**, *9*, 458. [CrossRef]

Article

Power Quality Disturbance Classification Using the S-Transform and Probabilistic Neural Network

Huihui Wang [1,2,*], **Ping Wang** [1] and **Tao Liu** [3]

[1]   School of Electrical Engineering and Automation, Tianjin University, Tianjin 300072, China;
      pingw@tju.edu.cn
[2]   School of Control and Mechanical Engineering, Tianjin Chengjian University, Tianjin 300384, China
[3]   School of Electrical Engineering and Automation, Tianjin Polytechnic University, Tianjin 300387, China;
      taoliu@tju.edu.cn
*   Correspondence: homeworldlt2014@gmail.com; Tel.: +86-135-122-71448

Academic Editor: Birgitte Bak-Jensen
Received: 19 September 2016; Accepted: 9 January 2017; Published: 17 January 2017

Abstract: This paper presents a transient power quality (PQ) disturbance classification approach based on a generalized S-transform and probabilistic neural network (PNN). Specifically, the width factor used in the generalized S-transform is feature oriented. Depending on the specific feature to be extracted from the S-transform amplitude matrix, a favorable value is determined for the width factor, with which the S-transform is performed and the corresponding feature is extracted. Four features obtained this way are used as the inputs of a PNN trained for performing the classification of 8 disturbance signals and one normal sinusoidal signal. The key work of this research includes studying the influence of the width factor on the S-transform results, investigating the impacts of the width factor on the distribution behavior of features selected for disturbance classification, determining the favorable value for the width factor by evaluating the classification accuracy of PNN. Simulation results tell that the proposed approach significantly enhances the separation of the disturbance signals, improves the accuracy and generalization ability of the PNN, and exhibits the robustness of the PNN against noises. The proposed algorithm also shows good performance in comparison with other reported studies.

Keywords: transient power quality; S-transform; width factor; feature extraction; probabilistic neural network (PNN)

---

1. Introduction

In recent years, with the development and increasing implementation of distributed generation and micro-grids, large numbers of high-speed switching devices and non-linear converters are integrated into the power system. They are sources of power quality (PQ) disturbances such as sag, swell, interruption, flicker, harmonic, and oscillatory transients, etc. These voltage disturbances degrade the end user's experience, and more importantly, it may damage modern precision devices and seriously affect the production efficiency of the manufacturing industry. Owing to the facts above, the problem of PQ disturbance detection and classification has received the attention of many scholars [1].

To understand the PQ disturbance, the time-frequency characteristics of disturbances signals need to be extracted and analyzed. Various signal processing algorithms, for example, Fourier transform (FT), short time Fourier transform (STFT) [2], wavelet transform (WT) [3,4], and Stockwell transform (S-transform) [5,6] have been utilized for this purpose. FT has the advantages of requiring a relatively less amount of calculation and broad applicability, however it has the disadvantages of spectrum leakage and fence effect, and it is limited to only detecting steady state signals. STFT transforms a signal into a two-dimensional complex function which reveals the frequency characteristics of the signal at each time instant, however its time window is relatively fixed and it is difficult to balance the frequency domain resolution and time domain resolution. WT can accurately detect the singular point and the time instant at which the PQ disturbance occurs, however it is time consuming and susceptible to noises. To overcome the disadvantages of the WT, Stockwell et al. [7] proposed the S-transform, which is conceptually a hybrid of STFT and WT. Compared with WT, S-transform can be treated as an extension of WT, on the other hand, S-transform can be treated as STFT with a variable window. S-transform has superior characteristics to WT and STFT, and it has good time-frequency focusing characteristics, satisfactory time-frequency resolution, and noise immunity. The features extracted with the S-transform of the disturbance signals are directly useable in intelligent algorithms for classification [8,9].

To make the S-transform more effective for different particular application focuses, various forms of S-transform have been proposed and examined. For example, to enable a flexible regulation of the time resolution and frequency resolution, a width factor is introduced into S-transform, which makes the width of Gaussian window adjustable. To have a better control over the scale and the shape of the analyzing window, a parameter P is introduced into the Gaussian window function [10–12]. Moreover, in [13], the frequency domain is segmented into domains of different frequencies, low, medium, and high. Within in different frequency domain, the width factor of the S-transform is different, which brings different time-frequency resolution the corresponding frequency domain, so the accuracy of PQ classification can be improved. In mechanical engineering, to analyze the vibration data, asymmetrical window function instead of the regular symmetrical one is used in S-transform [14]. To improve the ability for resolving signals whose frequency changes with time, a complex Gaussian window function is adopted in S-transform [15].

For PQ disturbance classification, it has been proven that S-transform-based approaches are effective, but efforts are needed to make them more favorable [16]. Conventionally, the S-transform used for this purpose takes its basic form—a symmetrical real Gaussian window function without a width factor or with a width factor but it takes a fixed value throughout. This paper studies how to make the S-transform more applicable to PQ disturbance classification by investigating its role in the S-transform-based classification approach. Specifically, an S-transform with a width factor is used; the determination of the value of the width factor is handled in a systemic view of the PQ disturbance classification problem itself. In other words, the width factor is treated as a variable. Its value is made to be feature oriented considering the fact that different features react to the time resolution or the frequency resolution differently.

As for the classification analysis, PNN is adopted in our study. The considerations are briefly given as follows. So far, various intelligent algorithms have been used to classify the PQ disturbances, for example, artificial neural network (ANN) [17], fuzzy logic [18], support vector machine (SVM) [19], etc. Artificial neural networks are mainly used for pattern matching, classification, function approximation, optimization, and data clustering. Neural networks have obvious advantages and have been widely used in PQ classification and fault diagnosis [20,21]. Multi-layer perceptron (MLP) and PNN are the typical forward neural network structures. Compared with MLP, PNN has many advantages such as simple structure, fast training rate, high accuracy, good generalization ability, and robustness, moreover, the training process of PNN is a single-pass network training stage without any iteration for weight adjustment, and it can be easily retrained to adapt to any network topology changes. In the comparison between PNN and the other two well-known neural networks i.e., feed forward multilayer (FFML) back propagation and learning vector quantization (LVQ), PNN provides better performance in terms of the classification results [16,19,22,23]. Based on the advantages of PNN and the characteristic of PQ disturbances, PNN is considered to be more favorable for PQ disturbances classification.

In summary, a PQ disturbance classification approach based on S-transform with a feature oriented regulatory and PNN is proposed and examined. Contents of sections below include: firstly, four features are selected through investigating the impacts of the width factor on the

S-transform-Amplitude matrix of eight disturbance signals and the behavior of signal separation based on the features extracted; secondly, favorable value of the width factor used to extract each of the four features from the S-transform-amplitude matrix of disturbance signals is determined by examining the impacts of the width factor on the performance of PNN used for disturbance classification, and finally the proposed approach is tested, showing that it has higher classification accuracy and performs well with and without the noises.

2. Power Quality Disturbance Classification Based on Generalized S-Transform and Probabilistic Neural Network

In this section, first, the PQ disturbance classification algorithm based on S-transform and PNN is briefly described. Then, details of each functional unit of the algorithm are given, including the definition of eight disturbance signal types and the normal sinusoidal signal used in this study, mathematic description of S-transform without and with a width factor, definition of signal features and corresponding extraction formulation, and PNN. Finally, work of the authors is briefly introduced. Specifically, it is the implementation of the S-transform with a feature oriented optimal width factor to realize effective classification of the nine types of signals with a PNN.

2.1. Method Overview

The PQ disturbance classification based on S-transform and PNN is illustrated in Figure 1. It contains three steps: (1) S-transform, by which the time and frequency information of the given PQ disturbance signal are obtained; (2) feature extraction, by which the customized feature vectors are further extracted from the time and frequency information obtained in step (1); (3) disturbance classification with PNN, the input of which is the feature vector of the PQ disturbance signal and the output of which is the disturbances classification result.

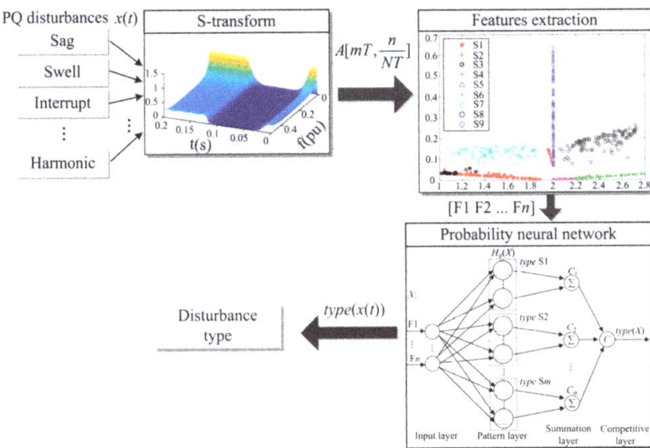

Figure 1. The schematic diagram of power quality (PQ) disturbance classification using S-transform and probabilistic neural network (PNN).

2.2. Disturbance Signal Types

IEEE Standard 1159 and the European standard EN50160 give definitions for various PQ disturbances and the corresponding suggested threshold settings [24,25]. For instance, Figure 2 shows the disturbance defined as a sag in the mentioned standards, which occurred to a normal sinusoidal voltage during the time approximately from 0.02 s to 0.14 s. The disturbance signal was captured on 21 March 2014 in Jurong East which located in the West Region of Singapore. The official source said it was due to a customer installation fault at Jurong Gateway Road. The PQ will be considered poor if the voltage magnitude is lower than the threshold set by the standard. In this specific case, the threshold should be 90% of the normal voltage magnitude.

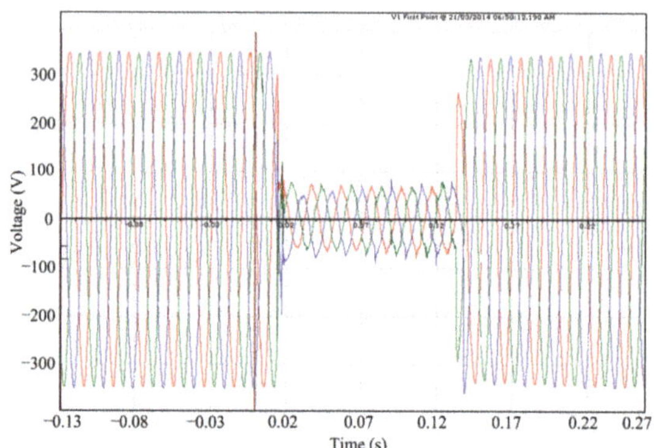

Figure 2. The waveform of sag captured from actual power system [26].

Table 1 lists the PQ disturbance types considered in this paper. Each disturbance type is described with an equation with parameters, with which the disturbance generated comply with the corresponding disturbance defined in IEEE standard 1159-1995 [24]. S1, S2, ... , and S8 denote sag, swell, interruption, flicker, oscillatory transient, harmonic, sag & harmonic, and swell & harmonic, respectively, S9 denotes normal signal. Where f = 50 Hz; ω = $2\pi f$; T = $1/f$. The parameters for the equations such as α, α_3, α_5, α_7, t_1, t_2, α_f, β, τ, f_n are determined randomly within the thresholds.

Table 1. Mathematical model of PQ disturbances.

Symbol	Type of Disturbance Signal	Equations	Parameters
S1	Sag	$y(t) = (1 - \alpha(u(t - t_1) - u(t - t_2)\,)\,)\sin(\omega t)$	$0.1 \leq \alpha \leq 0.9, T \leq t_2 - t_1 \leq 9T, t_1 \leq t_2$
S2	Swell	$y(t) = (\,1 + \alpha(\,u(t - t_1) - u(t - t_2)\,)\,)\sin(\omega t)$	$0.1 \leq \alpha \leq 0.8, T \leq t_2 - t_1 \leq 9T$ $u(t) = \begin{cases} 1 & t \geq 0 \\ 0 & t < 0 \end{cases}$
S3	Interruption	$y(t) = (\,1 - \alpha(\,u(t - t_1) - u(t - t_2)\,)\,)\sin(\omega t)$	$0.9 \leq \alpha \leq 1, T \leq t_2 - t_1 \leq 9T$
S4	Flicker	$y(t) = (\,1 + \alpha_f \sin(\beta \omega t)\,)\sin(\omega t)$	$0.1 \leq \alpha_f \leq 0.2, 5\text{Hz} \leq \beta \leq 20\text{Hz}$
S5	Oscillatory transient	$y(t) = \sin(\omega t) + \alpha e^{-(t - t_1)/\tau}\sin \omega_n$ $(t - t_1)(u(t_2) - u(t_1))$	$0.1 \leq \alpha \leq 0.8, 0.5T \leq t_2 - t_1 \leq 3T,$ $8\text{ms} \leq \tau \leq 40\text{ms}, 300\text{Hz} \leq f_n \leq 900\text{Hz}$
S6	Harmonic	$y(t) = \alpha_1 \sin(\omega t) + \alpha_3 \sin(3\omega t) +$ $\alpha_5 \sin(5\omega t) + \alpha_7 \sin(7\omega t)$	$0.05 \leq \alpha_3 \leq 0.15, 0.05 \leq \alpha_5 \leq 0.15,$ $0.05 \leq \alpha_7 \leq 0.15, \sum \alpha_i^2 = 1$
S7	Sag and harmonic	$y(t) = (1 - \alpha(u(t - t_1) - u(t - t_2)))$ $(\alpha_1 \sin(\omega t) + \alpha_3 \sin(3\omega t) + \alpha_5 \sin(5\omega t))$	$0.1 \leq \alpha \leq 0.9, T \leq t_2 - t_1 \leq 9T,$ $0.05 \leq \alpha_3 \leq 0.15, 0.05 \leq \alpha_5 \leq 0.15,$ $\sum \alpha_i^2 = 1$
S8	Swell and harmonic	$y(t) = (1 + \alpha(u(t - t_1) - u(t - t_2)))$ $(\alpha_1 \sin(\omega t) + \alpha_3 \sin(3\omega t) + \alpha_5 \sin(5\omega t))$	$0.1 \leq \alpha \leq 0.8, T \leq t_2 - t_1 \leq 9T,$ $0.05 \leq \alpha_3 \leq 0.15, 0.05 \leq \alpha_5 \leq 0.15,$ $\sum \alpha_i^2 = 1$
S9	Normal sine	$y(t) = \sin(\omega t)$	-

2.3. Generalized S-Transform and Feature Extraction

2.3.1. Generalized S-Transform

The continuous S-transform of signal $x(t)$ is described as:

$$S(\tau, f) = \int_{-\infty}^{\infty} x(t)w(\tau - t, f)e^{-i2\pi ft}\mathrm{d}t \tag{1}$$

where f is the frequency of signal $x(t)$. $w(\tau - t, f)$ is Gaussian window function, it is defined as:

$$w(\tau - t, f) = \frac{1}{\sigma\sqrt{2\pi}} e^{\frac{-(\tau - t)^2}{2\sigma^2}} \tag{2}$$

The S-transform matrix is a complex matrix whose rows pertain to frequency and columns to time. According to the Heisenberg principle, the time and frequency resolution cannot be improved at the same time. The intent of using the Gaussian window function in Equation (1) is to allow a better balancing between the time resolution and the frequency resolution. In Equation (2), τ is the point at the time axis at which the center of $w(\tau - t, f)$ sits. σ is a scale factor, it is defined as:

$$\sigma = 1/|f| \tag{3}$$

To make the Gaussian window function more effective, a width factor λ is introduced to Equation (3):

$$\sigma = \lambda/|f| \tag{4}$$

Changing the value of λ adjusts the value of scale factor σ and therefore makes the width of Gaussian window function adjustable. If a value greater than 1 is assigned to λ, for the same frequency point, Gaussian window will be wider than in the case where λ takes a value of 1. Thus, higher frequency resolution can be achieved in time-frequency domain. Inversely, if a value less than 1 is assigned to λ, for the same frequency point, Gaussian window will be narrower than in the case where λ takes a value of 1. Thus, higher time resolution can be achieved in time-frequency domain.

For instance, Figure 3 shows the effect of width factor on the time-frequency analysis of S-transform. The source signal, as shown in Figure 3a consists of two sinusoidal signals of different frequencies, 50 Hz for the signal from time 0 to 0.1 s, 100 Hz for the signal from 0.1 to 0.2 s. Width factors under examination are 0.2 and 2.0. In the case of $\lambda > 1.0$, the Gaussian window will be wider and cover more line frequency cycles, and this will help to improve the accuracy of frequency analysis of the signal. However, a wider Gaussian window will not make it easy to identify the time instant at which the frequency of the signal changes. In the case of $\lambda < 1.0$, the Gaussian window will be narrower and cover fewer line frequency cycles, this will decrease the accuracy of frequency analysis of the signal, but a narrower Gaussian window will make it easy to identify the time instant at which the frequency of the signal changes.

The result presented in Figure 3b shows that by using a width factor greater than 1.0, the frequency of two sine signals can be satisfactorily identified, as seen the two thin frequency bands in sharp yellow, however, this brings more uncertainty on the time of the signal changes because the changing part of the two frequency bands is not clear. This indicates a width factor greater than 1.0 helps to achieve the better frequency resolution, but the time resolution will get worse. On the contrary, the result presented in Figure 3c shows that using a width factor less than 1.0, the frequency of two sine signals cannot be clearly identified, as seen the two thick frequency bands in sharp yellow, however, the changing part of the two frequency bands is clearly identified. This indicates a width factor less than 1.0 results unsatisfactory frequency resolution, but the time resolution will get better.

Substituting Equation (4) into Equation (1), the continuous generalized S-transform takes the form:

$$S(\tau, f) = \int_{-\infty}^{\infty} x(t) \frac{|f|}{\lambda\sqrt{2\pi}} e^{\frac{-(t-\tau)^2 f^2}{2\lambda^2}} e^{-i2\pi ft} dt \tag{5}$$

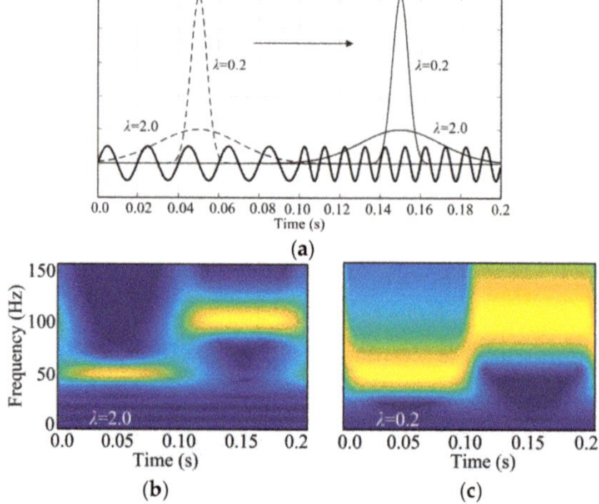

Figure 3. Time-frequency analysis results with different setting value of the width factor. (a) The comparison of the Gaussian windows with different width factor; (b) Time-frequency spectrum with λ = 2.0 and (c) Time-frequency spectrum with λ = 0.2.

Then, with the definition $\tau = mT$, $f = n/NT$, the discrete generalized S-transform expression can be obtained:

$$\begin{cases} S[mT, \frac{n}{NT}] = \sum_{k=0}^{N-1} X[\frac{k+n}{NT}]e^{\frac{-2\pi^2\lambda^2k^2}{n^2}}e^{\frac{i2\pi mk}{N}} & n \neq 0 \\ S[mT, 0] = \frac{1}{N}\sum_{k=0}^{N-1} X[\frac{k}{NT}] & n = 0 \end{cases} \tag{6}$$

where $k, m, n = 0, 1, \ldots, N - 1$, N is the total number of sampling points. T is the time interval between two consecutive sampling points. X is the discrete Fourier transform.

Further, for the purpose of extracting the features of disturbance signals, the S-transform-Amplitude (STA) matrix is calculated as follows:

$$A[mT, \tfrac{n}{NT}] = |S[mT, \tfrac{n}{NT}]| \quad m, n = 0, 1, \ldots, N - 1 \tag{7}$$

2.3.2. Feature Extraction

Feature extraction is one of the essential steps of PQ disturbance classification. Below are definitions of 10 features extracted from STA matrix.

F1: maximum amplitude of TmA-plot (time-maximum amplitude plot), and it has:

$$F1 = \max\{TmA(m)\} \tag{8}$$

where TmA-plot is maximum amplitude versus time by searching columns of STA matrix given in Equation (7) at every frequency, $TmA(m) = \max\{A[mT, \tfrac{n}{NT}]\}$.

F2: minimum amplitude of the TmA-plot:

$$F2 = \min\{TmA(m)\} \tag{9}$$

F3: mean value of the TmA-plot:

$$F3 = \frac{1}{N}\sum_{m=1}^{N} TmA(m) \tag{10}$$

F4: standard deviation of *TmA*-plot:

$$F4 = \sqrt{\frac{1}{N} \sum_{m=1}^{N} (TmA(m) - F3)^2} \tag{11}$$

F5: the summation of the maximum and minimum of the *TmA*-plot:

$$F5 = F1 + F2 \tag{12}$$

F6: the standard deviation of *FmA*-plot (frequency maximum amplitude plot, which is maximum amplitude versus frequency by searching the rows of STA matrix at each time instant) for frequencies above four times the line frequency. It has:

$$F6 = \sqrt{\frac{1}{N} \sum_{n=1}^{N} \left(FmA(n) - \overline{FmA(n)}\right)^2} \tag{13}$$

where $FmA(n) = \max\{A[mT, \frac{n}{NT}]\}$ with $\frac{n}{NT} > 200$, $\overline{FmA(n)}$ is the mean value of $FmA(n)$.

F7: the subtraction of the maximum and minimum of the $FmA(n)$:

$$F7 = \max(FmA(n)) - \min(FmA(n)) \tag{14}$$

F8: the Skewness of *FmA*-plot:

$$F8 = \frac{1}{(N-1)F6^3} \sum_{n=1}^{N} \left(FmA(n) - \overline{FmA(n)}\right)^3 \tag{15}$$

F9: the Kurtosis of *FmA*-plot:

$$F9 = \frac{1}{(N-1)F6^4} \sum_{n=1}^{N} \left(FmA(n) - \overline{FmA(n)}\right)^4 \tag{16}$$

F10: standard deviation of *mr* plot, where *mr* is time-amplitude plot determined by the frequency which has maximum amplitude in STA matrix above the frequency of 200 Hz, and it has

$$F10 = \sqrt{\frac{1}{N} \sum_{m=1}^{N} \left(mr(m) - \overline{mr(m)}\right)^2} \tag{17}$$

where $mr(m) = A[mT, \frac{n_{mr}}{NT}]$, and $n_{mr} = \underset{n}{\operatorname{argmax}}\{A[mT, \frac{n}{NT}]\}$ with $\frac{n}{NT} > 200$.

Apparently, when more features are used, a better classification effect may be expected. However, increasing the number of features results in a substantial increase of the calculation time as the scale of ANN used for disturbances classification will be increased. Therefore, the number of features used for the PQ disturbance classification should be as few as possible without obviously decreasing the classification accuracy. In this paper, four features are chosen according to the characteristic of disturbance and width factor, which are presented in Section 3.1.

2.4. Probabilistic Neural Network

As shown in Figure 1, a vector of features selected is used to be the input of PNN. The output of PNN will be the disturbance type. Figure 4 shows the schematic diagram of PNN. It contains four layers: input layer, pattern layer, summation layer, and competitive layer. The function of each layer is as follows.

Pattern layer calculates Euclidean distance between the feature vectors of PQ disturbance testing sample X and every PQ disturbance training sample X_{ij}, respectively. Below is the equation:

$$H_{ij}(X) = \frac{1}{(2\pi)^{n/2}\delta^n} e^{-\frac{\|X-X_{ij}\|^2}{2\delta^2}} \tag{18}$$

where $X = [F1\ F2\ F3\ \ldots\ Fn]^{\mathrm{T}}$ is the feature vector of an PQ disturbance testing sample, and F1, F2, F3, ... , Fn are the features as defined in Section 2.3; n is the dimension of the feature vector; X_{ij} is the feature vector of ith training sample of PQ disturbance type Sj, $Sj \in \{S1, S2, \ldots , S9\}$; δ is a smoothing parameter.

Summation layer makes summation of the results output from pattern layer, and the calculated conditional probability of X belonging to PQ disturbance type Sj is given below:

$$C_j(X) = \frac{1}{N_j} \sum_{i=1}^{N_j} H_{ij}(X) \tag{19}$$

where N_j is the number of training samples belonging to PQ disturbance type Sj.

In the competitive layer, X is assigned to the PQ disturbance type with maximum conditional probability. It has:

$$type(X) = \operatorname*{argmax}_{j}\{C_j(X)\} \tag{20}$$

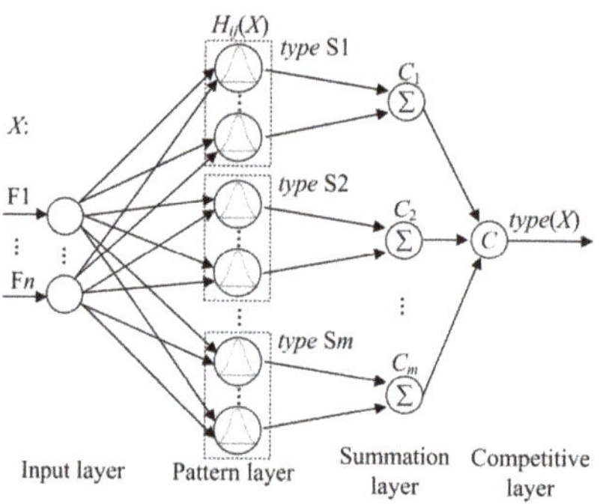

Figure 4. The schematic diagram of PNN.

3. Power Quality Disturbance Classification Based on Generalized S-transform with Feature Oriented Width Factor and Probabilistic Neural Network

The goal of this study is to realize an effective and efficient classification with high accuracy for eight types of disturbance signals plus the normal signal given in Section 2.2 using an approach based on the S-transform with a favorable width factor and PNN. Contents presented below include: selection of features used as inputs of PNN—four of ten are selected after examination; favorable value determination of the width factor for each of the four selected features by investigating the impacts of width factor on the PNN performance, and implementation of PNN for disturbance classification.

3.1. Method Proposed—Considering the Width Factor to Be Feature Oriented

In conventional PQ disturbance classification based on S-transform, the width factor λ is treated as a constant. In other words, all features interested are extracted from the S-transform matrix obtained with a width factor of the same value. Our studies, presented here, reveal that considering the width factor λ to be feature oriented renders a more satisfactory result. Implementation data and results presented below include: how does the STA-matrix of S-transform vary with λ, how does the effect of the feature separation behavior change with λ, and how does the classification accuracy of PNN vary with λ.

3.1.1. Effect of Width Factor on S-Transform-Amplitude-Matrix

Figure 5a–h shows the 3D plots of the STA-matrix of eight disturbance signals, listed in Table 1. For each disturbance signal, there are three 3D plots, which are graphical presentation of the STA-matrix obtained with three different values of the width factor λ, respectively. The STA-matrix is obtained per Equations (6) and (7); three values of λ are 0.1, 1.0, and 3.0. The values of λ, 0.1, 1.0, and 3.0 are selected per what was presented in Section 2.3.1: One less than 1.0, 0.1, which is used to examine its impacts on the low frequency domain resolution; one greater than 1.0, 3.0, which is used to examine its impacts on the high frequency domain resolution, and the value of 1.0, which is used as the base for comparison purpose.

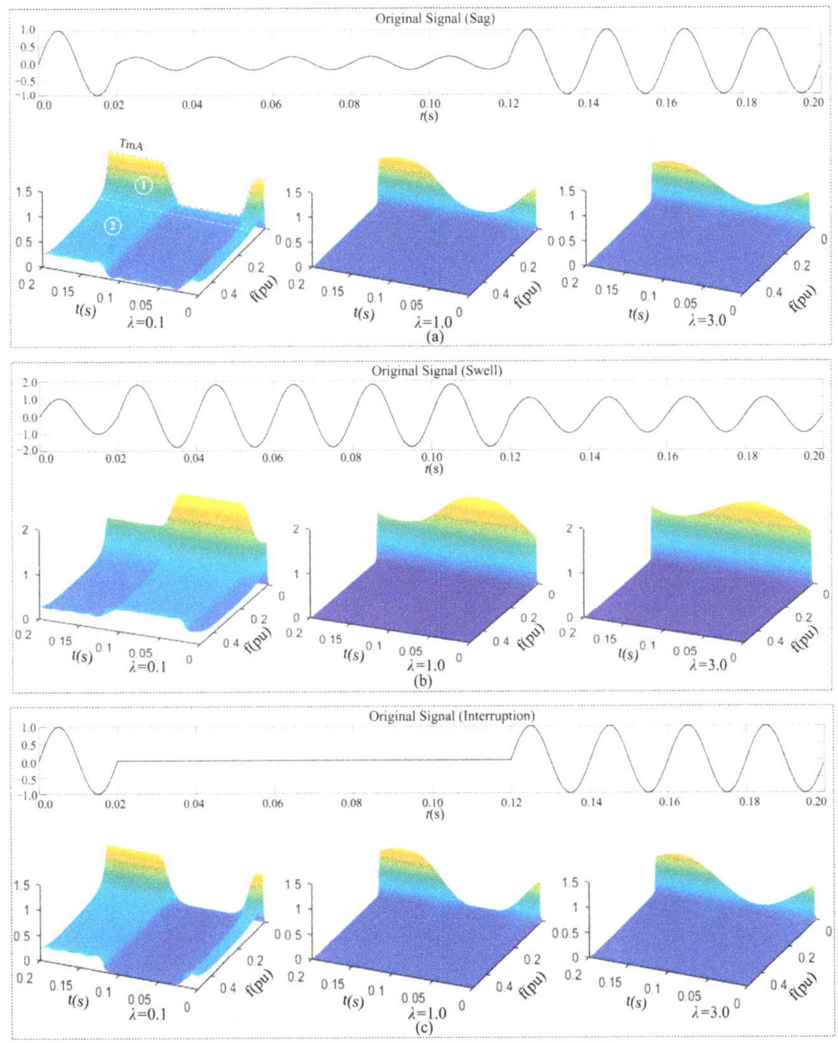

Figure 5. *Cont.*

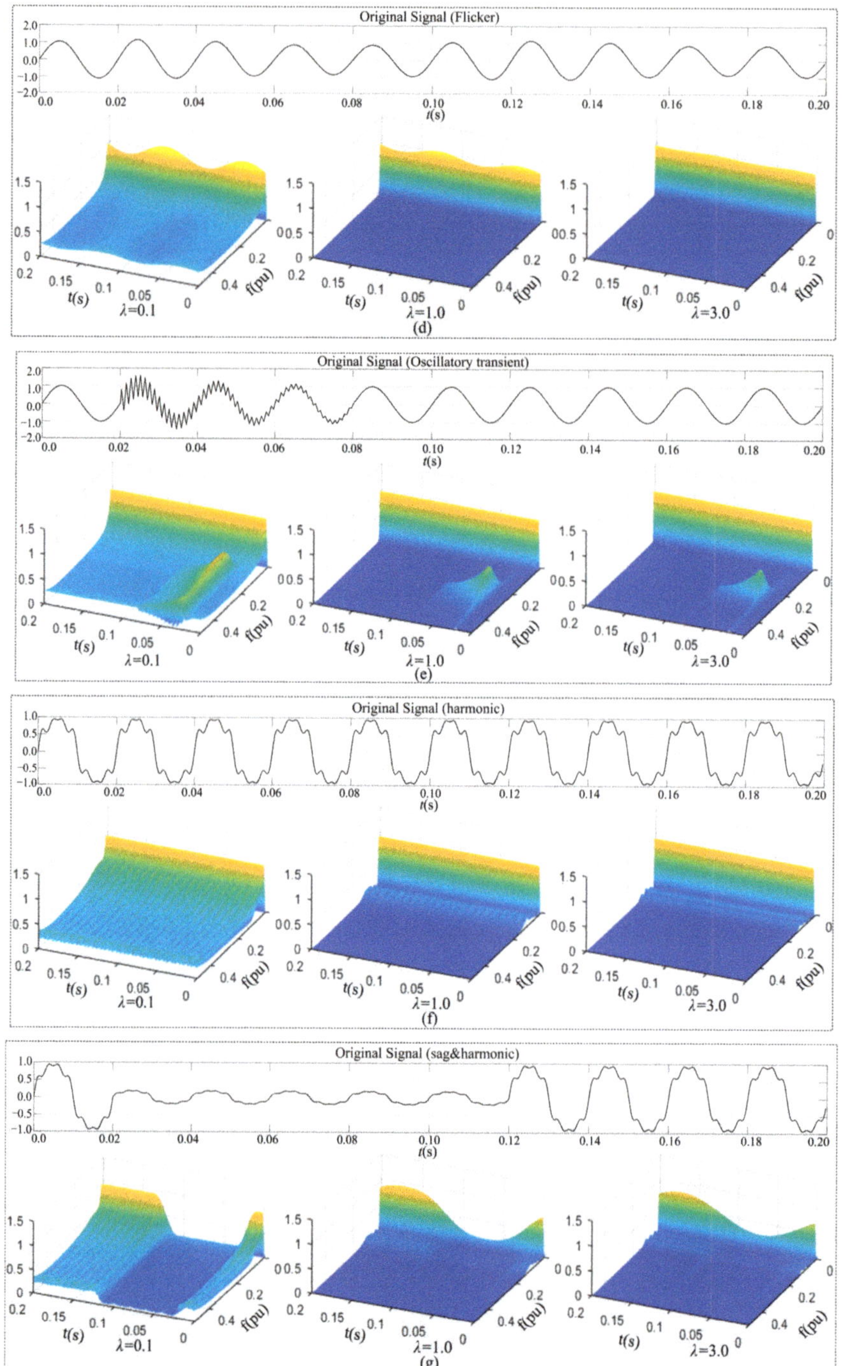

Figure 5. *Cont.*

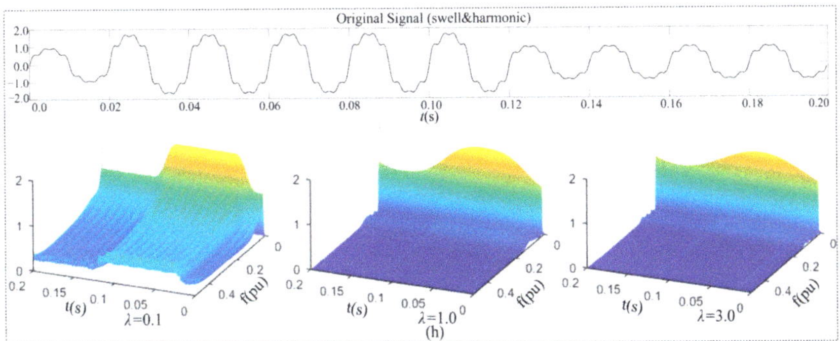

Figure 5. Disturbance signal and 3-D time-frequency magnitude spectrum. (**a**) sag; (**b**) swell; (**c**) interruption; (**d**) flicker; (**e**) oscillatory transient; (**f**) harmonic; (**g**) sag & harmonic; and (**h**) swell & harmonic.

3D plots in Figure 5 illustrate the behavior of the time-frequency characteristic of STA-matrix of each specific disturbance signal under the condition that different values of λ are used. This information helps to understand the advantages of introducing the width factor λ and the impacts of different λ values on distinguishing eight disturbance signals, S1 to S8.

Disturbance signals S1–S8 can be divided into three groups in terms of their characteristics: group 1 which has S1–S4, group 2 which has S5 and S6, and group 3 which has S7 and S8. The characteristic of group 1 is that the disturbances occur to the amplitude of the signals at the line frequency for a short period of time. Plots in Figure 5a, as an example, show the S-transform results of sag, and it can be seen that:

(1) When the width factor takes a value less than 1.0, the contour (*TmA* curve) represent the actual behavior of sag at the line frequency (shown by ①). This indicates that higher time resolution of the STA matrix is achieved. In other words, a less than 1.0 value of λ results in a satisfactory presentation of the behavior of sag at the line frequency; however, in the high-frequency domain (shown by ②), a less than 1.0 value of λ decreases the frequency resolution of the STA matrix, as one can find that obvious fluctuation, which is not the characteristic of the sag, appears in the high-frequency domain. That is, a less than 1.0 value of λ results in a wrong presentation of the behavior of sag in the high-frequency domain.

(2) When the width factor takes a value greater than 1.0, one can see that the high-frequency domain is almost flat without any high-frequency component, which is consistent with the actual behavior of sag in the high-frequency domain. This indicates that higher frequency resolution of the STA matrix is achieved. That is, a greater than 1.0 value of λ results a better presentation of the behavior of sag in the high-frequency domain. However, a greater than 1.0 value of λ decreases the time resolution of the STA matrix at the line frequency. It can be seen that the change of the *TmA* curve around line frequency becomes smoother, which may lead to a wrong identification of sag due to the inaccurate presentation of its amplitude variation with time. That is, a greater than 1.0 value of λ results an unsatisfactory presentation of the behavior of sag at line frequency. Similar conclusion can be obtained if examining plots in Figure 5b–d.

The characteristic of group 2 is that the disturbance appears to be the occurrence of harmonic components for a certain period of time or the harmonic components existing over the entire time range. The common feature of these two disturbance signal types is that the amplitude of the line frequency component of them stays unchanged. The difference between these two signals and the others are in the high-frequency domain. Plots in Figure 5e–f show the S-transform results of oscillatory transient and harmonic, and it can be seen that: (1) when the width factor takes a value less than 1.0, the frequency resolution of the STA matrix gets significantly reduced; High-frequency components which apparently do not belong to the original signals appear in the high-frequency domain; (2) these high frequency-components gradually fade away as λ increases. In the case when λ takes a value greater than 1.0, the high-frequency characteristics of the oscillatory transient and harmonic are well represented in the high-frequency domain.

Based on what are presented above, one can draw a conclusion that: A less than 1.0 value of λ results a better presentation of the behavior of the original signal at the line frequency, but distort the presentation of the behavior of the original signal in the high-frequency domain; A greater than 1.0 value of λ results a distorted presentation of the behavior of the original signal at the line frequency, but improves the presentation of the behavior of the original signal in the high-frequency domain. This indicates that combination of features is needed for the classification of different signals. As the signals of sag, swell, interruption, and flicker can be distinguished from others by the characteristic at the line frequency, a less than 1.0 value of λ is needed in the setting of the corresponding features; As the signals of oscillatory transient and harmonic can be distinguished from others by the characteristic in the high-frequency domain, a greater than 1.0 value of λ is needed in the setting of the corresponding features.

Characteristic of group 3 combines the characteristics of groups 1 and 2. Special attention needs to be paid to disturbance types S7 and S8, which are the combination of sag & harmonics and swell & harmonics, respectively. Investigation given above tells neither a value less than 1.0 nor a value greater than 1.0 ensures a successful separation of S7 or S8 form S1–S6, so in this condition, it needs combination of features to distinguish these disturbances from others.

In summary, to realize classification of the nine signal types defined in Table 1, a combination of features is needed; to make the classification more effective and efficient, making the width factor feature oriented may be considered.

3.1.2. Effect of Width Factor on Feature Distribution Behavior

Different features represent different characteristics of signals. Among the 10 features defined in Section 2.3.2, features F1–F5, give an intuitive description of disturbance signals S1–S9 at the line frequency. Differently, features F6–F10, give an intuitive description of signals in the high-frequency domain. Specifically, F6–F9 expresses the frequency characteristic of signals in the high-frequency domain and F10 expresses the time characteristic of signals in the high-frequency domain. For a successful distinction of PQ disturbances S1–S9, commonly experienced in power system, a combination of features is needed per what are presented above. The goal is to use the least number of features to realize the classification with high accuracy. Various combination possibilities were examined. Combination of F2, F5, F7, and F10 was found to be the most applicable to our purpose:

(1) F2 and F5 embody the line frequency characteristics of disturbance signals. F2 contributes to the separation of normal signal, sag, swell, and interruption by assigning a value less than 1.0 to width factor λ. Additionally, F5 is used to distinguish flicker from others by assigning a value less than 1.0 to width factor λ.

(2) F7 embodies the frequency characteristic characteristics of signals in the high-frequency domain, and it contributes to distinguishing S1–S4 (sag, swell, flicker and interruption) from S5–S6 (harmonic, transient oscillation) by assigning a greater than 1.0 value to width factor λ.

(3) F10 embodies the time characteristic of signals in the high-frequency domain, and it contributes to the separation of transient oscillation and harmonic by assigning a greater than 1.0 value to width factor λ.

Based on the features F2, F5, F7, F10 above, firstly, to get intuitionistic vision, the distributions of PQ disturbances are shown in plots with the features combination F5 & F7, F5 & F10, F2 & F10. By analyzing the separation degree of eight types of PQ disturbances and the normal signal with different setting of λ (λ_{F2}, λ_{F5}, λ_{F7}, λ_{F10}), the distributions of PQ disturbances are shown in Figures 6–8.

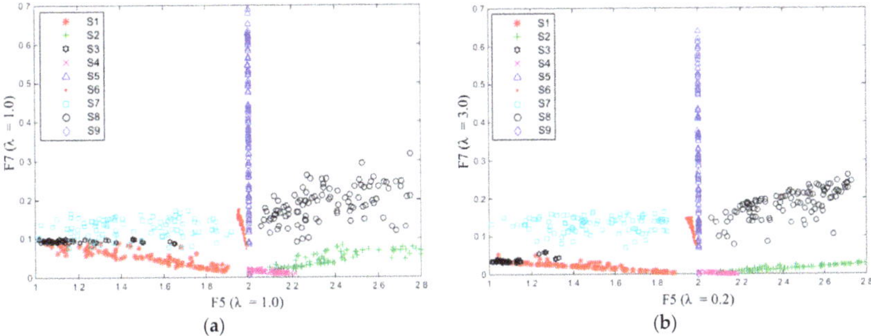

Figure 6. Comparison of F5 and F7 distribution behavior: (**a**) $\lambda_{F5} = 1.0$, $\lambda_{F7} = 1.0$; and (**b**) $\lambda_{F5} = 0.2$, $\lambda_{F7} = 3.0$.

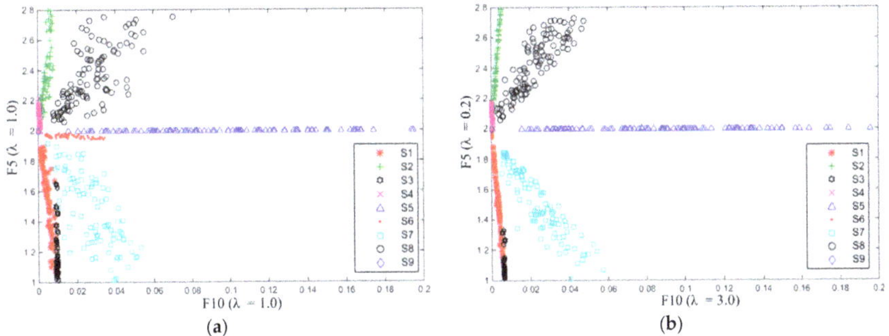

Figure 7. Comparison of F5 and F10 distribution behavior: (**a**) $\lambda_{F10} = 1.0$, $\lambda_{F5} = 1.0$; and (**b**) $\lambda_{F10} = 3.0$, $\lambda_{F5} = 0.2$.

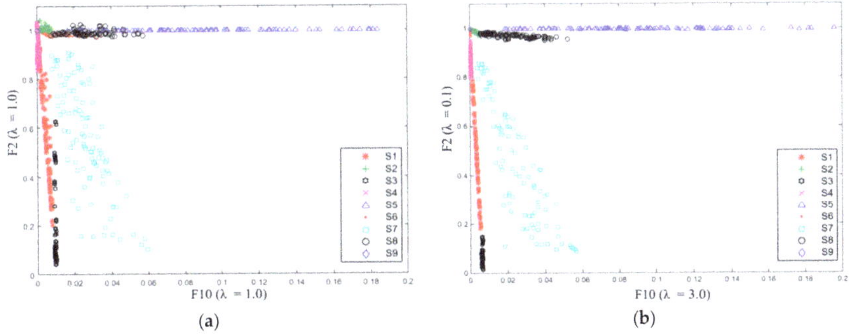

Figure 8. Comparison of F10 and F2 distribution behavior: (**a**) $\lambda_{F10} = 1.0$, $\lambda_{F2} = 1.0$; and (**b**) $\lambda_{F10} = 3.0$, $\lambda_{F2} = 0.1$.

Figure 6 shows distributions of F5 & F7 of signals S1–S9. Figure 6a is the distribution obtained with both λ_{F5} and λ_{F7} taking a value of 1.0, which is the case of the features extracted from traditional S-transform. It shows that there is obvious overlapping between areas taken by S1, S3, and S7, respectively. Also, the areas taken by S2 and S8 respectively are too close to be easily separated. This will reduce the accuracy of classification algorithm. For comparison, Figure 6b shows the results obtained with $\lambda_{F5} = 0.2$ and $\lambda_{F7} = 3.0$, determined per discussion above. It can be seen that areas taken by S1, S3, and S7 are nicely separated, and there appears a clear space between areas taken by S2 and

S8. Feature F7 is used to distinguish S1–S4 from S5–S6. Comparison between results obtained with $\lambda_{F7} = 1.0$ and $\lambda_{F7} = 3.0$, it can be seen the separation between S1–S4 and S5–S6 along the longitudinal axis (the direction of F7 axis) are much better in the case which λ_{F7} is equal to 3.0.

Figure 7 shows distributions of F10 & F5 of signals S1–S9. Figure 7a is the distribution obtained with both λ_{F10} and λ_{F5} taking a value of 1.0. It shows that there are obvious overlapping between areas taken by S1, S3 and S7; Figure 7b shows the results obtained with $\lambda_{F10} = 3.0$ and $\lambda_{F5} = 0.2$. It shows that areas taken by S1, S2, S3, S7, S8 are nicely separated. F10 is used to separate transient oscillation S5 and harmonic S6, F2 and F10 distribution map given in Figure 7a shows that there is obvious overlapping between S5 and S6 along the horizontal axis (the direction of F10 axis) with $\lambda_{F10} = 1.0$ while Figure 7b, there appears a clear space between areas taken by S5 and S6 with $\lambda_{F10} = 3.0$.

Figure 8 shows distributions of F10 & F2 of signals S1–S9. Feature F2 contributes to the separation of normal signal, sag, swell and interruption. The result in Figure 8a shows that, along the longitudinal axis (the direction of F2 axis), there is obvious overlapping between areas taken by S1 and S3, and also some overlapping between areas taken by S2 and S4 with $\lambda_{F10} = 1.0$; the result in Figure 8b shows that areas taken by S1 and S3, S2 and S4 are nicely separated with $\lambda_{F10} = 3.0$. Moreover, comparison between Figure 8a,b says the S7 can be nicely separated from other signals with $\lambda_{F10} = 3.0$, $\lambda_{F2} = 0.1$.

In summary, results presented in Figures 6–8 confirm that a combination of features is needed to separate S1 to S9. The results also verify that making λ feature oriented and using a favorable value instead of 1.0 for λ helps to achieve a satisfactory separation of S1–S9. In addition, for features which work better with λ greater than 1.0, λ values greater than 3.0 were tested and the result tells the improvement of the separation of signals S1 to S9 is insignificant. Similar results were obtained if values less than 0.1 are applied to λ for those features which work better with λ less than 1.0. Therefore, the variation range of λ used in the PNN section below is set to be from 0.1 to 3.0.

4. Determination of the Favorable Value of Feature Oriented Width Factor with the Use of Probabilistic Neural Network

This section presents the determination of favorable width factor set (λ_{F2}, λ_{F5}, λ_{F7}, and λ_{F10}) with PNN. Inputs of PNN are features F2, F5, F7, and F10 of the signal being classified. F2, F5, F7, and F10 are calculated from the S-transform matrix generated with width factor (λ_{F2}, λ_{F5}, λ_{F7}, and λ_{F10}); output of PNN is the classification result. The objective is to obtain the favorable value of (λ_{F2}, λ_{F5}, λ_{F7}, and λ_{F10}), with which and a trained PNN the classification of disturbances which falls into S1–S9 can be achieved with high accuracy. Steps for finding the favorable width factor (λ_{F2}, λ_{F5}, λ_{F7}, and λ_{F10}) are given in Figure 9.

$6^4 = 1296$ combinations of width factor λ_{F2} λ_{F5} λ_{F7} λ_{F10}, see the external loop of the flowchart in Figure 9, are examined, which are generated by assigning each element of [λ_{F2} λ_{F5} λ_{F7} λ_{F10}] with 0.1, 0.3, 0.6 1.0, 2.0, and 3.0, respectively. For each width factor combination set [λ_{F2} λ_{F5} λ_{F7} λ_{F10}]: (1) in PNN training, 900 samples of signals are used, which are obtained by randomly generating 100 signals for each of nine signal types; in PNN testing, similarly, 900 randomly generated samples of signals are used; (2) the feature vectors [F2 F5 F7 F10] are extracted from the 1800 samples of signals by using the S-transform with the corresponding width factor combination [λ_{F2} λ_{F5} λ_{F7} λ_{F10}]; (3) the PNN are trained and tested with the feature vectors [F2 F5 F7 F10] (900 for training; 900 for testing), and the classification error for the corresponding width factor combination [λ_{F2} λ_{F5} λ_{F7} λ_{F10}] is evaluated.

In the process explained above, the 200 signals (100 for PNN training; 100 for PNN testing) of each disturbance type are generated by randomly selecting the value of parameters used by each disturbance signal type, seen the parameter column of Table 1. To mitigate the possible randomicity-related classification error, the internal loop is repeated six times, $q = 6$, as shown in the flowchart in Figure 9. The classification errors of PNN are calculated six times, and the average of the obtained six classification errors is taken as testing error of the PNN. The obtained testing errors of PNN are presented in Figure 10 with spheres, the size of which is proportional to the value of the corresponding error.

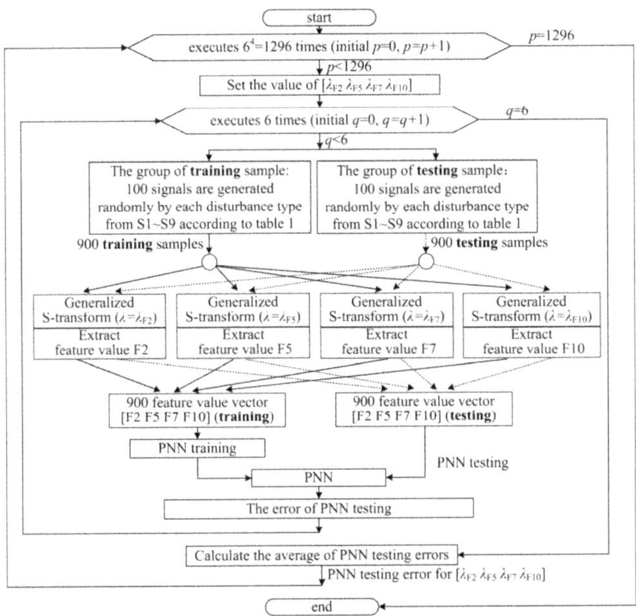

Figure 9. The flow chart for the PNN classification error.

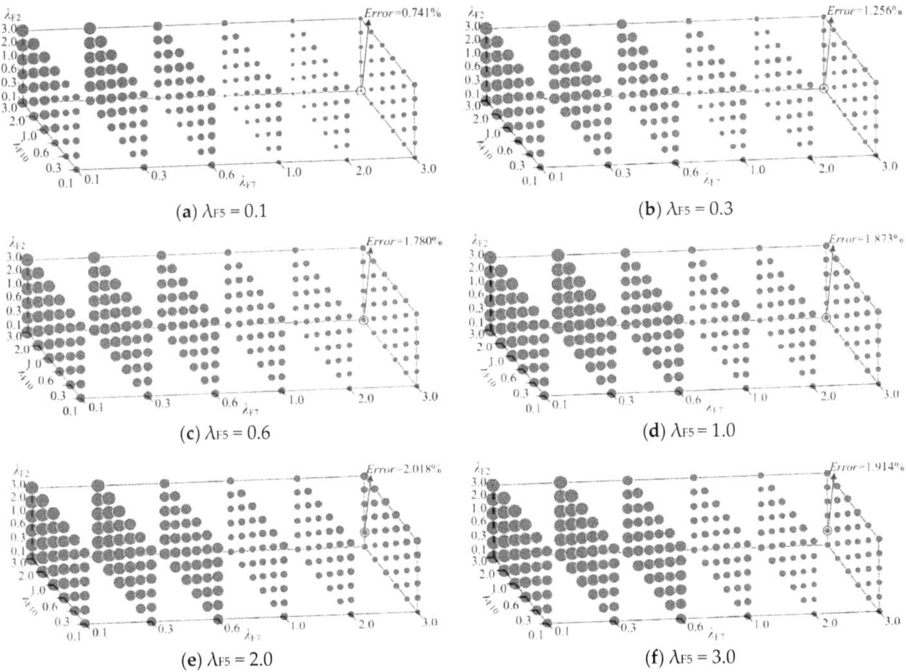

Figure 10. The error of PNN classification with the width factors combination $[\lambda_{F2}\ \lambda_{F5}\ \lambda_{F7}\ \lambda_{F10}]$ varying in the range of 0.1–3.0 (the smaller the radius of sphere, the smaller its corresponding classification error).

Firstly, take a look at Figure 10a. The size of spheres becomes smallest when λ_{F2} takes the smallest value 0.1, λ_{F7} and λ_{F10} takes their greatest value 3.0. Similarity exists in Figure 10b–f. In other words, the smallest spheres on Figure 10a–f all locate at the lower right corner in the back. This tells us that favorable values for λ_{F2}, λ_{F7} and λ_{F10} are 0.1, 3.0, and 3.0, respectively. Then, comparing the size of the smallest sphere of each individual subfigure, one can see that the sphere at the lower right corner in the back of Figure 10a has the smallest size. This tells us that the PNN classification error becomes the least if λ_{F5} takes a value of 0.1. Numbers in percentage format shown on the upper right corner in the back of each subfigure are the classification error of PNN corresponding to the sphere of the smallest size of each individual subfigure. It shows that the trained PNN will have the classification error be less than 1% (0.741% on Figure 10a) if λ_{F2}, λ_{F5}, λ_{F7}, and λ_{F10} take their values to be 0.1, 0.1, 3.0, and 3.0, respectively. The trained PNN, with features F2, F5, F7, F10 as inputs, may provide a satisfactory classification of PQ disturbance S1 to S8 and the normal sinusoidal signal if F2, F5, F7, F10 are extracted from S-transform matrix obtained with width factors λ_{F2}, λ_{F5}, λ_{F7}, λ_{F10} equal to 0.1, 0.1, 3.0, 3.0, respectively.

5. Accuracy of the Proposed Power Quality Disturbance Classification Approach

Without noise, classification results of PNN corresponding to the favorable width factor combination are shown in Table 2, and the classification accuracy is 99.259%. To analyze the effect of noise on the classification errors, different levels of noise are added to the nine types of signals. The results are listed in Table 3. The level of noise is expressed by the signal-to-noise ratio (SNR), and $SNR = 20\log_{10}(A_S/A_N)$, where A_S and A_N are the maximum amplitude of the signal and noise, respectively.

Table 2. PNN classification results with the favorable width factor combination.

Category	S1	S2	S3	S4	S5	S6	S7	S8	S9	Accuracy
S1	592	0	8	0	0	0	0	0	0	98.67%
S2	0	583	0	17	0	0	0	0	0	97.17%
S3	0	0	600	0	0	0	0	0	0	100%
S4	0	9	0	591	0	0	0	0	0	98.50%
S5	0	0	0	0	600	0	0	0	0	100%
S6	0	0	0	0	0	600	0	0	0	100%
S7	6	0	0	0	0	0	594	0	0	99.00%
S8	0	0	0	0	0	0	0	600	0	100%
S9	0	0	0	0	0	0	0	0	600	100%
PNN classification accuracy = 99.259%										

Table 3. PNN classification accuracy with noise conditions.

λ_{F2}, λ_{F5}, λ_{F7}, λ_{F10}	PNN Classification Accuracy (%)			
	Pure	40 dB	30 dB	20 dB
0.1, 0.1, 3.0, 3.0	99.26	99.13	98.63	98.38
1.0, 1.0, 1.0, 1.0	96.92	96.78	96.65	95.94
0.1, 2.0, 3.0, 0.6	97.42	96.80	96.62	96.17
2.0, 1.0, 0.1, 1.0	91.40	91.27	91.00	90.07
3.0, 1.0, 2.0, 1.0	97.04	95.98	95.87	95.68

As can be seen from Table 3, for the PQ disturbances without noise, the classification accuracy corresponding to the optimal width factor is 1.84%–7.86% higher than that of other width factor settings. The favorable width factor still maintains a good classification accuracy under noise (20 dB) and the classification accuracy is 2.21%–8.31% higher than the other width factor settings, including the traditional width factor settings ([λ_{F2} λ_{F5} λ_{F7} λ_{F10}] = [1.0, 1.0, 1.0, 1.0]).

6. Performance Comparison

In order to evaluate the effectiveness and feasibility of the proposed algorithm, Table 4 shows the comparison between the obtained results in this paper and the reported results by other studies [4,8,16,17].

Table 4. Performance comparison in terms of percentage of correct classification results.

Category	Classification Accuracy (%)				
	Proposed Method	[4]	[8]	[16]	[17]
S1	98.67	88	98	95	100
S2	97.17	96.5	92	91	100
S3	100	85.55	100	99	100
S4	98.50	-	98	98	94
S5	100	-	100	100	98
S6	100	100	92	96	98
S7	99.00	100	93	98	98
S8	100	100	90	98	97
S9	100	100	100	100	100
Average classification accuracy	99.26	95.71	95.5	97.22	98.33

In [8,16], the classification accuracy of each PQ disturbance is lower than that of the proposed algorithm. For [4], the classification accuracy of each PQ disturbance is lower than that of the proposed algorithm except S7. For [17], the classification accuracy of each PQ disturbance is lower than that of the proposed algorithm, except for S1 and S2. The average classification accuracy shows the ratio of correctly classified PQ disturbances to the total number of PQ disturbances, and the proposed method gives the best classification results for this case.

The classification accuracy comparison between the proposed algorithm and other reported studies is shown in Table 5. For the PQ disturbances without noise, the classification accuracy corresponding to the proposed algorithm is 99.26% higher than that of other algorithms. For the PQ disturbances with low level noise condition (40 dB), the classification accuracy corresponding to the proposed algorithm is 99.13% slightly lower than that of the algorithm in [27], but higher than that of other algorithms. For the PQ disturbances with high level noise condition (30 dB), the classification accuracy corresponding to the proposed algorithm is 98.63% higher than that of other algorithms.

Table 5. Comparison between the proposed algorithm and other algorithms with noise conditions.

PQ Classification Studies	Classification Accuracy (%)		
	Pure	40 dB	30 dB
[4]	95.71	93.64	91.85
[27]	-	99.7	98.5
[28]	98.5	96.875	93.625
[29]	-	98.8	97.49
Proposed method	99.26	99.13	98.63

7. Conclusions

This paper proposed a PQ disturbance classification approach based on S-transform with a feature-oriented width factor and PNN. By introducing a width factor into the conventional S-transform, the time resolution and the frequency resolution presented by the STA matrix of the signal being analyzed is made adjustable. In this way, the impact of the width factor on the 3D-STA time-frequency magnitude spectrum of eight disturbance signals are studied and the overall picture of how the regulator factor affects the description accuracy of the signal in the low frequency domain and the high frequency domain is obtained. On the basis of this and according to the joint consideration of the characteristics of eight disturbance types in frequency domain and the definition of 10 features, four out of 10 features are selected to be used for the disturbance classification. Three combinations of four selected features are investigated in terms of the 2D distribution behavior of their values for the eight disturbance signal types; the influence of the width factor on the separation of data points denoting the values of each feature combination is presented. From there, association between each feature and the width factor value favorable for the corresponding feature is established. Further, it is verified with PNN by examining the classification accuracy with a wide variation range of each width factor, from 0.1 to 3.0. Furthermore, a PNN satisfactorily trained is obtained. Simulation tells it renders high classification accuracy (less than 1% error) for 8 type disturbance signals by using only four features as inputs, which are extracted from the S-transform amplitude matrix with corresponding favorable width factor. In addition, the obtained PNN shows satisfactory robustness under various noise conditions. Finally, the proposed algorithm shows better performance in comparison with those presented in other research studies.

Acknowledgments: The authors are grateful to the financial support from the Specialized Research Fund for National Key Research and Development Plan (No.2016YFB0900204), and doctoral Program of Higher Education of China (No. 20120032110070).

Author Contributions: The paper was collaborative effort among the authors. The authors contributed collectively to the theoretical analysis and manuscript preparations.

Conflicts of Interest: The authors declare no conflict of interest.

References

1. Sainib, M.K.; Kapoora, R. Classification of quality disturbances event—A review. *Int. J. Electr. Power Energy Syst.* **2012**, *43*, 11–19. [CrossRef]
2. Gu, Y.H.; Bollen, M.H.J. Time-frequency and time scale domain analysis of voltage disturbance. *IEEE Trans. Power Deliv.* **2000**, *15*, 1279–1284. [CrossRef]
3. Biswal, B.; Biswal, M.; Hasanc, S. Nonstationary power signal time series data classification using LVQ classifier. *Appl. Soft Comput.* **2014**, *18*, 158–166. [CrossRef]
4. Uyara, M.; Yildirim, S.; Gencoglu, M.T. An effective wavelet-based feature extraction method for classification of power quality disturbance signals. *Electr. Power Syst. Res.* **2008**, *78*, 1747–1755. [CrossRef]
5. Uyara, M.; Yildirim, S.; Gencoglu, M.T. An expert system based on S-transform and neural network for automatic classification of power quality disturbances. *Expert Syst. Appl.* **2009**, *36*, 5962–5975. [CrossRef]
6. Ray, P.K.; Kishor, N.; Mohanty, S.R. Islanding and power quality disturbance detection in grid-connected hybrid power system using wavelet and S-transform. *IEEE Trans. Smart Grid* **2012**, *3*, 1082–1094. [CrossRef]
7. Stockwell, R.G.; Mansinha, L.; Lowe, R.P. Localization of the complex spectrum: The S-transform. *IEEE Trans. Signal Process.* **1996**, *44*, 998–1001. [CrossRef]
8. Bhende, C.N.; Mishra, S.; Panigrahi, B.K. Detection and classification of power quality disturbances using S-transform and modular neural network. *Electr. Power Syst. Res.* **2008**, *78*, 122–128. [CrossRef]
9. Dash, P.K.; Panigrahi, B.K.; Panda, G. Power quality analysis using S-transform. *IEEE Trans. Power Deliv.* **2003**, *18*, 406–411. [CrossRef]
10. Pinnegar, C.R.; Mansinha, L. The S-transform with windows of arbitrary and varying shape. *Geophysics* **2003**, *68*, 381–385. [CrossRef]
11. Djurović, I.; Sejdić, E.; Jiang, J. Frequency-based window width optimization for S-transform. *Int. J. Electr. Commun.* **2008**, *62*, 245–250. [CrossRef]
12. Sejdić, E.; Djurović, I.; Jiang, J. A window width optimized S-transform. *EURASIP J. Adv. Signal Process.* **2007**, *2008*, 672941. [CrossRef]
13. Huang, N.; Zhang, S.X.; Cai, G.W.; Xu, D.G. Power quality disturbances recognition based on a multiresolution generalized S-transform and a PSO-improved decision tree. *Energies* **2015**, *8*, 549–572. [CrossRef]
14. McFadden, P.D.; Cook, J.G.; Forster, L.M. Decomposition of gear vibration signals by the generalized S-transform. *Mech. Syst. Signal Process.* **1999**, *13*, 691–707. [CrossRef]

15. Pinnegar, C.R.; Mansinha, L. Time-local Fourier analysis with a scalable, phase-modulated analyzing function: The S-transform with a complex window. *Signal Process.* **2004**, *84*, 1167–1176. [CrossRef]

16. Mishra, S.; Bhende, C.N.; Panigrahi, B.K. Detection and classification of power quality disturbances using S-transform and probabilistic neural network. *IEEE Trans. Power Deliv.* **2008**, *23*, 280–287. [CrossRef]

17. Valtierra-Rodriguez, M.; Romero-Troncoso, R.D.J.; Osornio-Rios, R.A.; Garcia-Perez, A. Detection and classification of single and combined power quality disturbances using neural networks. *IEEE Trans. Ind. Electr.* **2014**, *61*, 2473–2481. [CrossRef]

18. Biswal, B.; Dash, P.K.; Panigrahi, B.K. Power quality disturbance classification using fuzzy c-means algorithm and adaptive particle swarm optimization. *IEEE Trans. Ind. Electr.* **2009**, *56*, 212–220. [CrossRef]

19. Mohanty, S.R.; Ray, P.K.; Kishor, N.; Panigrahic, B.K. Classification of disturbances in hybrid DG system using modular PNN and SVM. *Electr. Power Energy Syst.* **2013**, *44*, 764–777. [CrossRef]

20. Kumar, R.; Singh, B.; Shahani, D.T.; Chandra, A.; Al-Haddad, K. Recognition of power-quality disturbances using S-transform -based ANN classifier and rule-based decision tree. *IEEE Trans. Ind. Appl.* **2015**, *51*, 1249–1258. [CrossRef]

21. Meng, K.; Dong, Z.Y.; Wang, D.H.; Wong, K.P. A self-adaptive RBF neural network classifier for transformer fault analysis. *IEEE Trans. Power Syst.* **2010**, *25*, 1350–1360. [CrossRef]

22. Specht, D.F. Probabilistic neural networks and the polynomial Adeline as complementary techniques for classification. *IEEE Trans. Neural Netw.* **1990**, *1*, 111–121. [CrossRef] [PubMed]

23. Lin, W.M.; Lin, C.H.; Sun, Z.C. Adaptive multiple fault detection and alarm processing for loop system with probabilistic network. *IEEE Trans. Power Deliv.* **2004**, *19*, 64–69. [CrossRef]

24. *IEEE Recommended Practices for Monitoring Electric Power Quality*; IEEE Std 1159-1995; IEEE-SASB Coordinating Committees: New York, NY, USA, 1995.

25. *EN 50160: Voltage Characteristics of Electricity Supplied by Public Distribution System*; European Standards: Pilsen, Czech Republic, 2002.

26. Bohari, M.N. Voltage Dip 21/3/2014 0650hrs. Available online: http://powerquality.sg/wordpress/?p=194 (accessed on 21 March 2014).

27. Zhao, F.Z.; Yang, R.G. Power-quality disturbance recognition using S-transform. *IEEE Trans. Power Deliv.* **2007**, *22*, 944–950.

28. Hooshmand, R.; Enshaee, A. Detection and classification of single and combined power quality disturbances using fuzzy systems oriented by particle swarm optimization algorithm. *Electr. Power Syst. Res.* **2010**, *80*, 1552–1561.

29. Biswal, M.; Dash, P.K. Detection and characterization of multiple power quality disturbances with a fast S-transform and decision tree based classifier. *Digit. Signal Process.* **2013**, *23*, 1071–1083.

Article

A Wavelet-Modified ESPRIT Hybrid Method for Assessment of Spectral Components from 0 to 150 kHz

Luisa Alfieri [1,*], **Antonio Bracale** [1], **Guido Carpinelli** [2] **and Anders Larsson** [3]

[1]   Department of Engineering, University of Naples Parthenope, Centro Direzionale of Naples, Is C4, 80133 Naples, Italy; antonio.bracale@uniparthenope.it
[2]   Department of Electrical Engineering and Information Technology, University of Naples Federico II, Via Claudio, 21, 80138 Naples, Italy; guido.carpinelli@unina.it
[3]   Department of Engineering Sciences and Mathematics, Luleå University of Technology, 97187 Luleå, Sweden; anders.1.larsson@ltu.se
*   Correspondence: luisa.alfieri@uniparthenope.it; Tel.: +39-081-547-67-57

Academic Editor: Birgitte Bak-Jensen
Received: 11 November 2016; Accepted: 11 January 2017; Published: 14 January 2017

Abstract: Waveform distortions are an important issue in distribution systems. In particular, the assessment of very wide spectra, that include also components in the 2–150 kHz range, has recently become an issue of great interest. This is due to the increasing presence of high-spectral emission devices like end-user devices and distributed generation systems. This study proposed a new sliding-window wavelet-modified estimation of signal parameters by rotational invariance technique (ESPRIT) method, particularly suitable for the spectral analysis of waveforms that have very wide spectra. The method is very accurate and requires reduced computational effort. It can be applied successfully to detect spectral components in the range of 0–150 kHz introduced both by distributed power plants, such as wind and photovoltaic generation systems, and by end-user equipment connected to grids through static converters, such as fluorescent lamps.

Keywords: waveform distortion; high-frequency spectral components; wavelet transform; modified estimation of signal parameters by rotational invariance technique (ESPRIT); power quality

1. Introduction

Modern distribution systems are characterized by the simultaneous presence of renewable energy sources, storage systems and loads that actively contribute to the operation of such systems, along with an increasingly complex and performing information and communications technology relevant infrastructure. In this context, power quality (PQ) is one of the most important issue since adequate PQ levels guarantee the necessary compatibility among all equipment connected to the grid [1]. Among PQ disturbances, the waveform distortions are considered with much attention as a consequence of several factors such as: (1) the sensitivity of customers to such disturbances and (2) the growing penetration of electrical components responsible for new and significant waveform distortions.

Recently, renewable energy sources, especially wind and solar energy, and some high-efficiency end-user devices, e.g., fluorescent lamps powered by high-frequency ballasts, have attracted great interest as perturbing devices since they contribute significantly to increasing the distortion levels of voltage and current waveforms [2–4]. They have a spectrum that includes a wide range of frequencies that can exceed 2 kHz up to 150 kHz. This spectral content was initially indicated as "high-frequency distortion", but recently the term "supraharmonic" was introduced [5] and it is being used more frequently [2,6–8].

The presence of these high-frequency spectral components can cause different problems in the electrical power systems, such as: (i) the potential interferences with the power-line communication (PLC), that are included in the same range of frequency [7,8], or rather; (ii) the possible appearance of both series and parallel resonance phenomena [9–11] and (iii) the impact of this distortion on end-user equipment and on equipment in the grid [5,7–9]. Obviously, the aforesaid problems may be overcome with the exact knowledge of the high-frequency spectra in terms of both amplitude and frequency spectral components versus time, also when an active filtering has to be provided for their compensation [10]. However, these high-frequency spectral components are superimposed to low frequency spectral components with consequent conflicting needs in term of time window length (and frequency resolution) for their spectral analysis.

Moreover, such high-frequency disturbances still require not only appropriate analysis tools, but also adequate standardizations, which allow to define proper indices and their maximum thresholds for the evaluation and the limitation of the high-frequency spectral components. Note that the importance of covering the frequency range from 2 kHz to 150 kHz with adequate standards was also underlined in a European Committee for Electrotechnical Standardization (CENELEC) report, in the International Electrotechnical Commission (IEC) standard TS 62749, by different Conseil International des Grands Réseaux Électriques (CIGRE)/Congrès International des Réseaux Electriques de Distribution (CIRED)/Institute of Electrical and Electronic Engineers (IEEE) working group and in the application guide for the EN50160. Moreover, several international standard setting organizations are working on this topic [2,5–7].

As well known, currently, the IEC standards, which address the range from 0 kHz to 9 kHz, recommend for signal processing the use of the discrete Fourier transform over successive and rectangular time windows with a duration fixed as an integer multiple of the fundamental period, also called short time Fourier transform (STFT) [4,12]. The STFT method, together with the other STFT-based ones, has been proved to be an important tool for the global evaluation of waveform distortions, but it cannot provide more detailed information about each spectral component. This is due to the inaccuracies associated with the inherent fixed frequency resolution and spectral leakage problems. Moreover, in case of waveforms characterized by wide spectra, these problems became more significant due to the aforementioned different behaviours of low- and high-frequency spectral components and the increasing desynchronization of high-frequency components. As attempt to reduce STFT inaccuracies, IEC suggests the use of grouping not only for the low frequency range, but also for the spectral content from 2 kHz to 9 kHz, assuming high-frequency resolution to be unnecessary for this range, in opposition to the modern requirements [13,14]. Recently, in [14], some measurement methods have been also indicated to provide an overview of spectral content in the range from 2 kHz to 150 kHz, as opposed to detailed measurement methods used for low-frequency range. These measurement methods for the high-frequency spectral components are informative and not normative; they include the extension of the grouping from 9 kHz to 150 kHz using a 10 Hz frequency resolution and 200 Hz bands to group this spectral content [14] and, then, do not allow to obtain acceptable estimations of amplitudes and frequencies of high-frequency spectral components.

The IEC standard [14] suggests also the use of measurement methods according to Comité international spécial des perturbations radioélectriques CISPR 16-1-2 for the evaluation of the high-frequency spectral content; however, it states also that it is not always possible or practical to apply them. In fact, these measurement methods suffer of complex and expensive implementation; also, they provide an enormous amount of data to be stored as output. Moreover, IEC standard highlights that the measurement methods according to CISPR 16-1-2 focus only on emissions from devices under test, stating that they are *"not directly addressed to power quality investigations"*. Hence, the IEC standard suggests also the use of other techniques for the analysis of the high-frequency spectral content and advises about the possible inclusion of other methods in future editions of the standard.

However, with reference to the low frequency range, to solve the problems associated with the IEC standard method, many spectral analysis methods have been presented in the relevant literature for the assessment of time varying waveform distortions [4,12,14–17]. Among them, wavelet-based methods are introduced as an alternative tool, exploiting the wavelet insensitive behaviour to the frequency fluctuations. Moreover, several advanced parametric methods has also been proposed, such as the sliding-window estimation of signal parameters by rotational invariance technique (ESPRIT) method and the sliding-window Prony's method. Although these parametric methods are characterized by enhanced accuracy and great frequency resolution, their computational burden is excessive. In recent literature, hybrid sliding-window parametric-STFT methods and new modified sliding-window parametric methods have been proposed to reduce the computational effort [14–16,18,19]. With reference to the high-frequency range, in the relevant literature, only few methods were applied to the spectral analysis of waveforms with wide spectra [15,16]. In particular, the method proposed in [16] was not applied for the simultaneous spectral analysis of the entire range from 0 Hz to 150 kHz. It requires a high computational effort and a priori knowledge of the number of spectral components included in the analysed waveform, as well as the order of the correlation matrix. These two problems were overcome by the method proposed in [15], although it was not applied for the detection of high-frequency time varying spectral component, when large-spectrum waveforms are analysed.

Note also that low frequency spectral components and high frequency spectral components can be characterized by different behaviours in the frequency and time domain; low frequency components can be stationary while high frequency are usually not stationary.

Motivated by these issues, the new contribution of this study is in proposing a novel scheme to analyse electrical waveforms with spectral content up to 150 kHz that provides accurate estimation of parameters (mainly, frequencies, amplitudes and initial phases) of spectral components, while maintaining acceptable computational efforts. The proposed method allows to have an accurate contemporary knowledge of low and high frequency spectral components and their time behaviour; thus, it is useful for:

- the evaluation of supraharmonic emission;
- the definition of adequate power quality indices to introduce standard methods and limits;
- the study of supraharmonic propagation and impact in power systems;
- the definition of new models which can emulate supraharmonic injection;
- the use of proper adaptive active filters to reduce the emission of supraharmonics.

Based on the previously-cited main aspects of the spectral analysis methods offered by relevant literature, the proposed method applies a discrete wavelet transform (DWT) and a sliding-window modified ESPRIT method (SW MEM) [15] in two successive steps. The DWT divides the original signal into low-frequency and high-frequency waveforms. The SW MEM separately analyzes the two passbands for a separate estimation of the low-frequency and high-frequency spectral components.

The application of the proposed method has the following advantages:

- it allows a detailed estimation in time of each high-frequency spectral component;
- it has ability to provide the optimal time and frequency resolutions in each band to obtain an accurate time-frequency representation using the strategy of divide and conquer.

Eventually, we evidence that the main contribution of this paper refers to the application addressed (signal processing in power distribution systems in the entire frequency range) since the spectral analysis of waveforms in electrical distribution systems with wide spectra (including supraharmonics) is nowadays of the greatest interest. Moreover, even though the proposed scheme uses basic and well known methods, these methods are used in a novel and advantageous way.

This article is organized as follows: Section 2 introduces some considerations on the high-frequency components up to 150 kHz to outline their main spectral characteristics. In Section 3, the new sliding-window wavelet-modified ESPRIT-based method (SWWMEM) is described in detail.

Section 4 describes how the numerical validation of the proposed method is affected. The study conclusions are in Section 5.

2. Spectral Components in the Range from 2 kHz to 150 kHz

Nowadays, the main cause of waveform distortions in modern electrical power systems is the growing use of devices equipped with static converters. They can inject in the grid distorted currents with spectral components included in a wide frequency range (up to 150 kHz) [3,4,16,20]. In order to better underline how the problem of the waveform distortions that involve a wide frequency range is deeply rooted in modern electrical power systems, a detailed list of common supraharmonic emitters is provided in [21]. These are: (i) converters for industrial applications that mainly emit spectral components in the range from 9 kHz to 150 kHz; (ii) oscillation related to the electronic device commutation that mainly emit spectral components up to 10 kHz; (iii) street lamps that mainly emit spectral components up to 20 kHz; (iv) electro-vehicle chargers that mainly emit spectral components in the range from 15 kHz to 100 kHz; (v) photovoltaic and wind turbine inverters that mainly emit spectral components in the range from 4 kHz to 20 kHz; (vi) household devices, e.g., liquid-crystal display televisions or highly-efficient loads, that mainly emit spectral components in the range from 2 kHz to 150 kHz; (vii) PLC that mainly emit spectral components in the range from 9 kHz to 95 kHz.

Since, as seen, the supraharmonic emitters are very common the spectral components from 2 kHz to 150 kHz have recently received great interest in the relevant literature [2,7,16,20–27]. Many research activities have recently focused on the effects of the high-frequency sources on the power systems and their propagation [2,6,7]. These studies discerned between primary and secondary emissions. Primary emissions refer only to the distinctive disturbances of the considered load or production equipment. Secondary emissions refer to the disturbances caused by other sources near the considered equipment. Moreover, it was outlined that the frequencies from 2 kHz to 150 kHz are strictly linked to the switching technique adopted in the interfacing power-electronic converter. Several measurements taken by low power (up to 100 kW) equipment indicate that the most common switching frequencies are up to few tens of kHz [28]. Note that, sometimes, the secondary emission might also be due to PLC, they, as previously observed, generally work in the range from 9 kHz to 95 kHz and interferences could occur in the power system [8].

The main difference between the high-frequency emissions and low-frequency emissions is that, while the latter tend to propagate towards the distribution network, the former mainly flow within the installation, towards the other devices that offer a lower impedance than the network at so high frequencies (such as the electromagnetic compatibility (EMC) filter). Moreover, the presence of EMC filter as interface at the point of common coupling (PCC), also provides a low impedance for these high-frequency components coming from the distribution network. Therefore, high-frequency currents from both the installation and the network flow through the capacitor, causing high-frequency voltage distortion at the PCC [6].

Most of waveforms with spectral components from 0 kHz to 150 kHz are related to distributed generation (DG) and loads equipped with static converters.

With reference to DGs emission, photovoltaic systems (PVSs) and wind turbine systems (WTSs) are the more spread technologies and they generally are connected to the grid through power static converters [28]. Thus, high-frequency components are introduced at the PCC as primary emissions by the interfacing converters.

In PVSs, the high-frequency primary emission is linked to the adopted pulse-width modulation (PWM) technique of the inverter, although the presence of the electromagnetic interference (EMI) filter, used to reduce the harmonic emission, influences the amplitude of the high-frequency components with respect to an ideal PWM spectrum [7]. The high-frequency components can be detected in ideal operating conditions of the system, in correspondence of the frequencies $f_{k,m}^{PWM}$, given by:

$$f_{k,m}^{PWM} = [k \cdot f_{sw} \pm m \cdot f_0] \; \forall \, k \in \mathbb{N}, \; \forall \, m \in \mathbb{N}_0, \tag{1}$$

where f_{sw} is the switching frequency and f_0 is the power system fundamental frequency. They are sidebands centered on integer multiples of the switching frequency, which is generally a few tens of kHz.

Regarding the WTSs, the schemes equipped with a power static converter are the doubly fed induction generator (DFIG) and the full-scale power converter wind turbine [29]. For both these configurations, the power electronic converter is the main source of high-frequency primary emissions at the PCC.

The high-frequency secondary emissions at the PCC of both PVSs and WTSs are due to background voltage distortion and could increase significantly in the presence of resonance effects [20].

With reference to loads emission, several types of perturbing loads can introduce high-frequency components [7,8,30,31]. In particular, the propagation of high-frequency components as primary emissions of adjustable speed drives (ASDs) with a 6 kHz switching frequency in industrial networks has been analyzed in [31]. In [7] it was outlined that the electrical vehicles (EVs) produce a weak primary emission at low frequencies, while in the high-frequency range, the levels of current emission can be significant. Moreover, in the measured currents at the PCC of EV chargers, spectral components in a range from 10 kHz to 100 kHz have been detected [7].

The most diffuse disturbing loads are currently the fluorescent lamps powered by high-frequency ballast and the LED lamps [7,8]. Generally, an electronic ballast has a voltage stiff rectifier (which create harmonics) and also an inverter that is normally switching at constant frequency (typically from 30 kHz to 40 kHz). Components in both these frequency ranges can be measured at the grid side of the lamps as a emission in the current. In addition, since lamps larger than 25 W have to improve the harmonic emission, an active power factor correction (APFC) stage is placed directly after the rectifier. This is typically done with a DC to DC converter that make sure that the current drawn by the lamp is almost sinusoidal. However this means that there is an additional source of emission that changes amplitude and frequency over the fundamental cycle. The frequency emission from this part of the lamp is shifting from 30 to 40 kHz to 100 kHz over half of a fundamental cycle [32].

The high-frequency secondary emission at the PCC of the aforesaid loads are generally caused by background voltage or by the presence of several non-linear loads in the same installation.

3. The Proposed Method

In the most general case, a waveform in a power system can be characterized by both high-frequency and low frequency (up to 2 kHz) spectral components involving conflicting needs in term of time window length and frequency resolution for their spectral analysis.

The high-frequency spectrum is generally characterized by spectral components centred around frequencies not directly linked to the power system frequency and that are commonly defined as "asynchronous" components. The low frequency spectral components are mainly constituted by discrete components at frequencies that are linked to the power system frequency [6,7]. Moreover, high-frequency and low-frequency spectral components present different time behaviours, being high-frequency components often not stationary with fast dynamics, and consequently with frequencies and amplitudes that can rapidly vary vs. time. Finally, in the absence of resonance effects, the energy content of high-frequency components usually is very small if compared with the characteristic low-frequency spectral components.

Eventually, the spectral analysis of waveforms including both low-frequency and high-frequency spectral components requires an approach different from the traditional approach for the low-frequency components only. In particular, the method to be used should be characterized by the ability to:

- separate the original signal in different frequency bands;
- analyze different frequency bands with different time and frequency resolutions;
- analyze different frequency bands with adequate accuracy for different levels of distortions estimating both amplitude and frequency of each spectral component in time.

Motivated by above, the main features of the proposed method are (1) to isolate the different frequency bands of interest and (2) to separately analyse each band, taking into account the different behaviour and needs of low-frequency and high-frequency components (Figure 1).

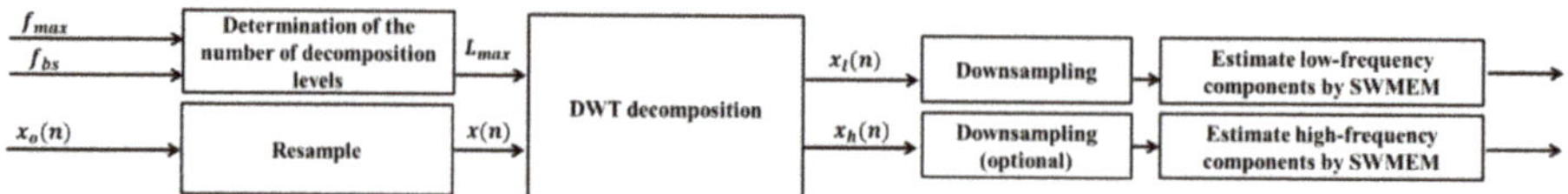

Figure 1. Scheme of the proposed method.

The proposed method involves a two-step procedure as shown in Figure 1. In particular, the selected values f_{bs} and f_{max} are the bands' separation frequency and the maximum frequency of interest, respectively. In the first step, the number of decomposition levels is calculated, the waveform is adequately resampled and then filtered by the decomposition of a discrete wavelet transform (DWT), which produces a low-frequency band and a high-frequency band of the original waveform. Then in the second step, the two parts of the waveform are resampled and analyzed separately by the sliding-window modified ESPRIT method (SW MEM).

In the first step, we take the advantage of wavelet suitability for studying signals in different frequency bands with different frequency resolution [33,34]. Moreover, the DWT, differently to a common low/band-pass filter, performs a waveform decomposition that guarantees no phase-shifting and no signal leakage among the decomposed bands of frequency. This allows that, if all of the reconstructed signals of the different bands of frequency are added up, the original waveform is again obtained. In the second step, we use ESPRIT method which is one of the most accurate parametric methods; in particular: (i) ESPRIT method performs better than Prony's method when the waveform to be analysed is corrupted by noise; (ii) DWT can be couplet better with ESPRIT method than with other parametric methods; in fact, if f_s is the waveform sampling rate, ESPRIT method, as well as DWT, is able to detect spectral components up to $f_s/2$. In the following subsections, the main features of each of these steps are described in detail.

3.1. First Step

The first step of the proposed hybrid method is based on the application of a DWT obtained by using a sampled mother wavelet with discrete scale and translation parameters, applied to a sampled waveform [12].

Specifically, given a sequence of samples $x(n)$ with $n = 0, 1, \ldots, N-1$ and the chosen continuous mother wavelet $\psi\left(\frac{t-b}{a}\right)$ with a and b continuous scale and translation parameters respectively, selecting $a = a_0^j$ and $b = ka_0^j b_0$, the sampled discrete mother wavelet $\psi_{j,k}(n)$ is:

$$\psi_{j,k}(n) = \frac{1}{\sqrt{a_0^j}}\psi\left(\frac{n - kb_0 a_0^j}{a_0^j}\right) \ \forall n = 0, 1, \ldots, N-1, \tag{2}$$

where k and j are integers and the values $a_0 > 1$, $b_0 > 0$ are fixed [12]. Hence, the DWT of $x(n)$ is provided by:

$$DWT(j,k) = \frac{1}{\sqrt{a_0^j}}\sum_{n=0}^{N-1}x(n)\psi^*\left(\frac{n - kb_0 a_0^j}{a_0^j}\right), \tag{3}$$

where the symbol * indicates the complex conjugate [12].

It is well known that the DWT achieves the decomposition of the waveforms on different levels. Two parts are obtained for each level and they represent the approximation a_j and the detail d_j of the waveform in progressively-halved bands of frequency. In particular, by defining the scaling function $\varphi\left(\frac{t-b}{a}\right)$ as the aggregation of wavelets with scale factor $a > 1$, and discretizing it as seen for the mother wavelet, at each level j, the approximation a_j and the detail d_j of the original waveform can be evaluated by the inverse DWT of the approximate $A_j(k)$ and detailed $D_j(k)$ coefficients, computed as follows:

$$
\begin{aligned}
D_j(k) &= \frac{1}{\sqrt{a_0^j}} \sum_{n=0}^{N-1} x(n)\psi^*\left(\frac{n-k}{a_0^j}\right) \\
A_j(k) &= \frac{1}{\sqrt{a_0^j}} \sum_{n=0}^{N-1} x(n)\varphi^*\left(\frac{n-k}{a_0^j}\right)
\end{aligned}
\tag{4}
$$

where k is related to the translation in time and j is related to the selected frequency band [12]. Figure 2 shows a representation of the DWT decomposition for a waveform sampled at $2f_{max}$ Hz, where f_{max} is the maximum frequency of interest (i.e., 150 kHz).

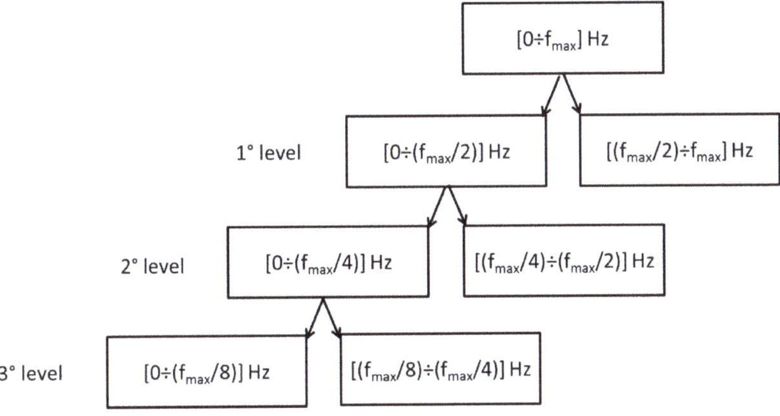

Figure 2. Discrete wavelet transform (DWT) decomposition scheme.

The DWT decomposition is often used as a filter bank, since it appears a sort of high and low pass band filters in cascade, with two bands of frequency obtained at each level. However, the bands are affected by overlap and attenuation at their edges. In the proposed method, by using a Meyer mother wavelet, multi-level DWT decomposition is achieved by assuming, as the number L_{max} of decomposition levels, a value that depends on f_{max} and on the bands' separation frequency f_{bs} (i.e., 2 kHz), i.e., $L_{max} = \left\lceil \log_2\left(\frac{f_{max}}{f_{bs}}\right) \right\rceil$, where the symbol $\lceil \cdot \rceil$ indicates the rounding up operation.

Note that as shown in Figure 1, the original waveform $x_o(n)$ to be analyzed can be pre-processed to adapt the sampling frequency to the maximum frequency of interest f_{max}, obtaining the DWT input waveform $x(n)$. This pre-processing consists in an upsampling of $x_o(n)$, in order to avoid filter bounds to be too close to the frequencies that must be detected.

Once the DWT decomposition is known, two different waveforms, $x_l(n)$ and $x_h(n)$, are obtained. They are the waveform $x_l(n)$ ("low-frequency waveform"), which includes only the approximation $a_{L_{max}-1}$ at the level $L_{max} - 1$ (Figure 2) and the waveform $x_h(n)$ ("high-frequency waveform"), which includes the sum of all of the details d_j (Figure 2). The level $L_{max} - 1$ instead of L_{max} was chosen to avoid the attenuation effects of the overlap introduced by the canonical DWT in the ranges of frequencies of interest.

Both the low-frequency waveform $x_l(n)$ and the high-frequency waveform $x_h(n)$ are separately analyzed in the second step of the proposed method (Figure 1).

3.2. Second Step

The second step of the proposed hybrid method is based on the application of the sliding-window modified ESPRIT method (SW MEM) described in [15]. This method is applied for the assessment of the low-frequency and high-frequency components included in $x_l(n)$ and $x_h(n)$, respectively.

Specifically, denoting with $x_i(n)$, of generic size N_i, either the sequence of the N_l-sized sampled data $x_l(n)$ or the sequence of the N_h-sized sampled data $x_h(n)$ (Figure 1), the ESPRIT model is used to approximate the waveform samples with a linear combination of M_i complex exponentials added to a white noise $r(n)$ [12]:

$$\hat{x}_i(n) = \sum_{k=1}^{M_i} A_k e^{j\psi_k} e^{(\alpha_k + j2\pi f_k)nT_s} + r(n), \; n = 0, 1, \ldots, N_i - 1, \tag{5}$$

where T_{si} is the sampling period and A_k, ψ_k, f_k, and α_k are the amplitude, initial phase, frequency and damping factor of the kth complex exponential, respectively. These are the unknown parameters to be estimated by using the transformation properties of the rotation matrix associated with the waveform samples. However in [15], it was shown that a reduction of the unknown parameters in the model (5) is possible, considering that the damping factors and the frequencies of spectral components in power system applications generally vary slightly vs. time, especially at low-frequencies. This is the basic principle of the SW MEM, where the estimation of frequencies and damping factors is realized only a reduced number of times along the whole waveform to be analyzed, with a great improvement in terms of computational effort [15]. In particular, the frequencies of the spectral components are initially evaluated in the first sliding window (also called "basis window"). Then the obtained values are assumed to be known quantities in the successive sliding windows (also called "no-basis windows"). The same handling is also performed for the damping factors.

Note that a check is conducted in each no-basis window to evaluate if the frequencies and damping factors obtained in the basis window can be still considered valid. In particular, the reconstruction error of the waveform can be checked; if the reconstruction error is greater than a fixed threshold, a new basis window is generated, the frequencies and the damping factors are updated and these new values are imposed in the following no-basis windows [15]. This check makes the SW MEM also suitable for the high-frequency components assessment, since they might significantly vary vs. time as previously mentioned.

The accuracy of the results and the computational effort of the SW MEM are influenced significantly by the number of exponentials, M_i, by the selected order N_{i1} of the correlation matrix and by the sampling frequency f_{si} [14].

In particular, it is worthwhile to outline the following considerations:

(1) Thanks to the DWT decomposition in the first step, both $x_l(n)$ and $x_h(n)$ include a reduced number of spectral components than the original sequence $x(n)$, so the analyses of these waveforms require smaller values of M_i than the analysis of the original sequence $x(n)$.

(2) Since $x_h(n)$ include only high-frequency components, the duration of the analysis window for this waveform can be chosen shorter than that for the spectral analysis of $x_l(n)$, consequently reducing the value of N_{i1}.

(3) Since the maximum frequency that the ESPRIT method and therefore the SW MEM are able to detect is half of the sampling frequency f_{si} [14]. The low frequency waveform $x_l(n)$ can be downsampled to only $2f_{bs}$ (two times the bands' separation frequency), reducing the number of samples in each window and, once again, the global computational burden.

Note also that to prevent the problem associated with the attenuation introduced by the DWT decomposition, only the spectral components over f_{bs} (i.e., 2 kHz) were stored when analyzing $x_h(n)$, whereas the spectral components up to f_{bs} were stored when analyzing $x_l(n)$.

Finally, it is important to observe that since the duration of window for the spectral analysis of $x_h(n)$ is shorter than that for the spectral analysis of $x_l(n)$, the proposed method could provide a number of high-frequency spectra greater than the number of spectra of the low-frequency waveform. This result is in accordance with the need for having a higher time resolution for the detection of the high-frequency components, since these components generally vary vs. time more than the low-frequency components.

4. Numerical Validation of the Proposed Method

Several numerical analysis of synthetic and measured waveforms in different operating conditions of typical generation systems and loads were performed. For sake of conciseness, only three case studies are shown in this Section. Specifically, the first two case studies analyse synthetic waveforms while the third case study is an actual waveform measured at the PCC of fluorescent lamps installation. In order to test the effectiveness of the proposed hybrid method (SWWMEM) in terms of both computational burden and accuracy, the same waveforms were analyzed also through the STFT method (STFTM), selecting a 5 Hz frequency resolution, and through the Sliding-Window ESPRIT method (SWEM) [12].

Moreover, also the spectrogram presented in [23] is included for all of the considered case studies, in order to underline the different approach of SWWMEM in respect to the currently available high-frequency signal processing technique.

All of the waveform analysis were performed in MATLAB environment. The MATLAB programs were developed and tested on a Windows PC with an Intel i7-3770 3.4 GHz and 16 GB of RAM.

4.1. Case Study 1

A synthetic 3 s waveform that emulates a current at the PCC of a PV system equipped with a full-bridge, unipolar inverter, is analyzed (Figure 3). Specifically, the waveform was assembled assuming a fundamental current of 40 A at 50.02 Hz and a frequency modulation index m_f of the inverter PWM technique equal to 200; then all odd harmonics up to the 27th order (low-frequency components) and a white noise with a standard deviation of 0.001 were added.

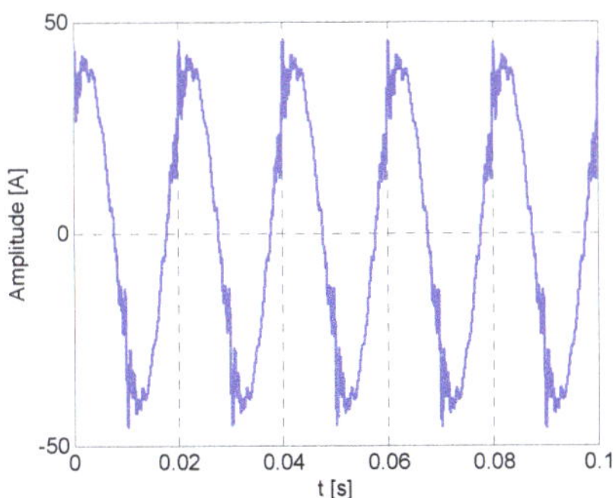

Figure 3. Case study 1: Synthetic current waveform.

The sampling frequency of the waveform was 50 kHz, in order to provide the most appropriate operating conditions for the parametric methods; this choice allowed the detection of the spectral components around the order $2m_f$, that are the most significant introduced by the inverter PWM and whose amplitudes were fixed up to 2% of the fundamental, in order to emulate the behavior of the PV system during high-irradiance conditions [35].

A resampling to 100 kHz was required in the first step of the proposed method, in order to guarantee an uncorrupted estimation of the spectral components of our interest. Moreover, fixing the bands' separation frequency f_{bs} equal to 2 kHz, the number L_{max} of decomposition levels was 5.

For the SWEM, the error threshold was fixed equal to 10^{-5}, and the window of analysis moves by 0.04 s without overlap. The same setting was chosen also for the spectral analysis of the low-frequency waveform (downsampled to 5 kHz) through the SWWMEM. For the analysis of the high-frequency range waveform, instead, the error threshold was fixed equal to 10^{-4}, and the window of analysis moves by 0.018 s without overlap.

Table 1 shows the average percentage errors of the frequencies (Table 1a) and of the amplitudes (Table 1b) for five selected spectral components. These spectral components are the fundamental, the 3rd and the 11th harmonic (low-frequency components) and two components ($2m_f + 1$ and $2m_f + 5$) linked to the inverter switching frequency (high-frequency components). The proposed method seems to outperform the STFTM, since it provides average percentage errors that are globally very similar to those obtained by the SWEM, both for amplitude and frequency estimation.

Table 1. Case study 1: (**a**) Average percentage errors of frequency; (**b**) Average percentage errors of amplitude.

(a)					
Average Errors of Frequency (%)					
	Fundamental	**3rd Harmonic**	**11th Harmonic**	**401st**	**405th**
SWEM	2.84×10^{-5}	6.55×10^{-4}	4.12×10^{-5}	3.36×10^{-6}	1.89×10^{-5}
SWWMEM	8.74×10^{-6}	2.59×10^{-4}	3.65×10^{-5}	1.38×10^{-7}	1.67×10^{-5}
STFTM	0.04	0.04	0.04	0.04	0.04

(b)					
Average Errors of Amplitude (%)					
	Fundamental	**3rd Harmonic**	**11th Harmonic**	**401st**	**405th**
SWEM	1.44×10^{-4}	0.017	0.0031	0.011	0.092
SWWMEM	4.59×10^{-4}	0.0029	0.0011	0.010	0.023
STFTM	0.06	0.70	0.25	82.02	87.63

In particular, for the low-frequency spectral components, the errors of SWWMEM are slightly lower than that of SWEM, while the proposed method provides average percentage errors on frequency and amplitude that are two and three orders of magnitude lower than those obtained through the STFTM.

For the high-frequency components, the average percentage errors on the amplitudes obtained by the proposed method show a significantly improved accuracy compared to the errors obtained through STFTM; in fact, STFTM errors are greater than 82% and are affected by spectral leakage problems that increase as the frequency increases. Also in this range of frequencies, the SWWMEM provides errors very close to the SWEM errors, and they do not exceed 0.09% in amplitude and 2.5×10^{-5}% in frequency. Similar mean errors were observed for all the spectral components of the whole waveform using both the SWWMEM and the SWEM. However, the proposed method generally seemed to produce slightly lower amplitude mean errors than SWEM.

Note that also the grouping, recommended by the IEC, both for the low-frequency components (using a 200 ms analysis window) and for the high-frequency components (using a 100 ms analysis window) was evaluated. This post-processing provides, for the low-frequency components, percentage errors on the amplitude practically coincident with the values obtained for the STFTM in Table 1. Conversely, for the high-frequency components, the grouping provides an aggregation of spectral components centered around fixed frequencies equal to 19.9 kHz and 20.1 kHz, as shown in Figure 4. It is clear that at high-frequency, this aggregation allows only to detect approximatively the allocation of the energetic content around the fixed frequencies.

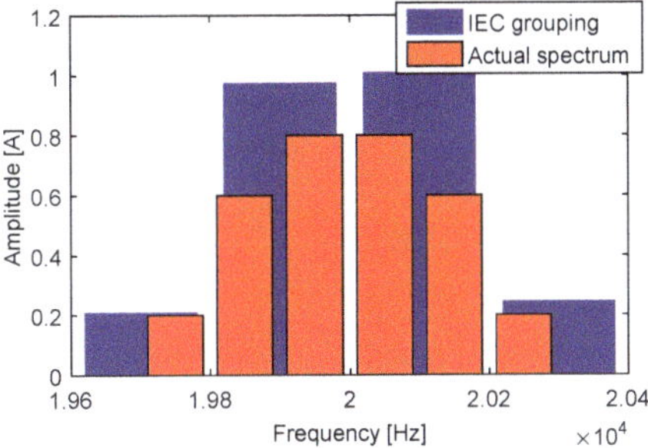

Figure 4. Case study 1: Comparison between high-frequency International Electrotechnical Commission (IEC) grouping and actual high-frequency spectrum.

In Figure 5 the spectrogram with 1 kHz frequency resolution and 1 ms time resolution is shown for the considered waveform, in order to underline the different approach of the spectrogram to the spectral analysis of wide-spectra waveforms in respect to SWWMEM.

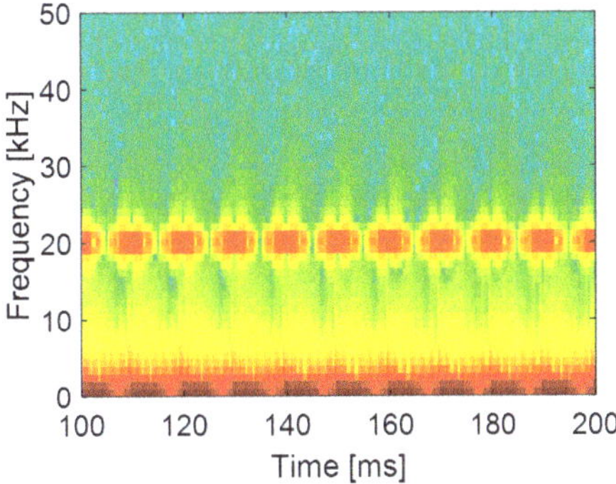

Figure 5. Case study 1: Time-frequency representation obtained by means the spectrogram with 1 kHz frequency resolution and 1 ms time resolution.

As shown in Figure 5, the spectrogram was able to detect the spectral content both at low-and high-frequency with reduced computational effort, even if a great spread of colour intensity around the real spectral components can be observed, cause of the spectral leakage and of the too large frequency resolution. In particular, the high-frequency spectral components related to the switching frequency appear concentrated around 20 kHz, but, unlike the proposed method, it is impossible to individuate both the correct number of spectral components included in that range and each specific frequency and/or amplitude. Moreover, also a fake periodical variability in time seems to be introduced by this spectral analysis method. Similar behaviour can be observed at low-frequency.

Finally, Table 2 shows the computational times required by all of the methods to analyze the 3 s waveform, per unit of computational time required by STFTM. It is evident that SWWMEM requires a computational time comparable to that required by STFTM and that SWEM requires a computational time that is about 22 times greater than SWWMEM, providing however results that are globally affected by similar errors. This is due to the presence of the DWT decomposition and of the resampling in the proposed method. In fact, they allow to model both the low-frequency and high-frequency waveforms with a reduced number of exponentials and with a reduced number of samples per analysis window than the SWEM. In this way, the computational burden of the SWWMEM is low although the method holds the accuracy of a parametric method.

Table 2. Case study 1: computational times in per unit of computational time required by STFTM.

	Computational Time (p.u.)
SWEM	221.40
SWWMEM	4.31
STFTM	1

4.2. Case Study 2

In this case study, a frequency-modulated, high-frequency spectral component was added to the synthetic waveform of the previous case study. This was made in order to both emulate the presence of secondary emission and to test the performance of the proposed method in the detection of time-varying spectral components, that are typical in the high-frequency range. The added spectral component $s_{tv}(t)$ was a tone at $f_{tv} = 17,598\ Hz$ with a sinusoidal modulation in frequency, according to Equation (6):

$$s_{tv}(t) = A_{tv}\cos(2\pi f_{tv}t + \varphi_{tv}(t)) \tag{6}$$

where:

$$\varphi_{tv}(t) = A_\varphi \sin(2\pi f_\varphi t) \tag{7}$$

and A_{tv} was fixed equal to the 0.5% of the fundamental amplitude, A_φ was 5 Hz and f_φ was 1 Hz. The spectral analyses of this waveform were performed through the same settings of the previous case study.

Figure 6 shows few ms of the actual modulated frequency and its estimations obtained through STFTM, SWEM, SWWMEM and IEC grouping. Specifically, Figure 6a shows that the IEC grouping associates the main part of the energetic content in correspondence of the group at 17.5 kHz. Figure 6b shows a focus on the graphical comparison among the actual modulated frequency, the STFTM, SWEM and SWWMEM estimations. Note that, since in this case the actual value of the spectral component was known "a priori", for STFTM it is assumed that a single component is around 17,598 Hz and the other components are spectral leakage. So the modulated component in Figure 5 is that with the maximum amplitude value in a range of 20 Hz around the actual mean value f_{tv}. Still, it is clear that STFTM fails the detection due to both the spectral leakage problems and the fixed frequency resolution, while SWEM and SWWMEM are able to clearly identify the modulation. However, the proposed method seemed to provide the best results for this estimation.

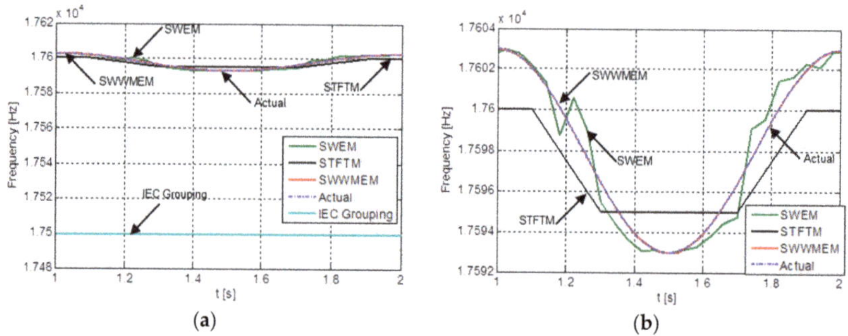

Figure 6. Case study 2: comparison among the actual time-varying frequency and its estimations obtained through: (**a**) IEC grouping, STFTM, SWEM and SWWMEM; (**b**) STFTM, SWEM and SWWMEM.

Since the average percentage errors on the estimation of the Table 1 spectral components slightly varied with respect to the results in case study 1, Table 3 shows only the average percentage frequency errors and the average percentage amplitude errors for the added spectral component. The results are coherent with the behaviors shown in Figure 4, since the lowest amplitude and frequency errors were provided by the proposed method. In particular, STFTM and SWEM amplitude errors were two order of magnitude higher than SWWMEM amplitude error. Moreover, the average percentage frequency error provided by SWWMEM was two and three orders of magnitude lower than the SWEM and STFTM frequency errors, respectively.

Table 3. Case study 2: Average percentage errors of frequency and amplitude for the frequency-modulated spectral component.

	Average Errors of Frequency (%)	Average Errors of Amplitude (%)
SWEM	0.0044	24.03
SWWMEM	2.72×10^{-5}	0.060
STFTM	0.010	19.18

In Figure 7 the spectrogram with 1 kHz frequency resolution and 1 ms time resolution is shown for the considered waveform, in order to underline once again the different approach of the spectrogram to the spectral analysis of wide-spectra waveforms in respect to SWWMEM.

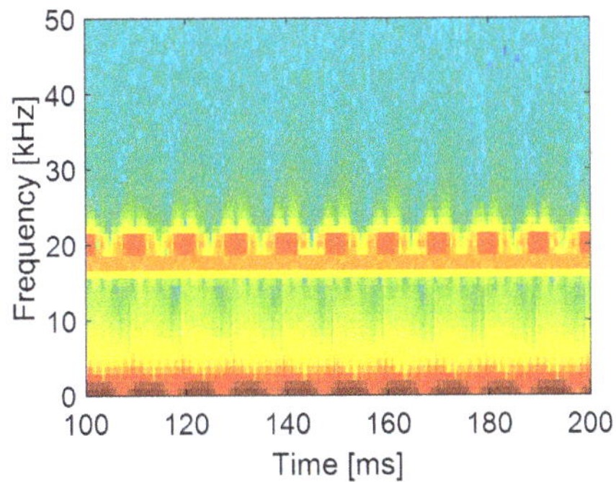

Figure 7. Case study 2: Time-frequency representation obtained by means the spectrogram with 1 kHz frequency resolution and 1 ms time resolution.

Similarly to the previous case study, in Figure 7 the spectrogram was able to detect the presence of spectral content both at low-and high-frequency, but, differently to SWWMEM, only rough and approximated information can be obtained. In particular, also the contribute of the spectral component around 17.958 kHz can be detected, but, in this case the related frequency appears time-invariant, since the real frequency modulation of this spectral component is not individuated, cause of the spectral leakage problems.

Finally, Table 4 shows the computational time required by the three methods to analyze the 3-s waveform, per unit of computational time required by STFTM. Note that the SWWMEM required a higher computational time than the previous case study, since the frequency variation of the high-frequency modulated component prevented to keep the estimated frequencies constant, requiring often an updating of their value (step 2 of proposed method). However, the time required by SWWMEM was still one order of magnitude lower than the time required by SWEM.

Table 4. Case study 2: Computational times in per unit of computational time required by STFTM.

	Computational Time (p.u.)
SWEM	217.76
SWWMEM	18.85
STFTM	1

4.3. Case Study 3

A 0.2 s-current waveform measured at the PCC of a single fluorescent lamp powered by high-frequency ballast was analyzed. The total active power consumption of the lamp was about 0.1 kW. The current waveform was measured with a Pearson current probe, model 411. The probe has its 3 dB cut-off frequencies at 1 Hz and 20 MHz with an amplitude accuracy of −0%, +1% in the frequency range of interest. The phase accuracy is less than 1 degree between 60 Hz to 333 kHz. The current was then sampled with 12 bits resolution and 10 MS/s sampling speed, and a low-pass filter with a cut-off frequency 1 MHz was used for anti-aliasing purpose. More details on the installation structure, instrument specification and error verification of measurement are available in [8,36].

The analyzed current is shown in Figure 8; the high-frequency components on the peak of the waveform are clearly evident, unlike the high-frequency distortion around the zero crossing.

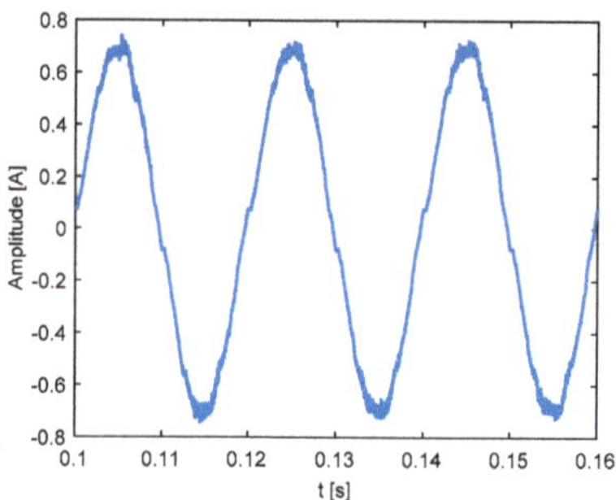

Figure 8. Case study 3: Actual current waveform.

However, according to the measurement procedure, based on 10 MHz sampling rate, the maximum frequency of interest cannot exceed 5 MHz. In this way, fixing the SWWMEM the bands' separation frequency f_{bs} to 2 kHz, the maximum number L_{max} of decomposition levels was 11. A resample to 10 kHz was performed for the low-frequency waveform while the high-frequency waveform was downsampled to 250 kHz, since our interest was the detection of the spectral components up to 150 kHz. Then, the two different waveforms, i.e., $x_l(n)$ and $x_h(n)$, were analyzed with a sliding window of 0.04 s and of 5×10^{-4} s, fixing an error threshold of 10^{-4} and 10^{-3}, respectively. The measured current was downsampled to 250 kHz also for the analyses through SWEM and STFTM, in order to guarantee a correct comparison of the methods in terms of computational burden. In particular, for the SWEM, the error threshold was fixed equal to 10^{-4}, and the sliding window duration was fixed to 0.04 s.

The spectra obtained through the SWWMEM and STFTM were almost similar at low-frequency. In fact, SWWMEM and STFTM individuated the same significant spectral components, though STFTM provided lower amplitudes than SWWMEM cause of the high spectral leakage. These low-frequency spectra included a fundamental component with a peak amplitude equal to about 0.7 A, and all of the other odd harmonics, whose amplitudes globally decrease as the harmonic order increases. These harmonic components reached a maximum amplitude equal to 3.6% of the fundamental in correspondence of the third harmonic, as shown in Figure 9a. Figure 9a shows the low-frequency spectrum obtained through SWWMEM, in percentage values of the fundamental amplitude. Finally the SWEM detected a reduced number of spectral components at low-frequency than the other two methods. Probably, this was due to a too high duration of the analysis window that shifted most of the detected spectral components to high-frequency.

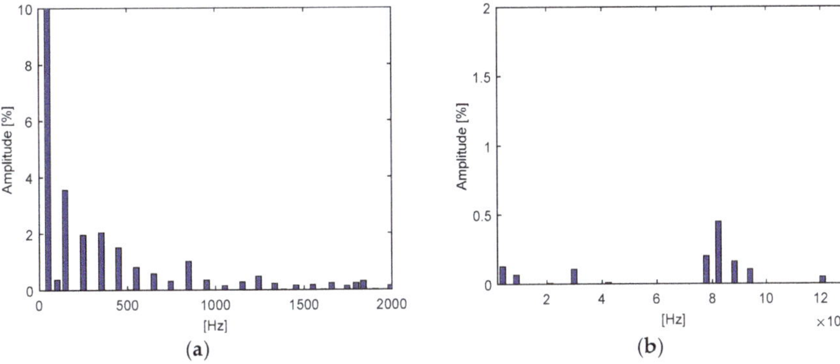

Figure 9. Case study 3: (**a**) Percentage low-frequency spectrum and (**b**) percentage high-frequency spectrum obtained through SWWMEM.

At high-frequency, the STFTM spectral leakage increased, spreading the spectral content and missing the detection of specific tones. In particular, a distributed spectral content from 50 kHz to 90 kHz was observed. The same high-frequency spectrum was also detected by SWEM. In both the cases, the amplitudes were decreasing from 50 kHz to 90 kHz, with a maximum equal to the 0.1% and 0.3% of the fundamental amplitude for STFTM and SWEM, respectively. Instead, the proposed method, due to the shorter analysis window, provided a high-frequency spectrum with defined single tones, as shown in Figure 9b.

The spectrum in Figure 9b is referred to the first 5×10^{-4} s of the waveform shown in Figure 8, but it is very interesting to note that in the following analysis windows the SWWMEM provided a periodic shift of the spectral components around 80 kHz, from 35 kHz to 95 kHz. This means that the proposed method is able to detect a time varying high-frequency spectra. Figure 10 shows the time trend of the spectral component with maximum amplitude detected in the aforesaid range of frequency. Specifically, Figure 10a shows the frequency variation in time, while Figure 10b shows the amplitude variation in time. This component could be related to the zero-crossing, since, as shown in Figure 10, both the frequency and the amplitude variations in time had a periodicity of about 10 ms [8]. Note also that this component reached the maximum amplitude (about 0.03 A) at about 50 kHz, namely when the positive and negative peaks of the current waveform shown in Figure 8 occurred.

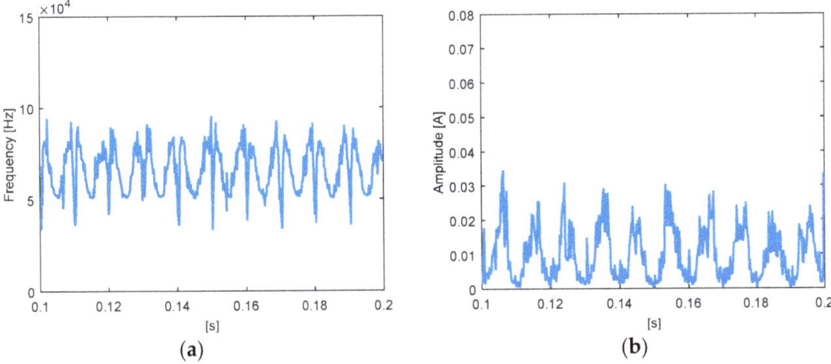

Figure 10. Case study 3: (**a**) Frequency and (**b**) amplitude of a time varying high-frequency component obtained by SWWMEM.

In Figure 11 the spectrogram with 1 kHz frequency resolution and 1 ms time resolution is shown also for the considered measured waveform, always in order to underline the different approach of the spectrogram to the spectral analysis of wide-spectra waveforms in respect to SWWMEM.

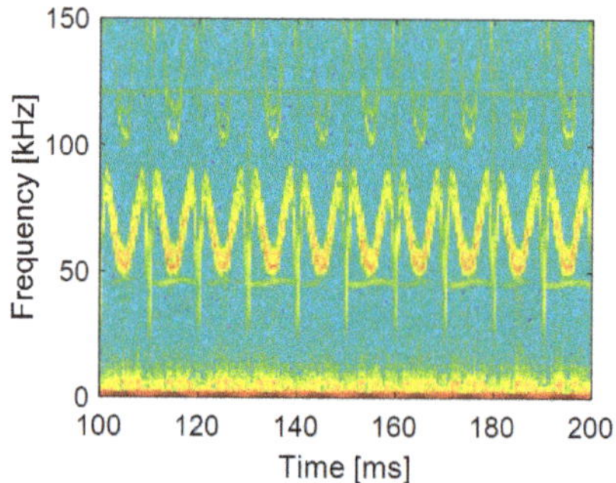

Figure 11. Case study 3: Time-frequency representation obtained by means the spectrogram with 1 kHz frequency resolution and 1 ms time resolution.

Figure 11 shows the time-varying spectral content in the range from 50 kHz to 100 kHz, so, the spectrogram appeared able to detect significant variations in time of the high-frequency spectral content. However, differently by SWWMEM, once again, only no-detailed information on the specific frequency and amplitude of that spectral component can be obtained.

Finally, Table 5 shows the computational times required by all of the methods to analyze the waveform, per unit of computational time required by STFTM: the results are coherent to what was theoretically expected, since the computational time required by the proposed method is about four time of that required by ICEM, but it is four order of magnitude lower than that required by the SWEM.

Table 5. Case study 3: computational times in per unit of computational time required by STFTM.

	Computational Time (p.u.)
SWEM	1.17×10^5
SWWMEM	4.35
STFTM	1

5. Conclusions

In this paper, we proposed a new sliding-window Wavelet-Modified ESPRIT scheme that seems to be particularly suitable for accurate analysis of waveforms that have very wide spectra; the scheme also guarantees a relatively low computational cost.

The method was based on the DWT decomposition of the original signal into low-frequency and high-frequency waveforms, separately analyzed by the SWWMEM in order to provide optimal time and frequency resolutions in each band and, then, an accurate time-frequency representation.

The proposed scheme was tested and evaluated on waveforms measured on an actual power system and on synthetically-generated data sequences, and was compared to the SWEM and the STFTM. These waveforms are characteristic emissions of dispersed generations (such as PVs and WTs) and modern loads (such as fluorescent lamps).

The numerical applications showed that, in presence of slightly time-varying components, the SWWMEM provided almost the same accuracy of the SWEM for both the low-frequency and the high-frequency components with a significantly reduced computational burden. In presence of highly time-varying high-frequency components, instead, the proposed method proved to overcome the limits of both the SWEM and STFTM for the detection of the high-frequency components, still with acceptable computational efforts.

The analyzed case studies give also great emphasis to the inability of the currently available high-frequency signal processing techniques, i.e., the spectrogram with 1 kHz frequency resolution and 1 ms time resolution, in a proper detection of each spectral component included in the wide-spectra (from 0 kHz to 150 kHz) waveforms. The sliding-window Wavelet-Modified ESPRIT scheme provides always detailed information in term of frequency, amplitude and time-variability for both the spectral components at low- and high-frequency.

Based on the aforesaid observations and on the possibility of properly select the duration of the SWWMEM analysis windows, in future works, the proposed method could reveal an adequate tool for the evaluation of new PQ indices in presence of wide spectra. This is due, i.e., to the opportunity of performing a time-aggregation of the spectral components in the range from 2 kHz to 150 kHz, in opposition or complementarily to the currently-proposed frequency-aggregation.

Acknowledgments: This work was supported by the University of Napoli Parthenope in the framework of "Bando per il Sostegno alla Ricerca individuale per il triennio 2015-17".

Author Contributions: Authors contributed in the same way all over the paper.

Conflicts of Interest: The authors declare no conflict of interest.

References

1. Hatziargyriou, N.; Dimeas, A.; Tomtsi, T.; Weidlich, A. *Energy-Efficient Computing and Networking*; Springer: Heidelberg, Germany, 2011.
2. Meyer, J.; Bollen, M.; Amaris, H.; Blanco, A.M.; Gil de Castro, A.; Desmet, J.; Klatt, M.; Kocewiak, Ł.; Rönnberg, S.; Yang, K. Future work on harmonics—Some expert opinions Part II—Supraharmonics, standards and measurements. In Proceedings of the 16th IEEE International Conference on Harmonics and Quality of Power (ICHQP), Bucharest, Romania, 25–28 May 2014.
3. International Council on Large Electrical Systems (CIGRE). *Impact of Increasing Contribution of Dispersed Generation on the Power System*; CIGRE WG 37-23; CIGRE: Paris, France, 1999.
4. Ribeiro, P. *Time-Varying Waveform Distortions in Power Systems*; John Wiley & Sons: New York, NY, USA, 2009.
5. Emanuel, A.; McEachern, A. Electric power definitions: A debate. In Proceedings of the IEEE Power & Energy Society (PES) General Meeting, Vancouver, BC, Canada, 21–25 July 2013.
6. Moreno-Munoz, A.; Gil-de-Castro, A.; Romero-Cavadal, E.; Rönnberg, S.; Bollen, M. Supraharmonics (2 to 150 kHz) and multi-level converters. In Proceedings of the IEEE 5th International Conference on Power Engineering, Energy and Electrical Drives (POWERENG), Riga, Latvia, 11–13 May 2015; pp. 37–41.
7. Renner, H.; Heimbach, B.; Desmet, J. Power quality and electromagnetic compatibility: Special report session 2. In Proceedings of the 23rd International Conference and Exhibition on Electricity CIRED, Lyon, France, 15–18 June 2015.
8. Larsson, E.O.A.; Bollen, M.H.J. Measurement result from 1 to 48 fluorescent lamps in the frequency range 2 to 150 kHz. In Proceedings of the 14th International Conference on Harmonics and Quality of Power (ICHQP), Bergamo, Italy, 26–29 September 2010; pp. 1–8.
9. European Committee for Electrotechnical Standardization (CENELEC). *Electromagnetic Interference between Electrical Equipment/Systems in the Frequency Range below 150 kHz*; CENELEC SC205A/Sec0339/R; CENELEC: Brussels, Belgium, 2013.
10. Gallo, D.; Langella, R.; Testa, A.; Hernandez, J.C.; Papic, I.; Blazic, B.; Meyer, J. Case studies on large PV plants: Harmonic distortion, unbalance and their effects. In Proceedings of the 2013 IEEE Power & Energy Society General Meeting, Vancouver, BC, Canada, 21–25 July 2013.
11. Luo, A.; Xie, N.; Shuai, Z.; Chen, Y.; Jin, G.; Ma, F.; Lv, Z. Large-scale photovoltaic plant harmonic transmission model and analysis on resonance characteristics. *IET Power Electr.* **2015**, *8*, 565–573. [CrossRef]
12. Caramia, P.; Carpinelli, G.; Verde, P. *Power Quality Indices in Liberalized Markets*; Wiley-IEEE Press: Chippenham, UK, 2009.

13. International Electrotechnical Commission (IEC). *IEC Standard 61000-4-7: General Guide on Harmonics and Interharmonics Measurements, for Power Supply Systems and Equipment Connected Thereto*; IEC: Geneva, Switzerland, 2010.

14. International Electrotechnical Commission (IEC). *IEC Standard 61000-4-30: Testing and Measurement Techniques—Power Quality Measurement Methods*; IEC: Geneva, Switzerland, 2015.

15. Alfieri, L.; Carpinelli, G.; Bracale, A. New ESPRIT-based method for an efficient assessment of waveform distortions in power systems. *Electr. Power Syst. Res.* **2015**, *122*, 130–139. [CrossRef]

16. Gu, I.Y.H.; Bollen, M.H.J. Estimating interharmonics by using sliding-window ESPRIT. *IEEE Tran. Power Deliv.* **2008**, *23*, 13–23. [CrossRef]

17. Alfieri, L. Some advanced parametric methods for assessing waveform distortion in a smart grid with renewable generation. *EURASIP J. Adv. Signal Process.* **2015**, 1–16. [CrossRef]

18. Chang, G.W.; Chen, C.-I. Measurement techniques for stationary and time-varying harmonics. In Proceedings of the IEEE Power & Energy Society (PES) General Meeting, Minneapolis, MN, USA, 25–29 July 2010.

19. Bracale, A.; Carpinelli, G.; Lauria, D.; Leonowicz, Z.; Lobos, T.; Rezmer, J. On some spectrum estimation methods for analysis of nonstationary signals in power systems. Part I. Theoretical aspects. In Proceedings of the 11th International Conference on Harmonics and Quality of Power, New York, NY, USA, 12–15 September 2004.

20. Barros, J.; de Apráiz, M.; Diego, R.I. Measurement of voltage distortion in the frequency range 2–9 kHz. In Proceedings of the IEEE International Workshop on Applied Measurements for Power Systems (AMPS), Aachen, Germany, 22–24 September 2010; pp. 70–73.

21. Rönnberg, S.; Bollen, M. Power quality issues in the electric power system of the future. *Electr. J.* **2016**, *29*, 49–61. [CrossRef]

22. Bollen, M.; Olofsson, M.; Larsson, A.; Rönnberg, S.; Lundmark, M. Standards for supraharmonics (2 to 150 kHz). *IEEE Electr. Comp. Mag.* **2014**, *3*, 114–119. [CrossRef]

23. Larsson, E.O.A.; Bollen, M.H.J.; Wahlberg, M.G.; Lundmark, C.M.; Rönnberg, S.K. Measurements of high frequency (2–150 kHz) distortion in low-voltage networks. *IEEE Trans. Power Deliv.* **2010**, *25*, 1749–1757. [CrossRef]

24. Bollen, M.; Wahlberg, W.; Ronnberg, S. Interaction between narrowband power-line communication and end-user equipments. *IEEE Trans. Power Deliv.* **2011**, *26*, 2034–2039.

25. Carrasco, J.M.; Franquelo, L.G.; Bialasiewicz, J.T.; Galvan, E.; Guisado, R.C.P.; Prats, M.A.M.; Leon, J.I.; Moreno-Alfonso, N. Power-electronic systems for the grid integration of renewable energy sources: A survey. *IEEE Trans. Ind. Electron.* **2006**, *53*, 1002–1016. [CrossRef]

26. Bartak, G.F.; Abart, A. EMI of emissions in the frequency range 2 kHz–150 kHz. In Proceedings of the 22nd International Conference and Exhibition on Electricity CIRED, Stockholm, Sweden, 10–13 June 2013.

27. International Electrotechnical Commission (IEC). *CISPR/1270/INF: International Special Committee on Radio Interference (CISPR) ISPR Guidance Document on EMC of Equipment Connected to the Smart Grid*; International Electrotechnical Commission (IEC): Geneva, Switzerland, 2014.

28. Chicco, G.; Russo, A.; Spertino, F. Supraharmonics: Concepts and experimental results on photovoltaic systems. In Proceedings of the International Symposium on Networks, Computers and Communications (ISNCC), Łagów, Poland, 15–18 June 2015; pp. 1–6.

29. Li, H.; Chen, Z. Overview of different wind generator systems and their comparisons. *IET Renew. Power Gener.* **2008**, *2*, 123–138. [CrossRef]

30. De Rosa, F.; Langella, T.; Sollazzo, A.; Testa, A. On the interharmonic components generated by adjustable speed drives. *IEEE Trans. Power Deliv.* **2005**, *20*, 2535–2543. [CrossRef]

31. Frey, D.; Schanen, J.L.; Quintana, S.; Bollen, M.; Conrath, C. Study of high frequency harmonics propagation in industrial networks. In Proceedings of the International Symposium on Electromagnetic Compatibility (EMC EUROPE), Rome, Italy, 17–21 September 2012; pp. 1–5.

32. International Electrotechnical Commission (IEC). *IEC 61000-3-2: Electromagnetic Compatibility (EMC)—Part 3-2: Limits—Limits for Harmonic Current Emissions (Equipment Input Current ≤ 16 A per Phase)*; IEC: Geneva, Switzerland, 2014.

33. Barros, J.; Diego, R.I.; de Apraiz, M. Applications of wavelet transform for analysis of harmonic distortion in power systems: A review. *IEEE Trans. Instrum. Meas.* **2012**, *61*, 2604–2611. [CrossRef]

34. Ribeiro, P.F. *Time-Varying Waveform Distortions in Power Systems*; Wiley-IEEE Press: Chippenham, UK, 2009.

35. Gallo, D.; Landi, C.; Luiso, M. AC and DC power quality of photovoltaic systems. In Proceedings of the IEEE International Instrumentation and Measurement Technology Conference (I2MTC), Graz, Austria, 13–16 May 2012; pp. 576–581.
36. Larsson, A. On High-Frequency Distortion in Low-Voltage Power Systems. Ph.D. Thesis, Department of Electric Power Engineering, Luleå University of Technology, Luleå, Sweden, 2011.

Article

Stability Analysis and Stability Enhancement Based on Virtual Harmonic Resistance for Meshed DC Distributed Power Systems with Constant Power Loads

Huiyong Hu, Xiaoming Wang, Yonggang Peng *, Yanghong Xia, Miao Yu and Wei Wei

College of Electrical Engineering, Zhejiang University, Hangzhou 310027, China; huhuiyong@zju.edu.cn (H.H.); wxm_15@zju.edu.cn (X.W.); royxiayh@zju.edu.cn (Y.X.); zjuyumiao@zju.edu.cn (M.Y.); wwei@zju.edu.cn (W.W.)
* Correspondence: pengyg@zju.edu.cn; Tel.: +86-571-8795-2653

Academic Editor: Birgitte Bak-Jensen
Received: 24 November 2016; Accepted: 2 January 2017; Published: 9 January 2017

Abstract: This paper addresses the stability issue of the meshed DC distributed power systems (DPS) with constant power loads (CPLs) and proposes a stability enhancement method based on virtual harmonic resistance. In previous researches, the network dynamics of the meshed DC DPS are often ignored, which affects the derivation of the equivalent system impendence. In addition, few of them have considered the meshed DC DPS including multiple sources with voltage-controlled converters and CPLs. To tackle the aforementioned challenge, this paper mainly makes the following efforts. The component connection method (CCM) is employed and expanded to derive the stability criterion of the meshed DC DPS with CPLs. This stability criterion can be simplified to relate only with the network node admittance matrix, the output impendences of the sources, and the input admittances of the CPLs. A virtual harmonic resistance through the second-order generalized integrator (SOGI) is added in the source with the voltage-controlled converter to lower the peak of the source output impendence, which can enhance the stability of the meshed DC DPS. The effectiveness of the proposed stability criterion and stability enhancement method are verified by nonlinear dynamic simulations.

Keywords: meshed DC distributed power systems; stability criterion; virtual harmonic resistance; stability enhancement

---

1. Introduction

With the increasing penetration of renewable energy generation into the modern electric girds, meshed DC distributed power systems (DPS) have been widely adopted [1–3]. The meshed DC DPS physically is composed of smaller power modules/subsystems. Usually, these smaller power modules/subsystems are designed only based on the stability requirement in its stand-alone operation, and thus it can operate well in the stand-alone application. However, the meshed DC DPS may become unstable due to the interaction among the modules/subsystems and the network [4]. Furthermore, the negative resistance characteristic of the constant power load (CPL) is an important unstable factor.

Many stability methods in previous researches are given to analyze the stability of the single-source-single-load system or the parallel-source-parallel-load system [4–6]. However, the network dynamics of the meshed DC DPS are often ignored, which affects the derivation of the equivalent system impendence. In addition, few of them have considered the meshed DC DPS including multiple sources with voltage-controlled converters and CPLs. A general approach for the stability analysis of the meshed DC DPS is to build its whole state-space model, and to identify the eigenvalues of the state matrix [7]. However, this method requires the detailed models of loads and network dynamics, and the formulation of the system matrices may be very complex. To overcome this problem, the component connection method (CCM) is introduced for the stability analysis [8]. In the CCM, the dynamics of network and power modules/subsystems are modeled separately by a set of two vector-matrix equations, which is beneficial to reduce the computation burden of formulating the system transfer matrices [9]. For the ac power-electronics-based power system, the CCM is employed to derivate the closed-loop response transfer matrix of the overall power which can predict the instability of the power system [10], but it does not separate the source impedance and load admittance with the network node admittance matrix. Thus, it is difficult to obtain the general stability criterion intuitionally. In this paper, the CCM is also employed and expanded to derive the stability criterion for the meshed DC DPS with the CPLs. The stability criterion can be simplified to relate only with the network node admittance matrix, the output impendence of the source, and the input admittance of the CPLs which can be separately obtained by theoretical calculation or impedance analyzer measurement.

Many stability enhancement methods have been given in previous researches. An adaptive active capacitor converter (AACC) is introduced to effectively stabilize the cascaded system [11], but this method needs additional hardware devices. Another approach is to add a virtual adaptive parallel impedance [12] or adaptive series-virtual-impedance [13] in the input of the load converter to improve the stability in the cascaded DC/DC converter system. However, in the meshed DC DPS, it is not very effective due to the existence of an electromagnetic interference (EMI) filter for the CPL with a switching-mode power converter. The input admittance of the CPLs is mainly determined by the EMI filter, and the virtual impedance is difficult to be added by the load converter control part. Another way is to adjust the source output impedance. Modifying the source output impedance to be zero [14] and adding virtual impedances in the source output filters [15] are both feasible strategies by the source converter control part, but both of them will influence the low and high frequency characteristics of the source output impedance. To overcome the problem, this paper proposes an effective and achievable method, which is to add a virtual harmonic resistance through the second-order generalized integrator (SOGI) in the source converter control part to lower the peak of the source output impendence.

The main contributions of this paper are follows. Firstly, the stability criterion for the meshed DC DPS is derived, which has been simplified to relate only with the network node admittance matrix, the output impendence of the source, and the input admittance of the CPLs. It is easy to implement, as the impedance and the admittance can both be separately obtained by the theoretical calculation or impedance analyzer measurement. Secondly, an effective stability enhancement method by adding a virtual harmonic resistance is proposed, with which only the small range middle frequency characteristic of the source output impedance is modified, while the low and high frequency characteristics of the source output impedance are still unmodified. The results of the nonlinear dynamic simulations verify the effectiveness of the proposed stability criterion and stability enhancement method.

The remainder of this paper is organized as follows. The stability criterion for the meshed DC DPS is given in Section 2. The modeling of the source with voltage-controlled converters and the CPLs are derived, and the proposed stability enhancement method and the stability analysis for the meshed DC DPS are given in Section 3. Then, the nonlinear dynamic simulations are conducted in Section 4. At last, conclusions are drawn in Section 5.

2. Stability Criterion for the Meshed DC DPS

The general form of DC DPS is given in [4], but it ignores the network dynamics. In [4], any converter in a DC DPS can be classified as either a bus voltage controlled converter (BVCC) or a bus current controlled converter (BCCC). In this study, a four node meshed DC DPS as Figure 1 is considered as an example, which is composed by two sources with voltage-controlled converters and two CPLs. The sources with voltage-controlled converters are controlled with constant voltage, hence they can be treated as BVCCs. The CPLs are controlled to absorb constant power from the meshed DC DPS, hence they can be treated as BCCCs.

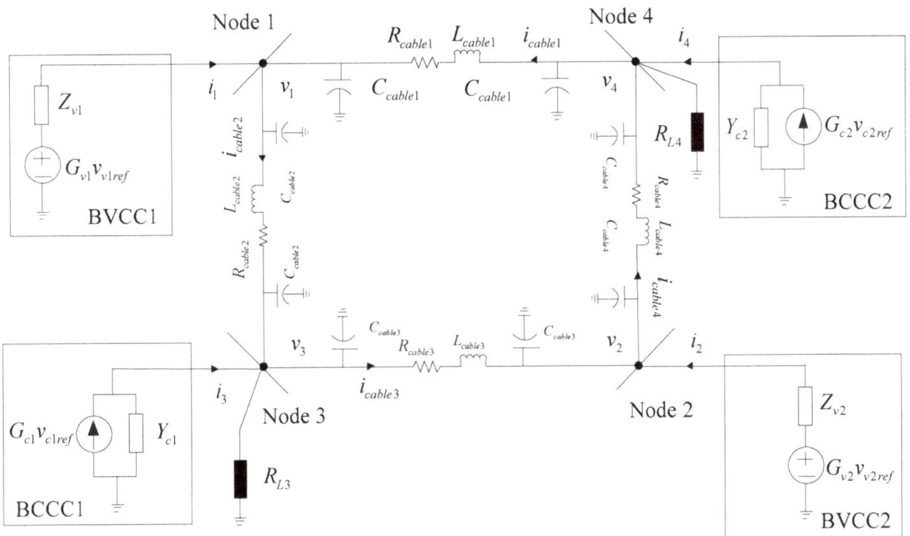

Figure 1. The block diagram of the component connection method (CCM) applied for the studied meshed DC distributed power systems (DPS).

Although all of the sources and CPLs are independently designed well, there will be complex interactions through the power cables in the meshed DC DPS. This consequently necessitates the use of CCM to analyze the stability of the meshed DC DPS.

The block diagram of the CCM applied for the studied meshed DC DPS is shown in Figure 1, where the CCM decomposes the overall DPS into four subsystems and the network. The two BVCCs are modeled by the Thevenin equivalent circuits [16], and the two BCCCs are modeled by the Norton equivalent circuits [17].

In Figure 1, Z_{v1} and Z_{v2} are the closed-loop output impedances of the sources, and Y_{c1} and Y_{c2} are the closed-loop input admittances of the CPLs. $\begin{bmatrix} v_1 & v_2 & v_3 & v_4 \end{bmatrix}^T$ is the node voltage column vector, and $\begin{bmatrix} i_1 & i_2 & i_3 & i_4 \end{bmatrix}^T$ is the node injection current column vector. $\begin{bmatrix} i_{cable1} & i_{cable2} & i_{cable3} & i_{cable4} \end{bmatrix}^T$ is the cable current column vector. R_{L3} and R_{L4} are the resistive loads connected to node 3 and node 4, respectively. $\begin{bmatrix} L_{cable1} & L_{cable2} & L_{cable3} & L_{cable4} \end{bmatrix}^T$ is the inductance column vector of cables, $\begin{bmatrix} R_{cable1} & R_{cable2} & R_{cable3} & R_{cable4} \end{bmatrix}^T$ is the parasitic resistance column vector of cables, and $\begin{bmatrix} C_{cable1} & C_{cable2} & C_{cable3} & C_{cable4} \end{bmatrix}^T$ is the capacitance column vector of cables. $\begin{bmatrix} G_{v1} & G_{v2} & G_{c1} & G_{c2} \end{bmatrix}^T$ is the voltage reference-to-output transfer function column vector of the source and CPLs. $\begin{bmatrix} v_{v1ref} & v_{v2ref} & v_{c1ref} & v_{c2ref} \end{bmatrix}^T$ is the reference voltage column vector of the source and CPLs, respectively.

Generally, the source is designed to be a voltage source that is stable when unloaded, and the CPL is designed to be stable when supplied by an ideal voltage source. That is, the unterminated behaviors of inverters $\begin{bmatrix} G_{v1} & G_{v2} & G_{c1} & G_{c2} \end{bmatrix}^T$ are stable, and $\begin{bmatrix} Z_{v1} & Z_{v2} & Y_{c1} & Y_{c2} \end{bmatrix}^T$ are stable, which means that there are no right-half plane poles.

To facilitate the formulation of the nodal admittance matrix, the Thevenin circuits of the BVCCs are converted to Norton circuits [10]. Then, the studied meshed DC DPS is represented as Figure 2.

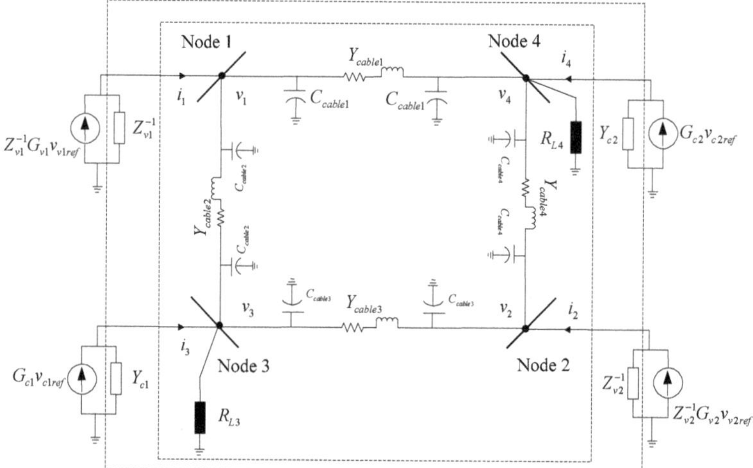

Figure 2. The equivalent Norton circuit of the studied meshed DC DPS.

Applying the node voltage equation to the external dashed box part in Figure 2, it can be derived as

$$
\begin{bmatrix} Z_{v1}^{-1}G_{v1}v_{v1ref} \\ Z_{v2}^{-1}G_{v2}v_{v2ref} \\ G_{c1}v_{c1ref} \\ G_{c2}v_{c2ref} \end{bmatrix} = \left(\begin{bmatrix} Z_{v1}^{-1} & & & \\ & Z_{v2}^{-1} & & \\ & & Y_{c1} & \\ & & & Y_{c2} \end{bmatrix} + Y_{sys} \right) \begin{bmatrix} v_1 \\ v_2 \\ v_3 \\ v_4 \end{bmatrix}
\tag{1}
$$

where Y_{sys} is the network nodal admittance matrix as

$$
Y_{sys} = \begin{bmatrix} C_{cable1}s + C_{cable1}s + Y_{cable1} + Y_{cable2} & 0 & -Y_{cable2} & -Y_{cable1} \\ 0 & C_{cable3}s + C_{cable4}s + Y_{cable3} + Y_{cable4} & -Y_{cable3} & -Y_{cable4} \\ -Y_{cable2} & -Y_{cable3} & Y_{L3} + C_{cable2}s + C_{cable3}s + Y_{cable2} + Y_{cable3} & 0 \\ -Y_{cable1} & -Y_{cable4} & 0 & Y_{L4} + C_{cable4}s + C_{cable1}s + Y_{cable4} + Y_{cable1} \end{bmatrix}
\tag{2}
$$

where Y_{L3} and Y_{L4} are the admittances for loads connected to node 3 and node 4, respectively. $Y_{cable1}\sim Y_{cable4}$ are the admittances for cables and $Y_{cablei} = 1/(R_{cablei} + L_{cablei}s)$, $i = 1\sim4$.

Then, the node voltage can be derived from (1) as follows:

$$
\begin{bmatrix} v_1 \\ v_2 \\ v_3 \\ v_4 \end{bmatrix} = \left(\begin{bmatrix} Z_{v1}^{-1} & & & \\ & Z_{v2}^{-1} & & \\ & & Y_{c1} & \\ & & & Y_{c2} \end{bmatrix} + Y_{sys} \right)^{-1} \begin{bmatrix} Z_{v1}^{-1} & & & \\ & Z_{v2}^{-1} & & \\ & & 1 & \\ & & & 1 \end{bmatrix} \begin{bmatrix} G_{v1}v_{v1ref} \\ G_{v2}v_{v2ref} \\ G_{c1}v_{c1ref} \\ G_{c2}v_{c2ref} \end{bmatrix}
\tag{3}
$$

$$
\begin{aligned}
v &= \left(\begin{bmatrix} Z_v^{-1} & \\ & Y_c \end{bmatrix} + Y_{sys} \right)^{-1} \begin{bmatrix} Z_v^{-1} & \\ & I_c \end{bmatrix} \begin{bmatrix} G_v & \\ & G_c \end{bmatrix} \begin{bmatrix} v_{vref} \\ v_{cref} \end{bmatrix} \\
&= \left(\begin{bmatrix} I_v & \\ & Y_c \end{bmatrix} + \begin{bmatrix} Z_v & \\ & I_c \end{bmatrix} Y_{sys} \right)^{-1} \begin{bmatrix} G_v & \\ & G_c \end{bmatrix} \begin{bmatrix} v_{vref} \\ v_{cref} \end{bmatrix}
\end{aligned}
\tag{4}
$$

where Z_v is the set of the closed-loop output impedances of the sources, and Y_c is the set of the closed-loop input admittances of the CPLs. v is the set of node voltages, and i is the set of the node injection currents. G_v is the the set of voltage reference-to-output transfer functions of the source, G_c is the set of voltage reference to output transfer functions of the CPLs, v_{vref} is the set of voltage references of the source, and v_{cref} is the set of voltage references of the CPLs. I_c is a unit diagonal matrix whose dimension is equal to Y_c. I_v is a unit diagonal matrix whose dimension is equal to Z_v. And they satisfy the following relations:

$$Z_v = \begin{bmatrix} Z_{v1} & \\ & Z_{v2} \end{bmatrix}, \; Y_c = \begin{bmatrix} Y_{c1} & \\ & Y_{c2} \end{bmatrix}, \; G_v = \begin{bmatrix} G_{v1} & \\ & G_{v2} \end{bmatrix}, \; G_c = \begin{bmatrix} G_{c1} & \\ & G_{c2} \end{bmatrix}, \; v_{vref} = \begin{bmatrix} v_{v1ref} \\ v_{v2ref} \end{bmatrix}, \; v_{cref} = \begin{bmatrix} v_{c1ref} \\ v_{c2ref} \end{bmatrix} \quad (5)$$

$$v = \begin{bmatrix} v_1 & v_2 & v_3 & v_4 \end{bmatrix}^T, \; i = \begin{bmatrix} i_1 & i_2 & i_3 & i_4 \end{bmatrix}^T \quad (6)$$

Applying the node voltage equation to the internal dashed box part in Figure 2, i can be derived as:

$$i = Y_{sys}v = Y_{sys}\left(\begin{bmatrix} I_v & \\ & Y_c \end{bmatrix} + \begin{bmatrix} Z_v & \\ & I_c \end{bmatrix} Y_{sys} \right)^{-1} \begin{bmatrix} G_v & \\ & G_c \end{bmatrix} \begin{bmatrix} v_{vref} \\ v_{cref} \end{bmatrix} \quad (7)$$

In the above equation, Y_{sys} is the network nodal admittance matrix for a real physical system, which means Y_{sys} is stable and has no right-half plane poles. G_v and G_c are both stable as in the above analysis. Therefore, the stability criterion of the meshed DC DPS T_m is

$$T_m = \left(\begin{bmatrix} I_v & \\ & Y_c \end{bmatrix} + \begin{bmatrix} Z_v & \\ & I_c \end{bmatrix} Y_{sys} \right)^{-1} \quad (8)$$

If the stability criterion T_m has right-half plane poles, the meshed DC DPS will be unstable. Through the analysis of the stability criterion T_m, it can be found that the stability of the meshed DC DPS is only related to the network node admittance matrix Y_{sys}, the output impendence of the sources Z_v, and the input admittance of the CPLs Y_c, which can be separately obtained by theoretical calculation or impedance analyzer measurement.

3. Modeling, Stability Enhancement Method and Stability Analysis

3.1. Modeling of the CPL

The modeling of the CPL connected to node 3 is similar to the CPL connected to node 4. Without loss of generality, taking the CPL connected to node 4 as an example. The CPL usually exploits a switching-mode power converter as the interface with the node, and the compliance with EMI standards usually requires the insertion of an EMI filter between the switching-mode power converter and the node [18–20]. The close-loop circuit of the CPL with an EMI filter is shown as Figure 3.

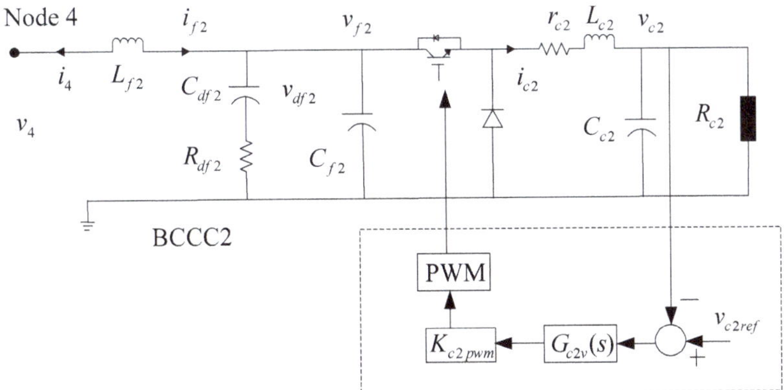

Figure 3. The close-loop circuit of the constant power load (CPL) with an electromagnetic interference (EMI) filter.

Applying switch period average, the small-signal model of the CPL with EMI filter can be achieved as Figure 4.

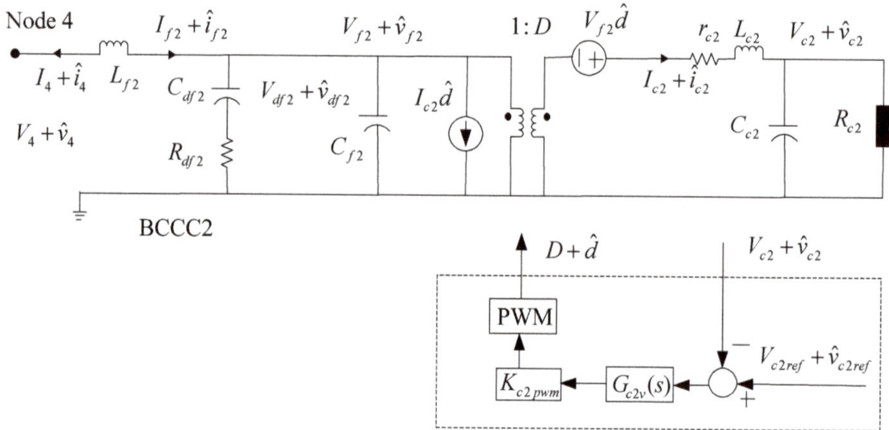

Figure 4. The small-signal model of the CPL with an EMI filter.

The variables showing in capital letters represent the steady-state values of duty cycle, inductor current, capacitor voltage, and output current, and the variables with "^" indicate the small-signal disturbance.

It can be derived from Figure 4 that the small-signal disturbance of the output current $\hat{i}_4$ can be expressed as:

$$\hat{i}_4 = -Y_{c2}\hat{v}_4 + G_{c2}\hat{v}_{c2ref} \tag{9}$$

$$Y_{c2} = \frac{a_4 s^4 + a_3 s^3 + a_2 s^2 + a_1 s^1 + a_0}{\Delta_{c2}} \tag{10}$$

$$G_{c2} = -\frac{c_3 s^3 + c_2 s^2 + c_1 s^1 + c_0}{\Delta_{c2}} \tag{11}$$

$$\Delta_{c2} = b_5 s^5 + b_4 s^4 + b_3 s^3 + b_2 s^2 + b_1 s^1 + b_0 \tag{12}$$

where G_{c2v} is the voltage controller, Y_{c2} is the closed-loop input admittance of the CPL, and G_{c2} is the voltage reference to output transfer function of the CPL. $a_0 \sim a_4$, $b_0 \sim b_5$, and $c_0 \sim c_3$ are given in Appendix A.

3.2. Modeling of the Source with Voltage-Controlled Converters

The modeling of the source with voltage-controlled converters connected to node 1 is similar to the source connected to node 2. Without loss of generality, taking the source connected to node 1 as an example, the close-loop circuit of the source with voltage-controlled converters is shown as Figure 5.

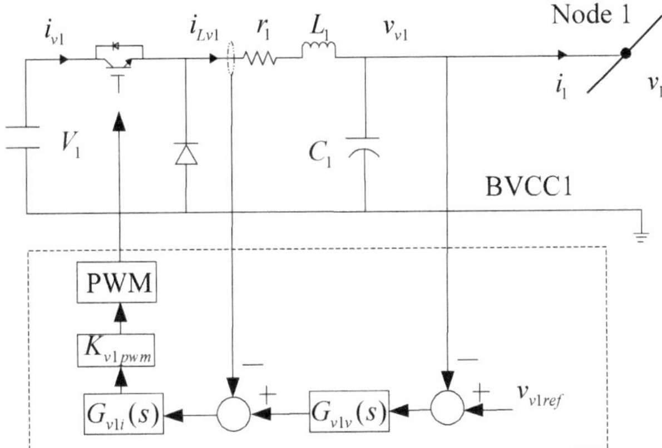

Figure 5. The close-loop circuit of the source with voltage-controlled converters (where G_{v1v} and G_{v1i} are the voltage and current controllers).

Figure 6 shows the small-signal block diagram of the multi-loop control system of the source with voltage-controlled converters.

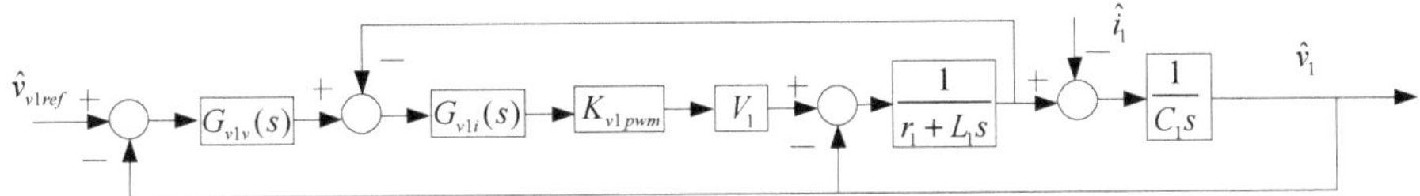

Figure 6. The small-signal block diagram of the multi-loop control system of the source.

Thus, the dynamic behavior of the close-loop system can be given by

$$\hat{v}_1 = -Z_{v1}\hat{i}_1 + G_{v1}\hat{v}_{v1ref} \tag{13}$$

where

$$Z_{v1} = \frac{G_{v1i}K_{v1pwm}V_1 + L_1 s + r_1}{L_1 C_1 s^2 + (C_1 G_{v1i}K_{v1pwm}V_1 + r_1 C_1)s + G_{v1i}G_{v1v}K_{v1pwm}V_1 + 1} \tag{14}$$

$$G_{v1} = \frac{G_{v1i}G_{v1v}K_{v1pwm}V_1}{L_1 C_1 s^2 + (C_1 G_{v1i}K_{v1pwm}V_1 + r_1 C_1)s + G_{v1i}G_{v1v}K_{v1pwm}V_1 + 1} \tag{15}$$

where Z_{v1} is the closed-loop output impedance of the source with voltage-controlled converters, and G_{v1} is the voltage reference-to-output transfer function of the source with voltage-controlled converters.

3.3. Stability Enhancement Method

The source output impedance has a peak at the resonant frequency. Based on the Nyquist criterion, instability between the source and the network may occur if the network input impedance becomes lower than the source output impedance at this frequency. Therefore, to ensure the system stability and to minimize the potential for inadvertent interactions with the source, it is important to lower the peak of the source output impedance. For this purpose, this paper proposes a method by adding a virtual harmonic resistance through the SOGI to be in parallel with the capacitor.

The transfer function $G_f(s)$ of SOGI can be expressed as

$$G_f(s) = \frac{s^2 + \omega^2}{s^2 + k\omega s + \omega^2} \tag{16}$$

where ω is the resonant frequency, and k is the frequency coefficient. The frequency-domain characteristic of $G_f(s)$ is shown as Figure 7. Around the resonant frequency, the amplitude of $G_f(s)$ is very small, and the amplitude of $G_f(s)$ in other frequency ranges is 0 dB.

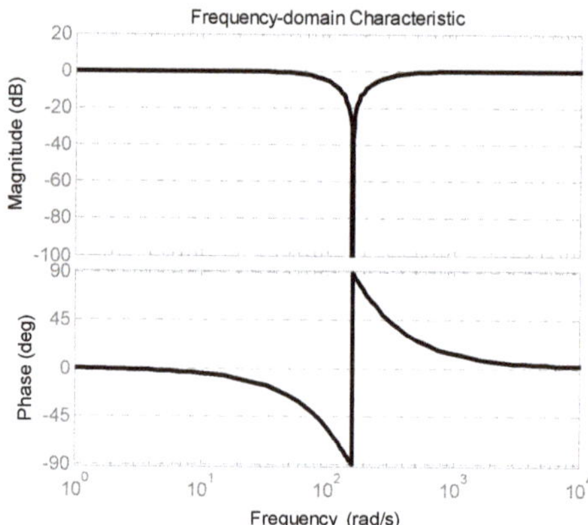

Figure 7. The frequency-domain characteristic of $G_f(s)$.

Figure 8 shows the small-signal control block of the source 1. If a virtual harmonic resistance R_{vh} is required to be added in parallel with the capacitor, it is intuitively to introduce R_{vh} to the block as a dot-dashed part in Figure 8, which is the basic idea of this control method. However, this method cannot be realized by control directly. Therefore, R_{vh} is moved to the output of $G_{v1v}(s)$ as the dashed part in Figure 8.

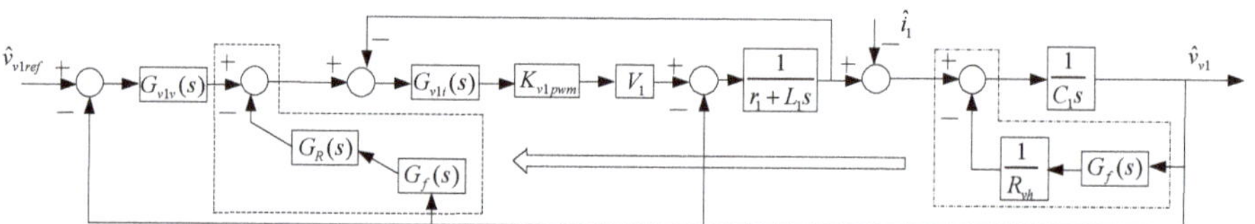

Figure 8. Adding a virtual harmonic resistance through the second-order generalized integrator (SOGI).

$G_R(s)$ is expressed as

$$G_R(s) = \frac{L_1 s + r_1 + G_{v1i} K_{v1pwm} V_1}{G_{v1i} K_{v1pwm} V_1 R_{vh}}. \tag{17}$$

With the parallel virtual harmonic resistance $R_{vh} = 1\ \Omega$, the frequency-domain characteristic of the source output impedance is shown as Figure 9.

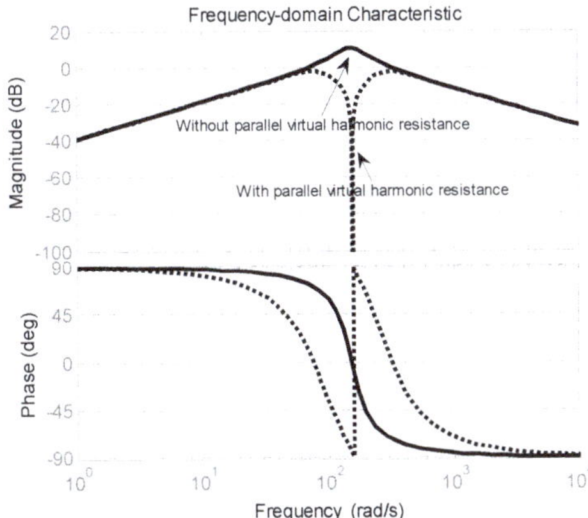

Figure 9. The source output impedance with/without parallel virtual harmonic resistance R_{vh}.

As can be seen, the peak of the source output impedance has been lower, and only the small range middle frequency characteristic of the source output impedance is modified, while the low and high frequency characteristics of the source output impedance are still unmodified.

3.4. Stability Analysis

All the main electrical parameters of the studied DC meshed DPS as shown in Figure 1 are listed in Appendix B.

The closed-loop output impedances of the sources Z_v can be represented by using impedance calculated in Equations (5) and (14), the closed-loop input admittances of the CPLs Y_c can be represented by using admittance calculated in Equations (5) and (10), and the network nodal admittance matrix Y_{sys} can be represented by using admittance calculated in Equation (2). Then, the the stability of the meshed DC DPS can be analyzed with the stability criterion T_m expressed in Equation (8). If the stability criterion T_m has right-half plane poles, the meshed DC DPS will be unstable.

At first, it is necessry to compare the proposed stability criterion T_m in this paper and the previous stability criterion T_{m2} given in [4] as follows:

$$T_{m2} = \frac{1}{1 + (Z_{v1}//Z_{v2})(Y_{c1} + Y_{c2} + Y_{L3} + Y_{L4})}. \tag{18}$$

The dominant poles of T_m ans T_{m2} are displayed as Figure 10, in which the cables are long $R_{cablei} = 0.05\ \Omega$, $L_{cablei} = 0.5$ mH, $C_{cablei} = 100\ \mu$F, and $i = 1 \sim 4$. As shown in Figure 10, all the dominant poles of T_{m2} are both in the left-half plane, which indicates that the meshed DC DPS is stable. However, there are two right-half plane poles $0.15 \pm 94.5i$ in the dominant poles of T_m, which indicates the meshed DC DPS is unstable and have an oscillation with approximately period 0.066 s. The results of the proposed stability criterion T_m and the previous stability criterion T_{m2} are not consistent, and the results of proposed stability criterion T_m will be confirmed to be right with the nonlinear dynamic simulations in Section 4. Moreover, many poles related to the network dynamics are also lost in T_{m2} compared to T_m.

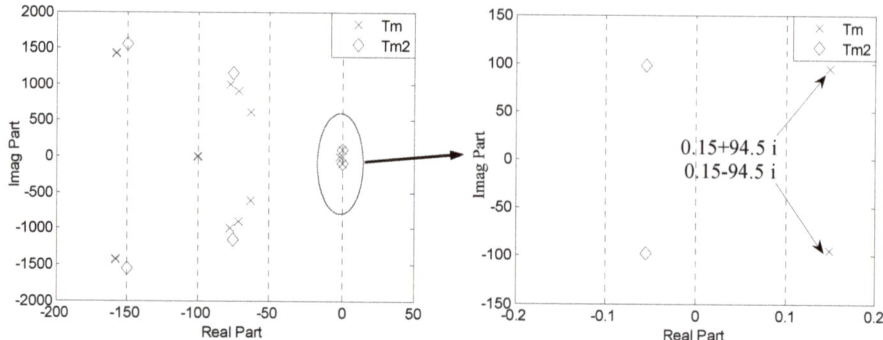

Figure 10. The dominant poles of T_m ans T_{m2} when the cables are long.

With the electrical parameters listed in Table B1, when the power of resistive load 4 P_{L4} is 50 kW, which means $R_{L4} = 5\,\Omega$, the dominant poles of T_m are displayed as Figure 11. When the power of resistive load 4 P_{L4} reduces from 50 kW to 10 kW, which means $R_{L4} = 25\,\Omega$, the dominant poles of T_m are displayed as Figure 12. Furthermore, when the power of resistive load 4 P_{L4} is 10 kW and the absorbed power of CPL2 P_{c2} reduces from 100 kW to 50 kW, which means $R_{L4} = 25\,\Omega$ and $R_{c2} = 1.25\,\Omega$, and the dominant poles of T_m are displayed as Figure 13.

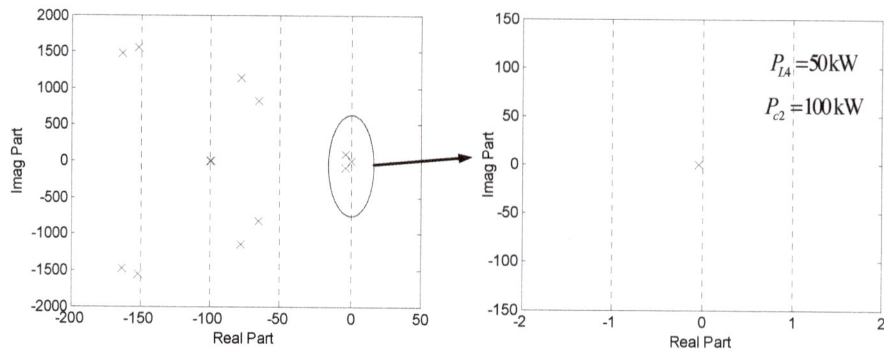

Figure 11. The dominant poles of T_m when $P_{L4} = 50$ kW, $P_{c2} = 100$ kW.

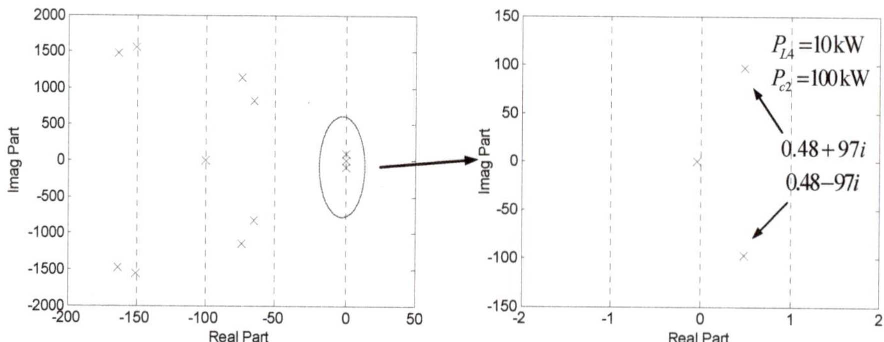

Figure 12. The dominant poles of T_m when $P_{L4} = 10$ kW, $P_{c2} = 100$ kW.

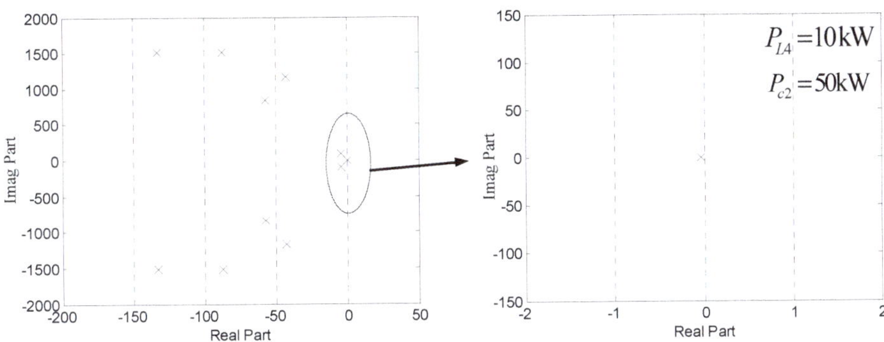

Figure 13. The dominant poles of T_m when $P_{L4} = 10$ kW, $P_{c2} = 50$ kW.

As can be seen from Figure 11, all of the poles of T_m are in the left half plane, which indicates that the meshed DC DPS is stable. But there are two right half plane poles $0.48 \pm 97i$ in the Figure 12, which indicates the meshed DC DPS is unstable and have an oscillation with approximately period 0.0647 s. Furthermore, with the power reduction of CPL2, the meshed DC DPS becomes stable as shown in Figure 13.

Applying the proposed stability enhancement method in the sources connected with node 1 and node 2 of the meshed DC DPS when P_{L4} is 10 kW and P_{c2} is 100 kW, the dominant poles of T_m are displayed as Figure 14. There are no right half plane poles, which means that the meshed DC DPS is stable. Compared to the poles in Figure 12, the stability of the meshed DC DPS is enhanced by the proposed stability enhancement method.

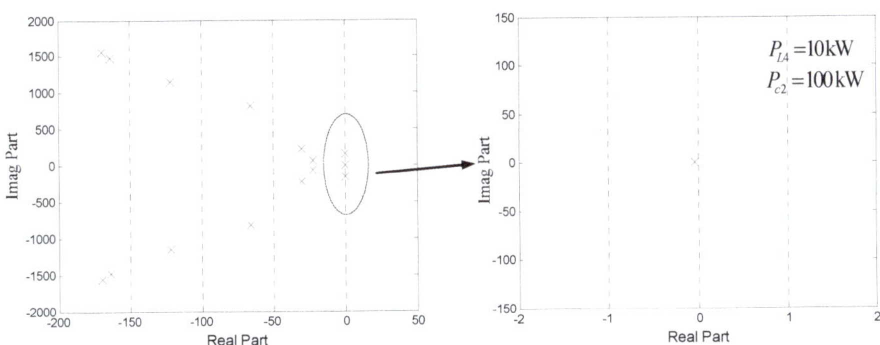

Figure 14. The dominant poles of T_m with the proposed stability enhancement method.

4. Nonlinear Dynamic Simulations

To validate the effectiveness of the proposed stability criterion and stability enhancement method, the meshed DC DPS in Figure 1 is built in the nonlinear time-domain simulation by using Matlab/Simulink. The electrical and controller parameters are all given in Appendix B.

To compare the proposed stability criterion T_m and the previous stability criterion T_{m2}, the cables are chosen to be long cables with parameters $R_{cablei} = 0.05 \, \Omega$, $L_{cablei} = 0.5$ mH, $C_{cablei} = 100$ µF, $i = 1{\sim}4$, the nonlinear dynamic simulation results of the node 1 voltage $v_1(t)$ and the cable 1 current $i_{cable1}(t)$ are shown in Figure 15. As can be seen in Figure 15, before 2.5 s, the meshed DC DPS gradually becomes to a stable state with $P_{L4} = 50$ kW, and once the power of resistive load 4 P_{L4} reduces to 15 kW at 2.5 s, the meshed DC DPS becomes to an oscillation state. The oscillation amplitude gradually grows, and the oscillation period is about 0.066 s, which satisfy the theoretical analysis in Figure 10. The results verify the effectiveness of the proposed stability criterion T_m compared to the previous stability criterion T_{m2} which does not take the network dynamics into account.

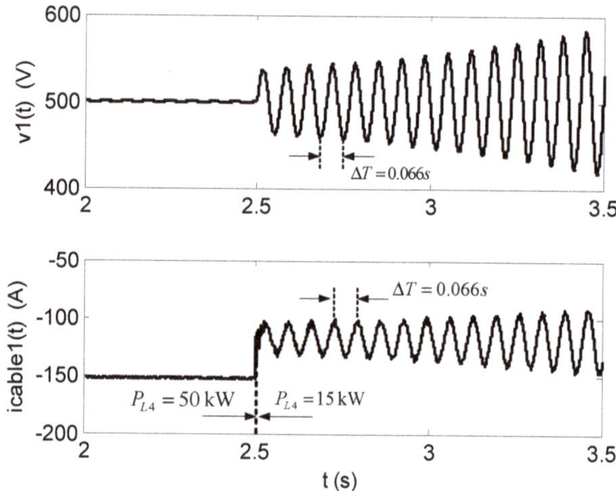

Figure 15. Comparison between T_m and T_{m2}.

To verify the effectiveness of the proposed stability criterion T_m with power changes of the resistive load and CPL, and the proposed stability enhancement method, the nonlinear dynamic simulation results are shown as the following Figures 16–18.

As can be seen in Figure 16, before 1.5 s, the meshed DC DPS gradually becomes to stable state with $P_{L4} = 50$ kW and $P_{c2} = 100$ kW, which confirms the theoretical analysis in Figure 11. Then, the power of resistive load 4 P_{L4} reduces to 10 kW at 1.5 s, the meshed DC DPS becomes to the oscillation state with about period 0.065 s, which confirms the theoretical analysis in Figure 12. The power of CPL2 reduces from 100 kW to 50 kW at 2 s, and the meshed DC DPS becomes to a stable state again, which confirms the theoretical analysis in Figure 13.

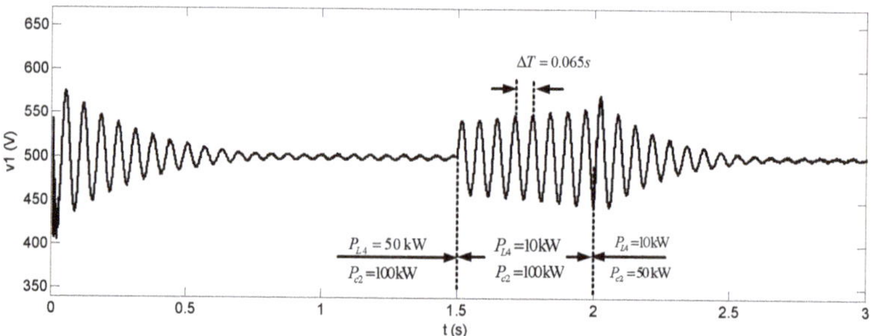

Figure 16. The node 1 voltage $v_1(t)$ with power changes of the resistive load and CPL.

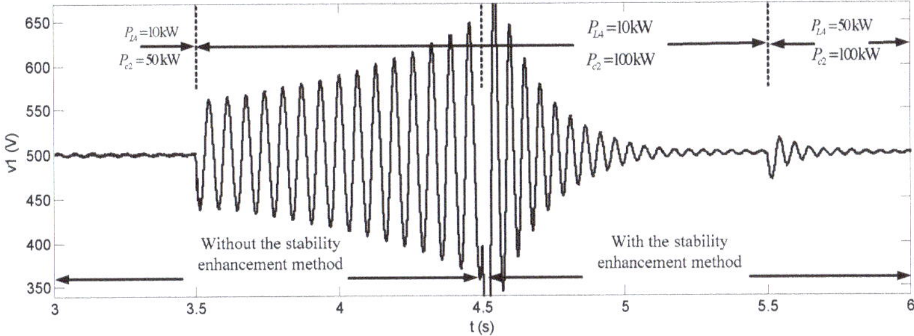

Figure 17. The node 1 voltage v_1 without/with the proposed stability enhancement method.

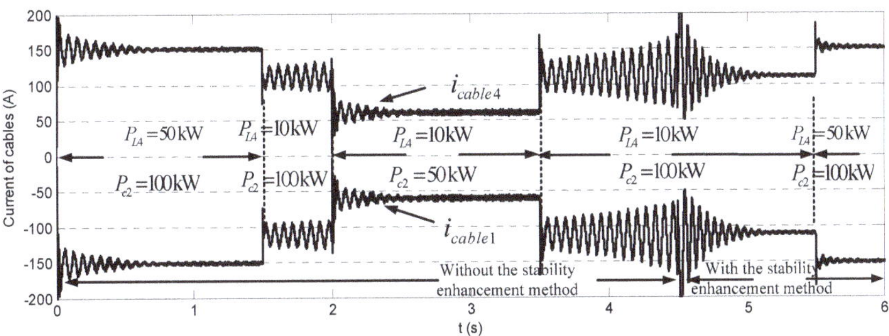

Figure 18. The currents of cable 1 i_{cable1} and cable 4 i_{cable4}.

As can be seen in Figure 17, when the power of CPL2 increases from 50 kW to 100 kW at 3.5 s, the meshed DC DPS becomes to the oscillation state again between 3.5–4.5 s. Then, the vitural harmonic resistances are added in the controllers of source 1 and source 2 at 4.5 s. With the vitural harmonic resistance, the meshed DC DPS becomes to the stable state again, and the node 1 voltage v_1 converges again to the steady value 500 V. The simulation results between 4.5–5.5 s confirm the theoretical analysis results in Figure 14. The power of resistive load 4 P_{L4} increases to 50 kW at 5.5 s, and the meshed DC DPS is maintained at a steady state.

The currents of cable 1 i_{cable1} and cable 4 i_{cable4} are displayed as Figure 18. The steady state and unstable state are similar to the node 1 voltage v_1 as shown in Figures 16 and 17. It is worth noting that the steady state current values in 0–1.5 s and 5.5–6 s are the same, which means that with or without the proposed stability enhancement method, the current distributions in the meshed DC DPS are not affected, which verifies that the low frequency characteristic of the source output impedance has not been modified.

5. Conclusions

This paper discusses the modeling and stability analysis of the meshed DC DPS including multiple sources with voltage-controlled converters and CPLs, and proposes a stability enhancement method. The stability criterion is found to relate only with the network node admittance matrix, the output impendence of the source, and the input admittance of the CPLs, and the three parts can be obtained separately. Virtual harmonic resistance through the second-order generalized integrator (SOGI) for the source proposed in this paper is a decoupling design which is only based on the characteristic of the source output impedance. Nonlinear dynamic time-domain simulation results verify the effectiveness of the proposed stability criterion and stability enhancement method.

Furthermore, the current control sources or loads can be treated as BCCCs, and the input admittances of current control sources or loads are related to the stability analysis. Multiple sources with droop control can be treated as BVCCs, and the output impendence of the droop control sources are related to the stability analysis. So the proposed stability criterion can also deal with the current control sources or loads and multiple sources with droop control. Related analyses are in the future researches.

Acknowledgments: The research leading to these results has received funding from the Nation "863" project of China (2014AA052001), the National Natural Science Foundation of China (51377142), the Zhejiang Provincial Natural Science Foundation of China (LY16E070002), and the Key Research and Development Project of Zhejiang Province (2016C01G2010726).

Author Contributions: All authors contributed equally in the presented research, and all authors approved the final manuscript.

Conflicts of Interest: The authors declare no conflict of interest.

Appendix A

$$
\begin{cases}
a_0 = D^2 - G_{c2v}I_{c2}K_{c2pwm}R_{c2}D \\
a_1 = -R_{c2}R_{df2}C_{df2}G_{c2v}I_{c2}K_{c2pwm}D + (C_{df2} + C_{f2})(G_{c2v}K_{c2pwm}R_{c2}V_{f2} + R_{c2} + r_{c2}) + (R_{c2}C_{c2} + R_{df2}C_{df2})D^2 \\
a_2 = R_{c2}R_{df2}C_{df2}(C_{f2}G_{c2v}K_{c2pwm}V_{f2} + C_{c2}D^2) + C_{df2}C_{f2}R_{df2}(R_{c2} + r_{c2}) + (L_{c2} + C_{c2}R_{c2}r_{c2})(C_{df2} + C_{f2}) \\
a_3 = (r_{c2}R_{c2}C_{c2} + L_{c2})R_{df2}C_{df2}C_{f2} + C_{c2}L_{c2}R_{c2}(C_{f2} + C_{df2}) \\
a_4 = R_{c2}C_{c2}R_{df2}C_{df2}L_{c2}C_{f2}
\end{cases}
\tag{A1}
$$

$$
\begin{cases}
b_0 = G_{c2v}K_{c2pwm}R_{c2}V_{f2} + R_{c2} + r_{c2} \\
b_1 = (R_{c2}R_{df2}C_{df2}V_{f2} - R_{c2}DI_{c2}L_{f2})G_{c2v}K_{c2pwm} + R_{c2}C_{c2}r_{c2} + R_{df2}C_{df2}(R_{c2} + r_{c2}) + D^2L_{f2} + L_{c2} \\
b_2 = -R_{c2}R_{df2}C_{df2}DI_{c2}G_{c2v}K_{c2pwm}L_{f2} + (C_{df2} + C_{f2})(R_{c2}V_{f2}G_{c2v}K_{c2pwm}L_{f2} + R_{c2}L_{f2} + r_{c2}L_{f2}) + (R_{c2}C_{c2} + R_{df2}C_{df2})(L_{f2}D^2 + L_{c2}) + R_{c2}C_{c2}R_{df2}C_{df2}r_{c2} \\
b_3 = R_{c2}R_{df2}C_{df2}C_{f2}G_{c2v}K_{c2pwm}L_{f2}V_{f2} + (D^2L_{f2} + L_{c2})R_{c2}C_{c2}R_{df2}C_{df2} + C_{df2}C_{f2}L_{f2}R_{df2}(R_{c2} + r_{c2}) + (C_{df2} + C_{f2})(C_{c2}L_{f2}R_{c2}r_{c2} + L_{c2}L_{f2}) \\
b_4 = C_{df2}C_{f2}L_{f2}R_{df2}(R_{c2}r_{c2}C_{c2} + L_{c2}) + (C_{df2} + C_{f2})C_{c2}L_{c2}L_{f2}R_{c2} \\
b_5 = C_{c2}C_{df2}C_{f2}L_{c2}L_{f2}R_{c2}R_{df2}
\end{cases}
\tag{A2}
$$

$$
\begin{cases}
c_0 = G_{c2v}K_{c2pwm}V_{f2}D + G_{c2v}I_{c2}K_{c2pwm}(R_{c2} + r_{c2}) \\
c_1 = (C_{c2}R_{c2} + C_{df2}R_{df2})G_{c2v}K_{c2pwm}V_{f2}D + \left(L_{c2} + R_{c2}R_{df2}C_{df2} + r_{c2}R_{df2}C_{df2} + r_{c2}R_{c2}C_{c2}\right)G_{c2v}K_{c2pwm}I_{c2} \\
c_2 = R_{c2}C_{c2}R_{df2}C_{df2}G_{c2v}K_{c2pwm}V_{f2}D + \left(R_{c2}C_{c2}L_{c2} + R_{df2}C_{df2}L_{c2} + R_{c2}C_{c2}R_{df2}C_{df2}r_{c2}\right)G_{c2v}I_{c2}K_{c2pwm} \\
c_3 = R_{c2}C_{c2}R_{df2}C_{df2}L_{c2}G_{c2v}I_{c2}K_{c2pwm}
\end{cases}
\tag{A3}
$$

Appendix B

This Appendix lists all the main electrical parameters of the studied DC meshed DPS as shown in Figure 1.

Table B1. Main electrical parameters of the studied DC meshed distributed power systems (DPS).

Parameters	Value	Parameters	Value
Parameters of CPLs		**Parameters of Sources**	
$R_{c1}\ R_{c2}$	0.625 Ω	$V_1\ V_1$	1000 V
$C_{c1}\ C_{c2}$	2700 μF	$r_1\ r_2$	0.001 Ω
$L_{c1}\ L_{c2}$	0.32 mH	$L_1\ L_1$	5 mH
$r_{c1}\ r_{c2}$	0.001 Ω	$C_1\ C_2$	4000 μF
$L_{f1}\ L_{f2}$	0.32 mH	$K_{v1pwm}\ K_{v2pwm}$	1/1000
$C_{f1}\ C_{f2}$	2700 μF	$G_{v1v}\ G_{v2v}$	0.24 + 89.39/s
$C_{df1}\ C_{df2}$	1350 μF	$G_{v1i}\ G_{v2i}$	9.6
$R_{df1}\ R_{df2}$	0.9373 Ω	**Parameters of Cables**	
$I_{c1}\ I_{c2}$	400 A	$R_{cable1} \sim R_{cable4}$	0.001 Ω
$V_{f1}\ V_{f2}$	500 V	$L_{cable1} \sim L_{cable4}$	0.01 mH
$K_{c1pwm}\ K_{c2pwm}$	1/500	$C_{cable1} \sim C_{cable4}$	10 μF
$G_{c1v}\ G_{c2v}$	0.5 + 500/s	**Parameters of Resistive Loads**	
$D_1\ D_2$	0.5	R_{L3}	12.5 Ω
$V_{c1ref}\ V_{c2ref}$	250 V	R_{L4}	16.67 Ω

References

1. Xu, C.D.; Cheng, K.W.E. A survey of distributed power system—AC versus DC distributed power system. In Proceedings of the 2011 4th International Conference on Power Electronics Systems and Applications (PESA), Hong Kong, China, 8–10 June 2011; pp. 1–12.
2. Lee, S.-W.; Cho, B.-H. Master–Slave Based Hierarchical Control for a Small Power DC-Distributed Microgrid System with a Storage Device. *Energies* **2016**, *9*. [CrossRef]
3. Liang, H.; Zhao, X.; Yu, X.; Gao, Y.; Yang, J. Study of Power Flow Algorithm of AC/DC Distribution System including VSC-MTDC. *Energies* **2015**, *8*, 8391–8405. [CrossRef]
4. Zhang, X.; Ruan, X.; Tse, C.K. Impedance-based local stability criterion for DC distributed power systems. *IEEE Trans. Circuits Syst. I* **2015**, *62*, 916–925. [CrossRef]
5. Sun, J. Impedance-based stability criterion for grid-connected inverters. *IEEE Trans. Power Electron.* **2011**, *26*, 3075–3078. [CrossRef]
6. Feng, X.; Liu, J.; Lee, F.C. Impedance specifications for stable DC distributed power systems. *IEEE Trans. Power Electron.* **2002**, *17*, 157–162. [CrossRef]
7. Anand, S.; Fernandes, B.G. Reduced-order model and stability analysis of low-voltage DC microgrid. *IEEE Trans. Ind. Electron.* **2013**, *60*, 5040–5049. [CrossRef]
8. Gaba, G.; Lefebvre, S.; Mukhedkar, D. Comparative analysis and study of the dynamic stability of AC/DC systems. *IEEE Trans. Power Syst.* **1988**, *3*, 978–985. [CrossRef]
9. Lefebvre, S. Tuning of stabilizers in multimachine power systems. *IEEE Power Eng. Rev.* **1983**, *PER-3*, 290–299.
10. Wang, X.; Blaabjerg, F.; Wu, W. Modeling and analysis of harmonic stability in an AC power-electronics-based power system. *IEEE Trans Power Electron.* **2014**, *29*, 6421–6432. [CrossRef]
11. Zhang, X.; Ruan, X.; Kim, H.; Tse, C.K. Adaptive active capacitor converter for improving stability of cascaded DC power supply system. *IEEE Trans. Power Electron.* **2013**, *28*, 1807–1816. [CrossRef]
12. Zhang, X.; Zhong, Q.-C.; Ming, W.-L. Stabilization of a cascaded DC converter system via adding a virtual adaptive parallel impedance to the input of the load converter. *IEEE Trans. Power Electron.* **2016**, *31*, 1826–1832. [CrossRef]
13. Zhang, X.; Zhong, Q.-C.; Ming, W.-L. Stabilization of Cascaded DC/DC Converters via Adaptive Series-Virtual-Impedance Control of the Load Converter. *IEEE Trans. Power Electron.* **2016**, *31*, 6057–6063. [CrossRef]
14. Karppanen, M.; Hankaniemi, M.; Suntio, T.; Sippola, M. Dynamical characterization of peak-current-mode-controlled buck converter with output-current feedforward. *IEEE Trans. Power Electron.* **2007**, *22*, 444–451. [CrossRef]
15. Lu, X.; Sun, K.; Guerrero, J.M.; Vasquez, J.C.; Huang, L.; Wang, J. Stability Enhancement Based on Virtual Impedance for DC Microgrids with Constant Power Loads. *IEEE Trans. Smart Grid* **2015**, *6*, 2770–2783. [CrossRef]
16. Wang, X.; Blaabjerg, F.; Chen, Z.; Wu, W. Resonance analysis in parallel voltage-controlled Distributed Generation inverters. In Proceedings of the 2013 Twenty-Eighth Annual IEEE Applied Power Electronics Conference and Exposition (APEC), Long Beach, CA, USA, 17–21 March 2013; pp. 2977–2983.
17. Wang, X.; Blaabjerg, F.; Liserre, M.; Chen, Z.; He, J.; Li, Y. An active damper for stabilizing power-electronics-based AC systems. *IEEE Trans. Power Electron.* **2014**, *29*, 3318–3329. [CrossRef]
18. Cespedes, M.; Xing, L.; Sun, J. Constant-power load system stabilization by passive damping. *IEEE Trans. Power Electron.* **2011**, *26*, 1832–1836. [CrossRef]
19. Xing, L.; Feng, F.; Sun, J. Optimal damping of EMI filter input impedance. *IEEE Trans. Ind. Appl.* **2011**, *47*, 1432–1440. [CrossRef]
20. Huangfu, Y.; Pang, S.; Nahid-Mobarakeh, B.; Rathore, A.; Gao, F.; Zhao, D. Analysis and Design of an Active Stabilizer for a Boost Power Converter System. *Energies* **2016**, *9*. [CrossRef]

MDPI

St. Alban-Anlage 66

4052 Basel

Switzerland

Tel. +41 61 683 77 34

Fax +41 61 302 89 18

www.mdpi.com

Energies Editorial Office

E-mail: energies@mdpi.com

www.mdpi.com/journal/energies